The Scaled Boundary Finite Element Method

The Scaled Boundary Finite Element Method

John P. Wolf
Swiss Federal Institute of Technology,
Lausanne, Switzerland

WILEY

Telephone (+44) 1243 779777

Email (for orders and customer service enquiries): cs-books@wiley.co.uk
Visit our Home Page on www.wileyeurope.com or www.wiley.com

Other Wiley Editorial Offices

John Wiley & Sons Inc.,
111 River Street, Hoboken, NJ 07030, USA

Jossey-Bass, 989 Market Street,
San Francisco, CA 94103–1741, USA

Wiley-VCH Verlag GmbH,
Boschstr. 12, D-69469 Weinheim, Germany

John Wiley & Sons Australia Ltd,
33 Park Road, Milton, Queensland 4064, Australia

John Wiley & Sons (Asia) Pte Ltd,
2 Clementi Loop #02–01, Jin Xing Distripark, Singapore 129809

John Wiley & Sons Canada Ltd,
22 Worcester Road, Etobicoke, Ontario, Canada M9W 1L1

Wiley also publishes its books in a variety of electronic formats. Some content that appears in print may not be available in electronic books.

Library of Congress Cataloging-in-Publication Data

Wolf, John P.
The scaled boundary finite-element method / John P. Wolf.
p. cm.
Includes bibliographical references and index.
ISBN 0–471–48682–5 (alk. paper)
1. Finite element method. 2. Boundary element methods. I. Title.
TA347. f5 W64 2003
620′.001′51535–dc21 2002029643

British Library Cataloguing in Publication Data

A catalogue record for this book is available from the British Library

ISBN 0 471 48682 5

Typeset in 10/12pt Times by Kolam Information Services Pvt. Ltd, Pondicherry, India.
Printed and bound in Great Britain by TJ International Ltd, Padstow, Cornwall.

This book is printed on acid-free paper responsibly manufactured from sustainable forestry in which at least two trees are planted for each one used for paper production.

Contents

Foreword

At the beginning of the 21st Century, the number of different types of solid mechanics problems, and in general continuum mechanics problems, that have been mathematically formulated is very large and quite comprehensive. There are models to represent the macromechanical behaviour of all sorts of materials under different loading conditions. Elastic, plastic, viscoelastic and other kinds of behaviour of isotropic and non-isotropic materials under static and dynamic loads are represented by systems of partial differential equations and/or integral equations. Formulations of problems related to new materials, fracture, contact, fatigue and other fields are well known. Nevertheless, the exact solution of these mathematical formulations is in most cases unknown. Only a few basic cases with simple geometry and boundary conditions can be solved analytically.

The only method of solving realistic solid mechanics problems has been discretisation. First, it was done by using representations of the solution in terms of a finite number of parameters which are coefficients of a simple series expansion, as occurs in classical methods for rectangular plate bending analysis. Later, it was done by using a set of selected shape functions extending over the domain and taking the amplitude of those functions as the problem unknowns, as in the Trefftz method. A new idea appeared with the Finite Differences Method: the field variable is represented in terms of its values at a number of points evenly distributed over the domain. The governing equations are enforced at each internal point with the derivatives being approximated by their expressions in terms of differences. The Finite Element Method (FEM) introduces a further step towards discretisation. Not only is the field variable represented in terms of its value at a series of nodes, but the domain itself is divided into elements. Local-based functions are used for interpolation.

The FEM has been developed for over forty years to become a general tool for the solution of almost any continuum mechanics problems for which there is a mathematical formulation. Recent ideas such as meshless formulations represent a further development of the domain discretisation concept. In spite of the fact that the FEM is a very general, robust and powerful tool that can be applied to almost any kind of solid mechanics problem, there are difficulties associated with its use in certain areas, such as: dynamic analysis of large or boundless domains; crack propagation; contact; and other stress concentration problems. These difficulties, together with the often tedious work associated with FE discretisation, gave rise to the appearance of the Boundary Element Method (BEM). In this case, the governing equations of the boundary value problem are expressed as boundary integral equations and, consequently, the discretisation of the field variable is restricted to the boundary of the domain. Elements, nodes, shape functions and other basic ingredients of the FEM are also part of the BEM, but are restricted to the boundary. The great advantage of reducing the dimension of the problem by one is clearly exploited for large and boundless domains. Problems related to crack propagation or contact mechanics also show the advantage of restricting the discretisation just to the boundary of the domain.

The fact that the FEM presents difficulties and is lacking for certain continuum mechanics fields has not only made the BEM a numerical tool widely used in those fields, but leaves room for the development of new numerical procedures. The novel method presented by John P. Wolf in this book is another step in the direction of producing numerical methods which overcome those difficulties of the FEM. The Scaled Boundary Finite Element Method (SBFEM) is very well suited to, and constitutes the most adequate numerical alternative for, the solution of certain types of continuum mechanics problems. It contains ingredients of both the FEM and BEM, and avoids some of their drawbacks. Obviously it has its own restrictions, which makes it more focused towards a limited yet large and important number of continuum mechanics problems. Researchers and engineers interested in the solution of problems such as soil-structure interaction, diffusion, fracture mechanics and other fields, should take advantage of the method and be ready to exploit its possibilities. The method can be applied to any continuum mechanics problem governed by linear partial differential equations. A scaling procedure in the radial direction is used in this semi-analytical approach to restrict the discretisation (in a finite element sense) to the circumferential directions.

There are not many books that present new methods for the solution of engineering problems. It has been possible to do so in this case, thanks to the long lasting hard work and extraordinary creativity of its author. Trying to obtain better approaches for dynamic soil-structure interaction (as shown in his book *Finite Element Modelling of Unbounded Media*, Wiley, 1996), the author has developed a method that has been extended to bounded media and to a large variety of engineering problems. Further progress in this promising novel computational procedure can be expected to occur in the future.

Jose Dominguez
University of Seville, Spain

Preface

The most widely used computational procedures in solid mechanics and many other fields of engineering and physics are the Finite Element Method and the Boundary Element Method. The unknown continuous solution is replaced, for instance in statics, by algebraic equations in terms of parameters defining the approximate solution. Both methods exhibit their specific features, advantages and disadvantages.

As this book will appeal to the numerical analyst with experience in a specific field, a more detailed technical description of the novel method, including the motivation to develop it, is appropriate than is normally found in a Preface. The beginner in numerical analysis for whom the book is equally well suited can essentially skip the following six paragraphs. It is important to realise that the novel approach, called the Scaled Boundary Finite Element Method, combines the advantages of the finite element and boundary element methods, which will both be explained, and also exhibits appealing features of its own.

In the *Finite Element Method* the domain is spatially discretised into non-overlapping elements. In each such finite element, shape functions in the form of polynomials interpolate, for instance, the displacements. Standard numerical integration of these regular functions leads to a simple approximation for the behaviour of each finite element, for instance, consisting of the (symmetric) static-stiffness and mass matrices. These local representations are then patched together enforcing compatibility and equilibrium. This leads to the global model. A great flexibility exists in representing the geometry and the material. For an unbounded domain, the finite element method cannot satisfy the boundary conditions at infinity exactly. The spatial discretisation is terminated on an artificial boundary where the truncated domain outside the boundary up to infinity can only be represented approximately.

In the *Boundary Element Method*, only the boundary is discretised spatially into elements, leading to a reduction of the spatial dimension by one. This diminishes the effort of data preparation and leads to fewer unknowns. However, a so-called 'fundamental solution' satisfying the governing differential equations in the domain must be available. This analytical solution is often very complicated, exhibiting singularities. In each boundary element, shape functions in the form of polynomials interpolate, for instance, the displacements and surface tractions. Special numerical integration of these polynomials with the fundamental solution involving singularities yields (non-symmetric) coefficient matrices. The boundary element method is well suited to model an unbounded domain, as the fundamental solution satisfies the boundary conditions at infinity exactly.

The finite element and boundary element methods converge, in general, to the exact solution for decreasing element size. The convergence is slow close to the point of stress singularities as occurring around the crack tip in fracture mechanics, because both methods use polynomials to interpolate the displacements. Special techniques and a large number of elements are necessary to achieve acceptable accuracy.

For most practical cases, the shape of the boundary, the variations of the material properties and of the prescribed loads in the domain and the boundary conditions preclude an analytical solution of the governing partial differential equations. However, if the physical problem is governed by ordinary differential equations, classical mathematical techniques can in important cases lead to an exact analytical solution in the single independent variable. As an example, in certain cases when symmetry exists leading to a one-dimensional problem, the governing ordinary differential equations in, for instance, the radial coordinate, can thus be solved exactly while the problem in three dimensions cannot be addressed analytically. To make use of this advantage also for the general case without any symmetry, a coordinate system consisting of the radial direction and two local circumferential directions (parallel to the boundary) is introduced. The governing partial differential equations are transformed to this coordinate system. In the circumferential directions the boundary is discretised with surface finite elements, reducing the governing partial differential equations to ordinary differential equations in the radial coordinate. The coefficients of the ordinary differential equations are determined by the finite-element approximation in the circumferential directions. These ordinary differential equations are then solved analytically in the radial direction. Thus a novel computational procedure called the *Scaled Boundary Finite Element Method* is developed which *combines the advantages of the analytical and numerical approaches*. The method is thus a *semi-analytical* procedure for solving (linear) partial differential equations. In the circumferential directions (parallel to the boundary), where the behaviour is, in general, smooth, the weighted-residual approximation of finite elements applies, leading to convergence in the finite-element sense. In the radial direction, the solution is exact. For a bounded domain the radial coordinate points from the boundary towards the interior, where in the presence of singularities the analytical solution permits the stress-intensity factors to be calculated directly based on their definition. For an unbounded domain the radial coordinate points from the boundary towards infinity, where the boundary conditions at infinity (radiation condition) can be incorporated exactly in the analytical solution.

The brief characterisation of the finite element and boundary element methods described above demonstrates that both methods exhibit disadvantages. Addressing these disadvantages leads to the scaled boundary finite element method, much as an oyster responds to a grain of sand. Or in other words, the scaled boundary finite element method *combines the advantages of the finite element and boundary element methods*. Most attractive features of both methods are kept: of the finite element method, that no fundamental solution is required – thus expanding the scope of application, for instance to anisotropic material without an increase in complexity – that singular integrals are avoided, and that symmetry of the results is automatically satisfied; of the boundary element method, that the spatial dimension is reduced by one as only the boundary is discretised with surface elements, reducing the data preparation and computational efforts, that the boundary conditions at infinity are satisfied exactly, and that no approximation other than that of the surface elements on the boundary is introduced. In addition, the scaled boundary finite element method presents *appealing features of its own*: an analytical solution inside the domain is achieved, permitting, for instance, accurate stress-intensity factors to be determined directly and, as will become apparent, no spatial discretisation of certain free and fixed boundaries and interfaces between different materials is required.

In a nutshell, the scaled boundary finite element method is a *semi-analytical fundamental solution-less boundary element method based on finite elements*. Paraphrasing the title of the famous paper by Professor Zienkiewicz, the best of both worlds (marriage *à la mode*) is achieved in two ways: with respect to the analytical and numerical methods, and with respect to the finite element and boundary element methods within the numerical procedures.

The scaled boundary finite element method was developed for bounded and unbounded domains over the past few years addressing elastodynamics and diffusion as examples of linear governing differential equations. Both the frequency and time domain solutions, as well as statics, were examined. Two derivations were presented. The mechanically-based derivation developed originally applies finite element assemblage and similarity, and lead to the name *consistent infinitesimal finite element cell method* for the novel approach. The so-called scaled boundary transformation-based derivation, developed later in an effort to streamline the procedure, starts from the governing partial differential equations. The weighted-residual technique is then applied, which is mathematically more appealing and is consistent with today's finite element technology. This work was published in specialised journals in the various areas as the development proceeded. Early results were also described in part of a book.

This book on the scaled boundary finite element method serves two goals, which can be pursued independently. In Part II, the derivations and solution procedures of the fundamental equations – including applications in the fields mentioned above for the general case in sufficient detail – are addressed, which is the first goal. As it would not be efficient for a student or engineer interested in one field, for instance structural mechanics, to be confronted with such a broad and necessarily abstract description, a primer, the second goal of the book, is provided in Part I. What is called for is an elementary text without any prerequisites, but which still covers all aspects of the method. The links to the finite element and boundary element methods must be made obvious. It should not be aimed at a specific and narrow area of application, and at the same time permit a physical interpretation. To attract students with different backgrounds the primer must be self-contained. The primer should provide a student with a concise introduction to numerical analysis and also permit a reader with some experience in finite elements or boundary elements to appreciate the salient features of the scaled boundary finite element method with a minimum effort. After consulting the clearly indicated specialised references in the literature, the reader can then master the novel approach in his or her area of expertise. A cookbook or a handbook containing a set of rules which, when followed, lead to an approximate numerical solution, is not called for. The famous order of the Charge of the Light Brigade ('Theirs not to reason why, theirs but to do and die', Alfred Lord Tennyson, 1809–1892) should not be followed. What is important is a description of the basic concepts on which the numerical method is founded. These will permit the reader to gain a deeper understanding, which can be expanded and applied to solving difficult problems in a correct and efficient manner.

To achieve these goals, as already mentioned, a *primer* is written, expanding two conference papers. A *model problem* is defined which is easy to follow. The *simplest wave propagation problem* is discussed, which still *contains all the essential features*: the scalar wave equation in two dimensions, which plays a dominant role in many fields of engineering and physics. A wedge as a bounded medium and a truncated semi-infinite wedge as an unbounded medium are used to derive the governing equations and solution procedures step-by-step in great detail. The reader can follow all aspects of the development and application to the model problem without using a pencil! This is described in Chapters 3 to 14 of Part I, which is the core of the primer. To make the book self-contained, standard fundamentals of numerical analysis are discussed in Chapter 1. The salient features of the scaled boundary finite element method are addressed in Chapter 2. At the end of Part I, three short appendices lead to deeper insight into certain aspects of the model problem, and provide the link to the generalisation to two- and three-dimensional statics, elastodynamics and diffusion of Part II. In Chapters 15 to 22 of Part II, the fundamental equations and solution procedures are derived rigorously for the general case in great detail. For the sake of illustration, Chapters 23 and 24 describe examples for bounded and unbounded media, emphasising the versatility and salient advantageous features. Based on a stress recovery technique error estimation and adaptivity are discussed in Chapter 25. Chapter 26 contains concluding remarks and addresses restrictive properties of the novel method (where there is light there is also shadow!), and suggestions for future research. Part II thus develops all aspects of the current state of the art of the scaled boundary finite element method, starting from scratch.

It is expected that most readers will study the derivation and solution procedure for the model problem in Part I in depth, gaining an overview, and will then turn to selected aspects in their area of interest in Part II. Others will examine Part I only and be content with this introduction covering all aspects of the method. The experienced numerical analyst can address directly Part II, consulting the model problem in Part I only when a simple illustration of a specific aspect is called for. To facilitate the switching between Parts I and II, the same chapter headings are used whenever possible.

The book is based on research and development performed from 1994–1998 with my colleague Dr. Chongmin Song. His significant creative contributions to key features and his leadership are gratefully acknowledged. He is also the co-author, in many cases the first author, of all papers written on the novel method during this period, and co-author of the book (*Finite-Element Modelling of Unbounded Media*, Wiley, 1996), which describes early results of the consistent infinitesimal finite element cell method, the name originally coined for the scaled boundary finite element method. Dr. Song would have been the ideal co-author of this book which would have lead to a more attractive publication of higher quality. Unfortunately, due to the cancellation of research funds, he decided to join a structural consulting firm to design with all his energy futuristic roofs of great complexity.

The cooperation from 2000–2002 with Dr. Andrew J. Deeks of the University of Western Australia led to the development of error estimators and adaptive procedures, which permit the computational efficiency of the scaled boundary finite element method to be proven. His other in-depth contributions to many aspects of the procedure are gratefully acknowledged.

As already mentioned, Part I with Chapters 1 and 2 can serve as a textbook of a first course in numerical analysis on the advanced undergraduate and first-year graduate levels. No prerequisites other than those taught at any university in engineering and science are required. In addition, the book permits an experienced numerical analyst in many fields of engineering such as engineering mechanics, acoustics, earthquake engineering, electromagnetism and computational mathematics to master the scaled boundary finite element method in a specific field efficiently making full use of previously acquired knowledge. Finally, the novel procedure will be a gratifying topic of research with many unexplored aspects, in particular in modelling problems with singularities and in analysing dynamic unbounded medium-structure interaction.

John P. Wolf
Swiss Federal Institute of Technology

1 Fundamentals of Numerical Analysis

This chapter provides an introduction to numerical analysis, in particular to the finite element and boundary element methods. Besides serving as an elementary description for the beginner, it should be helpful to the reader whose knowledge in computational procedures has become rusty over the years. Finally, the aim is to define and illustrate the nomenclature used throughout the book. The nomenclature applied to the simple examples in this introduction may seem unnecessarily complicated in certain instances. This is the price which has to be paid to ease the extension to more complicated cases in Parts I and II.

The reader familiar with the finite element method and to a lesser degree with the boundary element method in a specific field can skip Chapter 1 and move on to Chapter 2. Possibly, a glimpse at the scaled boundary finite element method presented in Section 1.6 to satisfy one's curiosity is appropriate.

Only the salient features of numerical analysis necessary to understand the concepts of the scaled boundary finite element method as developed for the so-called Model Problem in Part I are addressed. The exposure is restricted to the simplest forms of the finite element and boundary element methods, omitting important developments which partly eliminate certain disadvantages of the formulations.

In many places in this chapter, points of mathematical complexity are passed over lightly.

Many excellent elementary books describing the finite element method (e.g. [B2], [B3]) and the boundary element method (e.g. [B4], [P2]) exist. Reference [B4] contains an appealing discussion of the so-called weighted-residual formulation, which forms the basis for the finite element and boundary element methods. Section 1.3 is strongly influenced by this work.

1.1 PHYSICAL PROBLEM, MATHEMATICAL MODEL AND ITS NUMERICAL SOLUTION

To solve physical problems in analysis and design, numerical analysis is routinely used. The real *physical problem* present in nature consists of the actual structural configuration subjected to loads. As an example, a gravity dam embedded in soil excited by an earthquake is addressed (Figure 1.1). The dam with finite dimensions interacts dynamically with the soil which extends to infinity. The physical problem involves so-called

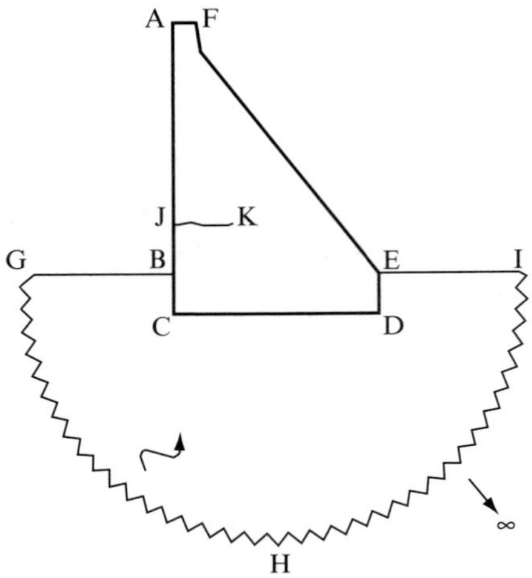

Figure 1.1 Dam embedded in soil

dynamic unbounded medium-structure interaction, with the structure consisting of the (finite) dam and the unbounded medium representing the (semi-infinite) soil (seismic soil-structure interaction). In addition, cracks in the dam are to be investigated which lead to *stress singularities* in fracture mechanics.

To be able to investigate the physical problem, certain idealisations leading to the *mathematical model* are introduced (see Chapter 1 of Reference [B2] for a thorough discussion). The choice of the mathematical model will depend on the response to be analysed, i.e. on the questions asked of nature. Far-reaching assumptions concerning the dimensions of the problem (one-, two- or three-dimensional), the kinematic linearity or non-linearity, the material behaviour, the geometry, the loading, the boundary conditions, etc. must be made, resulting in the governing differential equations. This step is decisive, as physical properties that are not present in the mathematical model cannot be predicted. For instance, for the structural configuration of Figure 1.1, a section leading to a two-dimensional linear model is selected. Three-dimensional effects will thus not be captured. The line B-C-D-E represents the interface between the two substructures, the dam and the soil. A crack runs from the edge J to the tip K. The mathematical model must also permit the important phenomena to be predicted effectively and reliably.

In general, the mathematical model cannot be solved exactly analytically in an efficient way. A *numerical solution* involving so-called discretisation is required which is only approximate. The corresponding accuracy must be addressed, and if not sufficient, the numerical procedure has to be repeated increasing the number of discretisation parameters. Again, the numerical solution should be performed effectively and reliably.

It is obvious that the numerical procedure only solves the selected mathematical model. The insight gained from the response analysed numerically will thus only cover those aspects of the real physical problem in nature, which are contained in the mathematical model. The selection of the mathematical model is the crucial step in engineering analysis!

1.2 LAPLACE EQUATION FOR OUT-OF-PLANE DISPLACEMENT OF SHEAR PLATE

To be specific, a very simple illustrative example is defined, which is used for demonstration throughout this chapter. A detailed numerical solution using the finite element and boundary element methods is developed. The example is the same as the model problem addressing a bounded medium, the wedge, and for which a novel computational procedure, called the scaled boundary finite element method, is discussed in Part I. The mathematical structure and approach in formulating its numerical solution are essentially the same as in more complex problems of greater significance.

The out-of-plane (anti-plane) motion of linear elasticity is addressed. Three-dimensional elasticity is decomposed with respect to a two-dimensional plane into in-plane motion (with two displacements, two normal stresses and a shear stress) and out-of-plane motion. The latter (Figure 1.2) caused by a load p acting in the direction of the z-axis perpendicular to the plane (x, y) will consist of a displacement u parallel to the z-axis and two shear stresses τ_x, τ_y corresponding to forces parallel to the z-axis. All variables are constant with respect to the z-coordinate. This displacement and stress state correspond to a shear plate under transverse loading. The coordinates of a point in the plane are denoted as x, y in the domain V. The shear stresses are shown acting on an infinitesimal element in Figure 1.2. Moving by $\mathrm{d}x$ from the negative face with τ_x to the corresponding positive face, the stress will change to $\tau_x(x, y) + \partial\tau_x(x, y)/\partial x \mathrm{d}x = \tau_x + \tau_{x,x}\mathrm{d}x$. Formulating equilibrium for statics of the forces parallel to the z-axis acting on the infinitesimal element with thickness $\mathrm{d}z$

$$-\tau_x \mathrm{d}y\mathrm{d}z + (\tau_x + \tau_{x,x}\mathrm{d}x)\mathrm{d}y\mathrm{d}z - \tau_y \mathrm{d}x\mathrm{d}z + (\tau_y + \tau_{y,y}\mathrm{d}y)\mathrm{d}x\mathrm{d}z + p\mathrm{d}x\mathrm{d}y\mathrm{d}z = 0 \tag{1.1}$$

yields

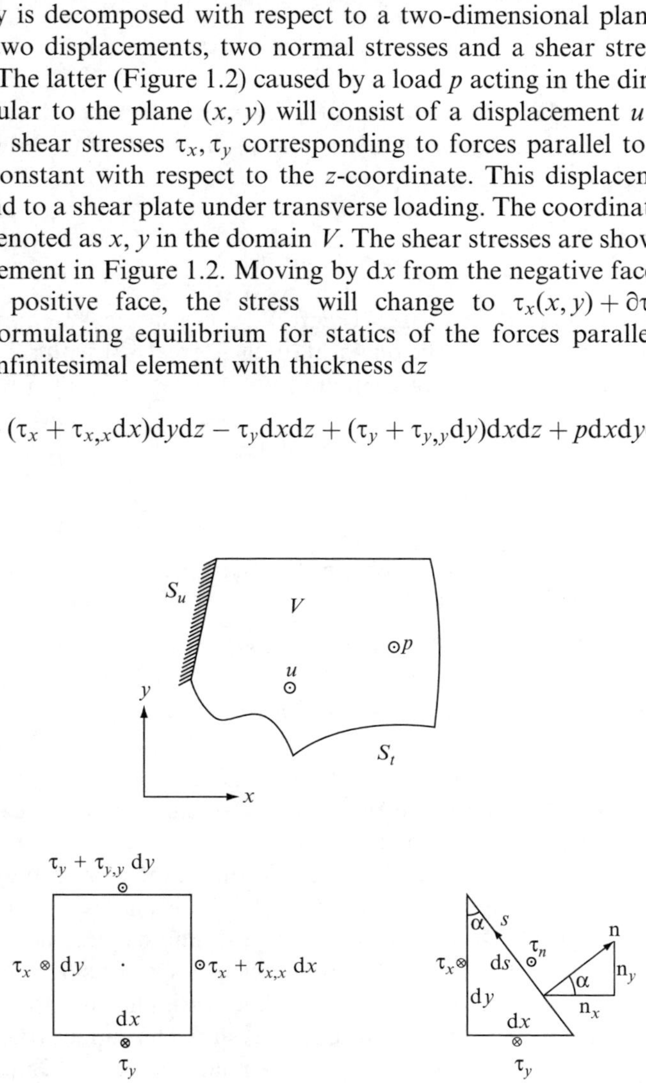

Figure 1.2 Two-dimensional out-of plane motion

$$\tau_{x,x} + \tau_{y,y} + p = 0 \tag{1.2}$$

with the body load p per unit volume (dimension force per length^3). The constitutive law expressed in derivatives of the displacement for a homogeneous isotropic material equals

$$\{\tau\} = \begin{Bmatrix} \tau_x \\ \tau_y \end{Bmatrix} = G\begin{Bmatrix} u,_x \\ u,_y \end{Bmatrix} \tag{1.3}$$

with the shear modulus G. Substituting Equation (1.3) into Equation (1.2) leads to the governing differential equation in displacement for statics in the domain V

$$G(u,_{xx} + u,_{yy}) + p = 0 \tag{1.4}$$

On the boundary S, either the displacement u (essential boundary condition on S_u) or the surface traction τ_n (natural boundary condition on S_t) are enforced

$$u = \bar{u} \text{ on } S_u \tag{1.5a}$$

$$\tau_n = \bar{\tau}_n \text{ on } S_t \tag{1.5b}$$

with a bar indicating a prescribed value (Figure 1.2). The surface traction represents the shear stress acting on a face with the outward (unit) normal $\{n\}$ with components n_x, n_y (subscript n for normal). Formulating equilibrium of the infinitesimal wedge

$$-\tau_x \mathrm{d}y\mathrm{d}z - \tau_y \mathrm{d}x\mathrm{d}z + \tau_n \mathrm{d}s\mathrm{d}z = 0 \tag{1.6}$$

results in

$$\tau_n = \frac{\mathrm{d}y}{\mathrm{d}s}\tau_x + \frac{\mathrm{d}x}{\mathrm{d}s}\tau_y = n_x\tau_x + n_y\tau_y = \{n\}^T\{\tau\} \tag{1.7}$$

Substituting Equation (1.3) into Equation (1.7) yields

$$\tau_n = G(n_x u,_x + n_y u,_y) = Gu,_n \tag{1.8}$$

The boundary value problem consists of solving the governing partial differential equation of Equation (1.4) subject to the boundary conditions of Equation (1.5) with the surface traction expressed in Equation (1.8).

The illustrative example consists of a wedge in the shape of a triangle (Figure 1.3). The three vertices are the origin 0 and points 1 with coordinates $x_1 = r_0, y_1 = -r_0/\sqrt{5}$ and 2 with $x_2 = r_0, y_2 = 0$. On the sides 0-1 and 0-2, the displacement $\bar{u} = 0$ and the surface traction $\bar{\tau}_n = \bar{\tau}_y = 0$, respectively, are prescribed. On the side 1-2, the prescribed displacement perpendicular to the plane (dashed line) varies linearly from 0 at point 1 to $\bar{u}$ at point 2. A uniform distributed load p acts. As results, the surface traction $\tau_n = \tau_x$ on 1-2 and the displacement u at point P with coordinates $0.5r_0, -0.25r_0/\sqrt{5}$ are calculated. It is the goal to demonstrate all steps of the finite element and boundary element methods in great detail, and not to determine accurate results.

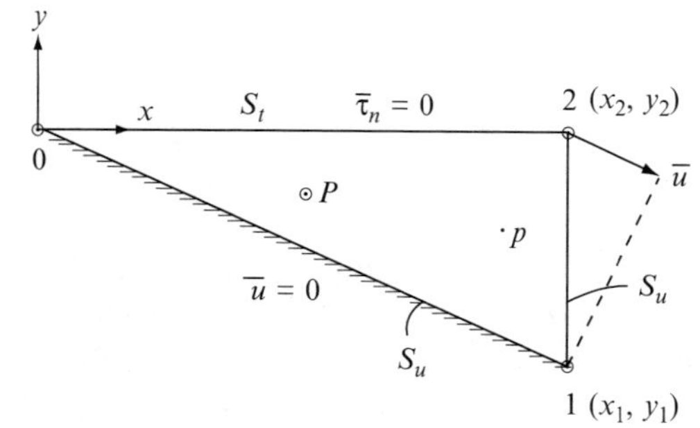

Figure 1.3 Wedge as an illustrative example

For $p \neq 0$ the differential equation is sometimes called Poisson's equation. For conciseness, the expression Lagrange equation will be used also for this case.

The Laplace equation (Equation (1.4)) involving a function (in structural mechanics displacement) together with boundary conditions (Equations (1.5) with Equation (1.8)) specified in the function and the normal flux (in structural mechanics surface traction, which can be expressed in stresses) represents many problems in engineering and physics. In a constitutive law (corresponding to Equation (1.3)), the flux is related to the spatial derivative of the function involving a material modulus. The function and the flux will of course have different meanings depending on the problem. Heat conduction, ideal fluid flow, electrostatics and magnetostatics can be mentioned as examples outside structural mechanics. A conservation law (corresponding to the equilibrium condition in Equation (1.2)) enforces that the net flux of an infinitesimal element is zero. The flux is present at the faces of the infinitesimal element, and can be generated within by a source (corresponding to the load).

The reader who does not master structural mechanics is urged to re-formulate the equations of the illustrative example in terms of his or her speciality. This will permit all results to be discussed in this chapter and in Part I to be 'translated' from structural mechanics to the other field.

The extension from statics to dynamics, where all variables become a function of time t, proceeds as follows. The only modification consists of including the inertial load when formulating equilibrium of the infinitesimal element. The term mass times acceleration, acting in the negative direction, $-\rho dx\, dy\, \ddot{u}(t)$ is added in Equation (1.2) yielding

$$\tau_{x,x}(t) + \tau_{y,y}(t) + p(t) - \rho\ddot{u}(t) = 0 \tag{1.9}$$

with the mass density ρ. Substituting the same constitutive law (Equation (1.3)) leads to the differential equation of motion

$$G(u_{,xx}(t) + u_{,yy}(t)) - \rho\ddot{u}(t) + p(t) = 0 \tag{1.10}$$

which can also be written as the (two-dimensional) scalar wave equation

$$u_{,xx}(t) + u_{,yy}(t) - \frac{1}{c_s^2}\ddot{u}(t) + \frac{p(t)}{G} = 0 \tag{1.11}$$

with the shear-wave velocity

$$c_s = \sqrt{\frac{G}{\rho}} \tag{1.12}$$

The problem consists in solving the governing partial differential equation of Equation (1.10) subject to the same (in general time-dependent) boundary condition of Equations (1.5) with Equation (1.8), and enforcing the initial conditions throughout the domain

$$u(t = 0) = \bar{u}_0 \tag{1.13a}$$

$$\dot{u}(t = 0) = \bar{\dot{u}}_0 \tag{1.13b}$$

1.3 METHOD OF WEIGHTED RESIDUALS

The shape of the boundary S, the boundary conditions on S_u and S_t and the load p (Figure 1.2) preclude in most practical problems the use of classical mathematical techniques to solve the boundary value problem (Equations (1.4) and 1.5), and thus the possibility to calculate an exact analytical solution. This even applies to the simple illustrative example of a wedge (Figure 1.3). The continuous exact solution for the function (displacement) u corresponds to an infinite number of unknowns (degrees of freedom), as the governing differential equation has to be satisfied in all points of the domain and the boundary conditions in all points on the boundary. To construct an approximate numerical solution, the infinite number of degrees of freedom of the continuous solution is reduced to a finite number n of the discrete solution. The approximate solution for u is formulated as

$$u(x, y) = \alpha_1\varphi_1(x, y) + \alpha_2\varphi_2(x, y) + \ldots = \sum_{i=1}^{n} \alpha_i\varphi_i(x, y) \tag{1.14}$$

where α_i are unknown coefficients and φ_i are a set of known linearly-independent functions. Often in finite element and boundary element methods, the α_i correspond to nodal values of the function u or its spatial derivative (normal flux). In this case, the φ_i are assumed interpolation functions, so-called *shape functions*. In many cases, polynomials in a local coordinate system are used (see also Appendix A).

As Equation (1.14) does, in general, not represent the exact solution, substituting in the governing differential equation (Equation (1.4)) leads to an error or *residual function* $R = R(x, y)$

$$R = G(u,_{xx} + u,_{yy}) + p = G\left(\sum_{i=1}^{n} \alpha_i(\varphi_{i,xx} + \varphi_{i,yy})\right) + p \neq 0 \text{ in } V \tag{1.15}$$

Analogously, the boundary conditions (Equations (1.5) with Equation (1.8)) are not identically satisfied, yielding residual functions on the boundary

$$R_1 = u - \bar{u} = \sum_{i=1}^{n} \alpha_i \varphi_i - \bar{u} \neq 0 \text{ on } S_u \tag{1.16a}$$

$$R_2 = \tau_n - \bar{\tau}_n = G \sum_{i=1}^{n} \alpha_i \varphi_{i,n} - \bar{\tau}_n \neq 0 \text{ on } S_t \tag{1.16b}$$

These residual functions will be made as small as possible over the domain and on the boundary by the various numerical techniques. To do so, the residuals are *distributed* in a certain manner, which produces different types of approximate methods. The residuals multiplied by a certain function, called the *weighting function*, and integrated over the domain and/or the boundary will be forced to be zero. This leads to the name *weighted-residual method.*

When the assumed approximate solution (Equation (1.14)) satisfies the governing differential equation (Equation (1.4)), leading to $R = 0$ (Equation (1.15)), but not the boundary conditions (Equation (1.5)), yielding $R_1 \neq 0, R_2 \neq 0$ (Equation (1.16)), only integrals of residuals multiplied by the weighting function over the boundary will occur. A *boundary method* thus results. When the approximate solution satisfies the boundary conditions leading to $R_1 = R_2 = 0$, but not the governing differential equation yielding $R \neq 0$, only integrals of residuals multiplied by the weighting function over the domain will occur. This leads to a *domain method.*

The weighted-residual formulation for a domain method is addressed. The functions φ_i will thus satisfy exactly the boundary conditions (Equation (1.5)). The weighting function w is defined as

$$w(x, y) = \beta_1 \psi_1(x, y) + \beta_2 \psi_2(x, y) + \ldots = \sum_{j=1}^{n} \beta_j \psi_j(x, y) \tag{1.17}$$

where β_j are coefficients and ψ_j are a set of known linearly-independent functions. The residual R of the governing differential equation (Equation (1.15)) multiplied by w is integrated over the domain V and set equal to zero

$$\int_V R w \mathrm{d}V = 0 \tag{1.18}$$

In this two-dimensional case, $\mathrm{d}V = \mathrm{d}x\mathrm{d}y$ holds. As the β_j are arbitrary, Equation (1.18) corresponds to a system of equations

$$\int R \psi_j \mathrm{d}V = 0 \quad j = 1, 2, \ldots, n \tag{1.19}$$

Substituting Equation (1.15) into Equation (1.19) leads to

$$\int \left(G \sum_{i=1}^{n} \alpha_i (\varphi_{i,xx} + \varphi_{i,yy}) + p \right) \psi_j \mathrm{d}V = 0 \qquad j = 1, 2, \ldots, n \tag{1.20}$$

This represents a system of n equations to determine the unknown n coefficients α_i. Thus, a numerical solution of a differential equation in an unknown continuous function (Equation (1.4)) leads to a system of algebraic equations (Equation (1.20)) in terms of the coefficients defining the approximate solution (Equation (1.14)).

The weighted-residual formulation provides a common basis for all fundamental numerical techniques, in particular the finite element and boundary element methods. It also permits the required continuity of the functions used to construct the approximate solution and the weighting function to be identified. Starting from the weighted-residual formulation of a domain method, it is possible to derive the weighted-residual equation for the more general case, where the boundary conditions are not satisfied exactly by the approximate solution. The approximations involved can be clearly identified.

This derivation proceeds as follows. The weighted-residual formulation of a domain method with domain V and boundary S (Figure 1.4) equals (Equation (1.18))

$$\iint (G(u_{,xx} + u_{,yy}) + p)w\mathrm{d}x\mathrm{d}y = 0 \tag{1.21}$$

The derivative terms are integrated by parts as follows. The first term on the right-hand side of the identity

$$\begin{aligned}\iint (u_{,xx} + u_{,yy})w\mathrm{d}x\mathrm{d}y = &\iint (u_{,x}w)_{,x}\mathrm{d}x\mathrm{d}y - \iint u_{,x}w_{,x}\mathrm{d}x\mathrm{d}y \\ &+ \iint (u_{,y}w)_{,y}\mathrm{d}x\mathrm{d}y - \iint u_{,y}w_{,y}\mathrm{d}x\mathrm{d}y\end{aligned} \tag{1.22}$$

is integrated by parts in the x-direction

$$\int \left(\int (u_{,x}w)_{,x}\mathrm{d}x \right) \mathrm{d}y = \int ((u_{,x}w)^{+} - (u_{,x}w)^{-})\mathrm{d}y \tag{1.23}$$

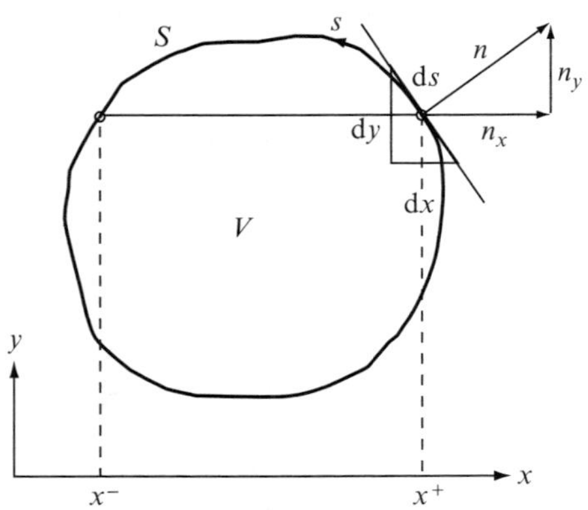

Figure 1.4 Integration anti-clockwise on the boundary with the domain on the left with outward normal

with the + and − signs referring to the coordinates x^+ and x^- on the boundary. Analogously, the third term is formulated as

$$\int\left(\int (u_{,y}w)_{,y}\mathrm{d}y\right)\mathrm{d}x = \int((u_{,y}w)^+ - (u_{,y}w)^-)\mathrm{d}x \tag{1.24}$$

Integrating anti-clockwise in the s-direction on the boundary S with the domain V on the left, with the outward (unit) normal $\{n\}$ with components n_x, n_y and substituting $\mathrm{d}x = n_y\mathrm{d}s, \mathrm{d}y = n_x\mathrm{d}s$ (Figure 1.4) in Equations (1.23) and (1.24) yields

$$\iint (u_{,x}w)_{,x}\mathrm{d}x\mathrm{d}y + \iint (u_{,y}w)_{,y}\mathrm{d}x\mathrm{d}y = \int (u_{,x}n_x + u_{,y}n_y)\, w\mathrm{d}s = \int u_{,n}w\mathrm{d}s \tag{1.25}$$

Substituting Equation (1.25) into Equation (1.22) leads to the *integration-by-parts formula*

$$\iint (u_{,xx} + u_{,yy})w\mathrm{d}x\mathrm{d}y = \int u_{,n}w\mathrm{d}s - \iint (u_{,x}w_{,x} + u_{,y}w_{,y})\mathrm{d}x\mathrm{d}y \tag{1.26}$$

Subjecting the second term on the right-hand side to integration by parts yields with

$$\iint u_{,x}w_{,x}\mathrm{d}x\mathrm{d}y = \iint (uw_{,x})_{,x}\mathrm{d}x\mathrm{d}y - \iint uw_{,xx}\mathrm{d}x\mathrm{d}y \tag{1.27}$$

$$\int\left(\int (uw_{,x})_{,x}\mathrm{d}x\right)\mathrm{d}y = \int((uw_{,x})^+ - (uw_{,x})^-)\mathrm{d}y = \int uw_{,x}n_x\mathrm{d}s \tag{1.28}$$

and analogously with

$$\iint u_{,y}w_{,y}\mathrm{d}x\mathrm{d}y = \int uw_{,y}n_y\mathrm{d}s - \iint uw_{,yy}\mathrm{d}x\mathrm{d}y \tag{1.29}$$

$$\iint (u_{,x}w_{,x} + u_{,y}w_{,y})\mathrm{d}x\mathrm{d}y = \int uw_{,n}\mathrm{d}s - \iint u(w_{,xx} + w_{,yy})\mathrm{d}x\mathrm{d}y \tag{1.30}$$

Substituting Equation (1.30) into Equation (1.26) results in *Green's theorem*

$$\iint (u_{,xx} + u_{,yy})w\mathrm{d}x\mathrm{d}y = \int (w_{,xx} + w_{,yy})u\mathrm{d}x\mathrm{d}y + \int u_{,n}w\mathrm{d}s - \int uw_{,n}\mathrm{d}s \tag{1.31}$$

The nomenclature is slightly modified. Integrations over the domain V occur and integrations over the boundary S decomposed into S_u with prescribed $u = \bar{u}$ and S_t with prescribed $\tau_n = Gu,_n = \bar{\tau}_n$ (Equation (1.5) with Equation (1.8)). Substituting Equation (1.31) into Equation (1.21) yields

$$G\int_V (w,_{xx} + w,_{yy})u\mathrm{d}V + \int_V pw\mathrm{d}V = G\int_S uw,_n\mathrm{d}S - \int_S Gu,_n w\mathrm{d}S \tag{1.32}$$

The prescribed boundary conditions are introduced, and using Equation (1.8), results in

$$G\int_V (w,_{xx} + w,_{yy})u\mathrm{d}V + \int_V pw\mathrm{d}V = G\int_{S_u} \bar{u}w,_n\mathrm{d}S + G\int_{S_t} uw,_n\mathrm{d}S - \int_{S_u} \tau_n w\mathrm{d}S - \int_{S_t} \bar{\tau}_n w\mathrm{d}S \tag{1.33}$$

Note that, replacing u by $\bar{u}$ on S_u and τ_n by $\bar{\tau}_n$ on S_t, introduces a modification.

To retrieve the original weighted-residual formulation, integration by parts is performed backwards twice. Substituting the last term in Equation (1.30) into Equation (1.33) leads to

$$G\int_V (u,_x w,_x + u,_y w,_y)\mathrm{d}V - \int_V pw\mathrm{d}V = G\int_{S_u} (u - \bar{u})w,_n\mathrm{d}S + \int_{S_u} \tau_n w\mathrm{d}S + \int_{S_t} \bar{\tau}_n w\mathrm{d}S \tag{1.34}$$

Substituting the last term in Equation (1.26) into Equation (1.34) yields

$$G\int_V (u,_{xx} + u,_{yy})w\mathrm{d}V + \int_V pw\mathrm{d}V = -G\int_{S_u} (u - \bar{u})w,_n\mathrm{d}S + \int_{S_t} (\tau_n - \bar{\tau}_n)w\mathrm{d}S \tag{1.35}$$

Introducing the residual functions (Equations (1.15), (1.16)) results in

$$\int_V Rw\mathrm{d}V + G\int_{S_u} R_1 w,_n\mathrm{d}S - \int_{S_t} R_2 w\mathrm{d}S = 0 \tag{1.36}$$

Thus, introducing the prescribed boundary conditions into Equation (1.33) leads to a generalised weighted-residual formulation with residual functions in the differential equation in the domain, in the displacement boundary condition and the surface traction boundary condition on the boundary. Although in Equation (1.33), $u = \bar{u}$ on S_u and $\tau_n = \bar{\tau}_n$ on S_t are enforced, residual (error) functions on the boundary still exist ($R_1 \neq 0, R_2 \neq 0$ in Equation (1.36)). The requirement for the approximate solution u with respect to the boundary conditions is relaxed in Equation (1.35). u does not need to satisfy the boundary conditions in this weighted-residual formulation.

The weighted-residual formulation corresponding to Equation (1.35) can be constructed for any boundary value problem. In structural mechanics, Equation (1.34) can be interpreted as a virtual work statement selecting $u = \bar{u}$ on S_u. The weighting function represents the virtual displacement, which does not have to satisfy any boundary conditions. For a virtual displacement, the first term is equal to the internal work (elastic strain energy), the second term to the external work of the load and the right-hand side to the external work on the boundary (unknown and prescribed surface tractions multiplied by virtual displacement).

The three key weighted-residual formulations in Equations (1.35), (1.34) and (1.33) also supply some insight on the continuity requirements of the functions. In Equations (1.33) and (1.35), continuity up to the second derivative is required, i.e. at least up to quadratic terms for polynomials. In Equation (1.34), continuity up to the first derivative is sufficient, i.e. up to linear terms is acceptable. The continuity requirement is thus 'weakened'. The weighted-residual statement of Equation (1.34) which is an approximation, is called the *weak form*. (The governing differential equation, Equation (1.4) is called the strong form.)

In section 1.4 it will be shown that the weighted-residual formulation in Equation (1.34) is the starting point of the finite element method. Essentially, for the weighting function w, the same function as for u is selected, but vanishing on $S_u(w = 0)$. The approximate function u is also chosen to satisfy the displacement boundary condition on $S_u(u = \bar{u})$. This results in

$$G\int_V (u,_x w,_x + u,_y w,_y)\mathrm{d}V = \int_V pw\mathrm{d}V + \int_{S_t} \bar{\tau}_n w\mathrm{d}S \tag{1.37}$$

In section 1.5 it is demonstrated that the weighted-residual formulation in Equation (1.33) forms the basis of the boundary element method. The weighting function w is selected in such a way that the first integral results in a term on the boundary only, leading to the so-called boundary-integral equation. With the exception of the term addressing the load p, only integrals on the boundary thus occur.

The extension of the weighted-residual formulations from statics to dynamics is straightforward. The inertial term $-\rho\ddot{u}(t)$ present in the differential equation of motion (Equation (1.10)) is multiplied by the weighting function w and is added. For instance, Equation (1.37) is transformed into

$$G\int_V (u,_x w,_x + u,_y w,_y)\mathrm{d}V + \rho\int_V \ddot{u}w\mathrm{d}V = \int_V pw\mathrm{d}V + \int_{S_t} \bar{\tau}_n w\mathrm{d}S \tag{1.38}$$

1.4 FINITE ELEMENT METHOD

The essential features of the finite element method can be summarised as follows. The domain is subdivided into non-overlapping sub-domains called finite elements. For each finite element an approximate simple solution reflecting the physical behaviour is determined. These local representations are then patched together to form the global model of the total problem, which can be solved. These three activities are strongly interrelated. Thus, the finite element method is not just based on the concept 'Divide and Conquer' but can be described by the citation from Goethe (1749–1832): "Divide and command, a wise maxim; unite and guide, a better one".

The basic steps of the finite element method are explained addressing the illustrative example of Laplace's equation described in section 1.2. The function is equal to the out-of-plane displacement u and the flux corresponds to the shear stresses τ_x, τ_y. The boundary value problem consists in solving the governing partial differential equation of Equation (1.4) subject to the boundary conditions expressed in displacement in Equation (1.5a) and in surface traction (stress) in Equation (1.5b). The specific problem, a wedge in the form of a triangle, is presented in Figure 1.3.

For statics, the following six steps can be identified.

Step 1: Spatial Discretisation of Domain

The domain is subdivided into a number of non-overlapping sub-domains, each of simple geometry, which are the finite elements. For each finite element, nodes are assigned with global coordinates. The geometry of the finite element can be described by interpolating its nodal values by so-called *mapping functions*. This aspect called solid modelling is discussed in Appendix A, and is further addressed in Step 2.

The wedge 0-1-2 (Figure 1.5) is divided into four triangles with nodes at the vertices. The coordinates of the nodes are specified in section 1.2, with node 3 coinciding with point P in Figure 1.3.

Step 2: Displacement within Finite Element

The approximate solution of the displacement is defined *piecewise* in each finite element. The nodal displacements are selected as the unknowns. The displacement within the individual finite element is interpolated in terms of the nodal values by so-called *shape functions* in a local coordinate system. Polynomials are well suited as differentiation and integration are simple to perform. The degree of the polynomial follows from the number of nodes of the finite element and the continuity requirement of the displacement along its sides. In most cases the same interpolation functions are selected for the mapping functions describing the geometry and for the shape functions determining the displacement.

Over each triangle, a linear variation of the displacement is assumed as an approximation (Equation (1.14))

$$u(x,y) = \alpha_1 + \alpha_2 x + \alpha_3 y \tag{1.39}$$

with the three unknown coefficients $\alpha_1, \alpha_2, \alpha_3$. To determine the three unknowns, the displacements in three nodes are required, which leads to selecting the nodes at the vertices 1, 2, 3 shown in Figure 1.6. (This is a local numbering scheme applicable to each finite element.) Formulating

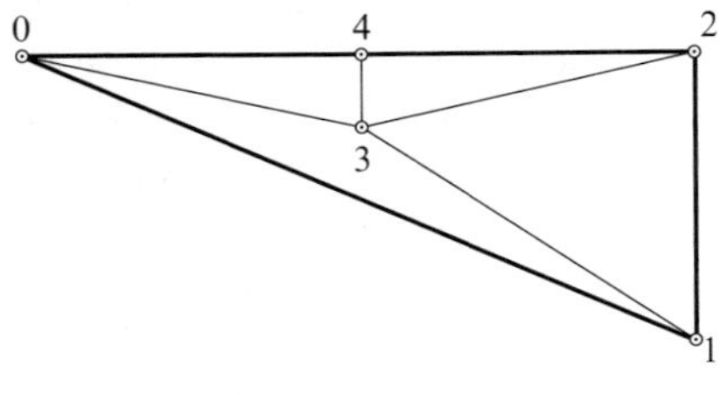

Figure 1.5 Division of wedge into four triangular finite elements with nodes at vertices

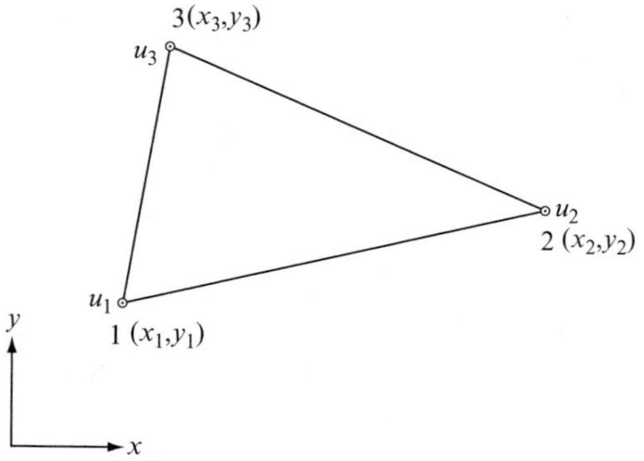

Figure 1.6 Triangular finite element with nodes at vertices

$$\begin{aligned} u_1 &= u(x_1, y_1) = \alpha_1 + \alpha_2 x_1 + \alpha_3 y_1 \\ u_2 &= u(x_2, y_2) = \alpha_1 + \alpha_2 x_2 + \alpha_3 y_2 \\ u_3 &= u(x_3, y_3) = \alpha_1 + \alpha_2 x_3 + \alpha_3 y_3 \end{aligned} \tag{1.40}$$

and solving yields the unknowns as a function of the displacements and coordinates of the nodes

$$\begin{aligned} \alpha_1 &= \frac{1}{2\Delta}[(x_2 y_3 - x_3 y_2)\,u_1 + (x_3 y_1 - x_1 y_3)\,u_2 + (x_1 y_2 - x_2 y_1)\,u_3] \\ \alpha_2 &= \frac{1}{2\Delta}[(y_2 - y_3)\,u_1 + (y_3 - y_1)\,u_2 + (y_1 - y_2)\,u_3] \\ \alpha_3 &= \frac{1}{2\Delta}[(x_3 - x_2)\,u_1 + (x_1 - x_3)\,u_2 + (x_2 - x_1)\,u_3] \end{aligned} \tag{1.41}$$

with the area Δ of the triangle

$$\Delta = \frac{1}{2}(x_2 y_3 + x_1 y_2 + x_3 y_1 - y_1 x_2 - y_2 x_3 - y_3 x_1) \tag{1.42}$$

Substituting Equation (1.41) into Equation (1.39) leads to

$$u(x, y) = N_1(x, y)\,u_1 + N_2(x, y)\,u_2 + N_3(x, y)\,u_3 \tag{1.43}$$

with the shape functions

$$\begin{aligned} N_1(x, y) &= \frac{1}{2\Delta}[x_2 y_3 - x_3 y_2 + (y_2 - y_3)x + (x_3 - x_2)y] \\ N_2(x, y) &= \frac{1}{2\Delta}[x_3 y_1 - x_1 y_3 + (y_3 - y_1)x + (x_1 - x_3)y] \\ N_3(x, y) &= \frac{1}{2\Delta}[x_1 y_2 - x_2 y_1 + (y_1 - y_2)x + (x_2 - x_1)y] \end{aligned} \tag{1.44}$$

Note that $N_i(x_j, y_j) = 1$ for $j = i$ and $= 0$ for $j \neq i$. The nodal interpolation in Equation (1.43) can also be formulated as

$$u(x, y) = [N(x, y)]\{u\} \tag{1.45}$$

with the nodal displacements

$$\{u\} = \begin{Bmatrix} u_1 \\ u_2 \\ u_3 \end{Bmatrix} \tag{1.46a}$$

and

$$[N(x, y)] = [N_1(x, y), N_2(x, y), N_3(x, y)] \tag{1.46b}$$

It is important to stress that the shape functions are linear. Along a specific side of a finite element, the linear function of the displacement is determined by the two nodal displacements. As the adjacent finite element with the same side shares the same two nodes with identical nodal values, the displacements along the common side calculated in the two finite elements will be identical, and thus the displacement will be continuous across the side. Addressing the global approximate solution of the wedge, the shape functions of adjacent finite elements are simply patched together, forming local pyramids. For instance, the global shape functions corresponding to node 2 in Figure 1.5 will merely decrease with the distances from this node to the opposite sides 4-3 and 3-1 of the two finite elements 2-4-3 and 2-3-1, respectively, which are present in this node 2. Disregarding the boundary conditions, the approximate displacement, piecewise linear over each finite element, is a function of the displacements in the nodes 0 to 4. As already mentioned, the approximate displacement is continuous across the sides of the finite elements. No displacement gaps occur throughout the wedge.

In Step 1, triangular finite elements are selected, where the coordinates along a side are linear functions of those of the nodes at the two vertices located at the ends of the side. This is the same interpolation as for the displacement. A natural correspondence thus exists. The same interpolation described in Equation (1.45) for the displacement will also apply for the coordinates of an individual finite element

$$\begin{aligned} x &= [N(x, y)]\{x\} \\ y &= [N(x, y)]\{y\} \end{aligned} \tag{1.47}$$

with the nodal coordinates (Figure 1.6)

$$\{x\} = \begin{Bmatrix} x_1 \\ x_2 \\ x_3 \end{Bmatrix} \qquad \{y\} = \begin{Bmatrix} y_1 \\ y_2 \\ y_3 \end{Bmatrix} \tag{1.48}$$

and the mapping functions $[N(x, y)]$ specified in Equations (1.46b) and (1.44).

In this simple example, the shape functions in Equation (1.44) are described in a global Cartesian coordinate system, and not in a local system associated with a finite element. As a preparation for the extension in the Model Problem of Part I, a local coordinate system ξ, η defined in a so-called parent element is introduced, which is then mapped to a general finite element. This procedure for the triangle is analogous to that for the quadrilateral described in Appendix A.

The general triangle is shown in the upper part of Figure 1.7 with the global coordinates of the nodes at the vertices. The parent triangle with a right angle and two sides of unit length (right-isosceles) and the local coordinates ξ, η scaled appropriately are presented in the lower part. A point ξ, η of the parent element is mapped to a point x, y of the general triangle with the parameters ξ, η. For instance, the side 1-2 of the parent element ($0 \leq \xi \leq 1, \eta = 0$) is mapped to the skew side 1-2 of the general triangle with the same values of ξ, η. By inspection, the mapping functions for the three nodes defined in the parent element equal

$$\begin{aligned} N_1(\xi, \eta) &= 1 - \xi - \eta \\ N_2(\xi, \eta) &= \xi \\ N_3(\xi, \eta) &= \eta \end{aligned} \tag{1.49}$$

Interpolating the global coordinates of the nodes $(x_i, y_i), i = 1, 2, 3$ in the local coordinates ξ, η leads to the map

$$\begin{aligned} x(\xi, \eta) &= [N(\xi, \eta)]\{x\} \\ y(\xi, \eta) &= [N(\xi, \eta)]\{y\} \end{aligned} \tag{1.50}$$

with the nodal coordinates defined in Equation (1.48), and with

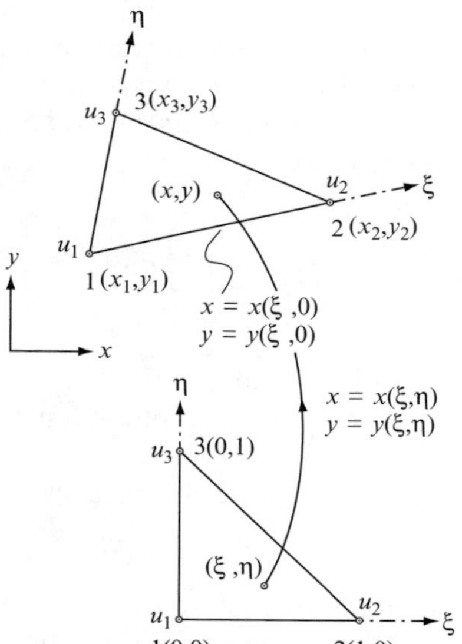

Figure 1.7 Generation of general triangular finite element by mapping right-isosceles triangular parent element

$$[N(\xi, \eta)] = [1 - \xi - \eta, \xi, \eta] \tag{1.51}$$

For this linear transformation the inverse map can easily be constructed. Substituting Equation (1.51) into Equation (1.50) and solving for ξ, η yields these variables as a function of x, y. Proceeding further, substituting these relations into Equation (1.50) yields the transformation in global coordinates of Equation (1.47).

Analogously, the displacement in a finite element can be expressed as

$$u(\xi, \eta) = [N(\xi, \eta)]\{u\} \tag{1.52}$$

with the shape functions specified in Equation (1.49).

Step 3: Properties of Finite Element (Static Stiffness and Nodal Load)

Based on a weighted-residual formulation, the properties of the individual finite element are determined. Although a finite element will have an infinite number of interconnection points on the sides with the adjacent elements, a discretised representation of these properties corresponding to the nodal values is required. In statics, the static-stiffness matrix relating the nodal displacements to the concentrated (integrated) forces in the same nodes is obtained. For a distributed load, acting on a finite element and for prescribed surface tractions, a nodal load vector is also calculated. With these two properties the nodal force-nodal displacement relationship of the individual finite element can be formulated. By selecting essentially the same shape functions for the weighting function as for the approximate displacement, a *symmetric* formulation follows. In particular, the static-stiffness matrix is symmetric.

As already mentioned at the end of section 1.3, the weak form of the weighted-residual statement is the starting point of the derivation of the finite element method. In Equation (1.37) the approximate displacement u satisfies the prescribed displacement $\bar{u}$ on $S_u(u = \bar{u})$ and the weighting function w vanishes on $S_u(w = 0)$. This virtual work statement is repeated for easy reference

$$G \int_V (u_{,x} w_{,x} + u_{,y} w_{,y}) \mathrm{d}V = \int_V p w \mathrm{d}V + \int_{S_t} \bar{\tau}_n w \mathrm{d}S \tag{1.53}$$

As the selected displacement u is a linear function, $u_{,x}$ and $u_{,y}$ are constants in each element. Across inter-element boundaries (sides of the finite elements), u is continuous, but these partial derivatives of u are step functions. As essentially w is the same function as u, the same properties also apply to the weighting function. Thus, the products of the partial derivatives of u and w present in Equation (1.53) exist, and are square integrable over the individual element.

The integrations of Equation (1.53) apply to the total domain. It can, however, be decomposed into the individual finite element's contributions, which are then summed. To simplify the nomenclature, the same symbols are used for quantities corresponding to an individual finite element and to the total domain consisting of the assemblage of all finite elements. In the following, an individual finite element is addressed.

The *static-stiffness matrix* of the triangle of Figure 1.6 with the shape functions specified in global coordinates (Equation (1.45)) is calculated. The left-hand side of Equation (1.53) for the triangle is rewritten as

$$G\int (w,_x^T u,_x + w,_y^T u,_y)\,\mathrm{d}x\mathrm{d}y \tag{1.54}$$

Based on Equation (1.45),

$$\begin{aligned} u,_x &= [N,_x]\{u\} \\ u,_y &= [N,_y]\{u\} \end{aligned} \tag{1.55}$$

follows with (Equation (1.44))

$$\begin{aligned} [N,_x] &= \frac{1}{2\Delta}[y_2 - y_3, y_3 - y_1, y_1 - y_2] \\ [N,_y] &= \frac{1}{2\Delta}[x_3 - x_2, x_1 - x_3, x_2 - x_1] \end{aligned} \tag{1.56}$$

For the weighting function formulated as

$$w = [N(x, y)]\{w\} = [N]\{w\} \tag{1.57}$$

with the weighting function in the nodes

$$\{w\} = \begin{Bmatrix} w_1 \\ w_2 \\ w_3 \end{Bmatrix} \tag{1.58}$$

analogous expressions result. Substituted into Equation (1.54) yields

$$\{w\}^T G\int ([N,_x]^T[N,_x] + [N,_y]^T[N,_y])\mathrm{d}x\mathrm{d}y\{u\} \tag{1.59}$$

As $\{w\}^T$ is arbitrary, the remaining vector representing the nodal forces $\{P\}$ in Equation (1.59) can be addressed

$$\{P\} = G\int ([N,_x]^T[N,_x] + [N,_y]^T[N,_y])\mathrm{d}x\mathrm{d}y\{u\} \tag{1.60}$$

The nodal force-nodal displacement relationship equals for vanishing load

$$\{P\} = [K]\{u\} \tag{1.61}$$

Comparing with Equation (1.60), the static-stiffness matrix is calculated as

$$[K] = G\int([N,_x]^T[N,_x] + [N,_y]^T[N,_y])\mathrm{d}x\mathrm{d}y \tag{1.62}$$

Substituting Equation (1.56) yields

$$[K] = \frac{G}{4\Delta}([K_y] + [K_x]) \tag{1.63}$$

where

$$[K_y] = \begin{bmatrix} (y_2 - y_3)^2 & (y_2 - y_3)(y_3 - y_1) & (y_2 - y_3)(y_1 - y_2) \\ (y_3 - y_1)(y_2 - y_3) & (y_3 - y_1)^2 & (y_3 - y_1)(y_1 - y_2) \\ (y_1 - y_2)(y_2 - y_3) & (y_1 - y_2)(y_3 - y_1) & (y_1 - y_2)^2 \end{bmatrix} \tag{1.64}$$

and $[K_x]$ is constructed from $[K_y]$ by replacing y_i by $x_i (i = 1, 2, 3)$. Note that $[K]$ is indeed symmetric.

The static-stiffness matrix can also be calculated based on a parent element with the local coordinates ξ, η (Figure 1.7). Equation (1.54) still applies with the partial derivatives with respect to the global coordinates x,y expressed in Equation (A.7) by those with respect to ξ, η

$$\begin{Bmatrix} \dfrac{\partial}{\partial x} \\ \dfrac{\partial}{\partial y} \end{Bmatrix} = \frac{1}{|J|}\begin{bmatrix} y,_\eta & -y,_\xi \\ -x,_\eta & x,_\xi \end{bmatrix}\begin{Bmatrix} \dfrac{\partial}{\partial \xi} \\ \dfrac{\partial}{\partial \eta} \end{Bmatrix} \tag{1.65}$$

with the Jacobian (Equation (A.8))

$$|J| = x,_\xi y,_\eta - x,_\eta y,_\xi \tag{1.66}$$

Based on Equations (1.50) and (1.51), the partial derivatives equal

$$\begin{aligned} x,_\xi &= [N(\xi,\eta),_\xi]\{x\} = -x_1 + x_2 \\ x,_\eta &= [N(\xi,\eta),_\eta]\{x\} = -x_1 + x_3 \\ y,_\xi &= -y_1 + y_2 \\ y,_\eta &= -y_1 + y_3 \end{aligned} \tag{1.67}$$

It follows that $|J|$ in Equation (1.66) is equal to twice the area Δ of the triangle (Equation (1.42)).

The displacement u in a finite element expressed in ξ, η is specified in Equation (1.52). The partial derivatives are formulated as

$$\begin{Bmatrix} \dfrac{\partial u}{\partial \xi} \\ \dfrac{\partial u}{\partial \eta} \end{Bmatrix} = \begin{bmatrix} [N(\xi,\eta),_\xi] \\ [N(\xi,\eta),_\eta] \end{bmatrix}\{u\} = \begin{bmatrix} -1 & 1 & 0 \\ -1 & 0 & 1 \end{bmatrix}\{u\} \tag{1.68}$$

Formulating Equation (1.65) for the displacement u and substituting Equations (1.66) into (1.68) leads to the strain-nodal displacement matrix $[B]$, as $\partial u/\partial x = u,_x$ and $\partial u/\partial y = u,_y$ represent the strains.

$$\left\{ \begin{array}{c} u_{,x} \\ u_{,y} \end{array} \right\} = [B]\{u\} \tag{1.69}$$

with

$$[B] = \frac{1}{2\Delta} \begin{bmatrix} y_2 - y_3 & -y_1 + y_3 & y_1 - y_2 \\ -x_2 + x_3 & x_1 - x_3 & -x_1 + x_2 \end{bmatrix} \tag{1.70}$$

For the weighting function

$$w(\xi, \eta) = [N(\xi, \eta)]\{w\} \tag{1.71}$$

analogous expressions for the derivatives follow. The infinitesimal element dxdy formulated in the $\xi-, \eta-$ coordinates of the parent element equals (Equation (A.9))

$$\mathrm{d}x\mathrm{d}y = |J|\mathrm{d}\xi\, \mathrm{d}\eta \tag{1.72}$$

Substituting Equation (1.69) with (1.70) and Equations (1.71) and (1.72) in Equation (1.54) yields

$$\{w\}^T G \int (\left\{ \begin{array}{c} y_2 - y_3 \\ -y_1 + y_3 \\ y_1 - y_2 \end{array} \right\} [y_2 - y_3, -y_1 + y_3, y_1 - y_2]$$

$$+ \left\{ \begin{array}{c} x_2 - x_3 \\ -x_1 + x_3 \\ x_1 - x_2 \end{array} \right\} [x_2 - x_3, -x_1 + x_3, x_1 - x_2]) \frac{1}{2\Delta} \mathrm{d}\xi \mathrm{d}\eta \{u\} \tag{1.73}$$

The integral over the parent element equals ½ (Figure 1.7). Again, $\{w\}^T$ is arbitrary, and the remaining vector is equal to the product of the static-stiffness matrix $[K]$ and $\{u\}$. This leads to the same result for $[K]$ as specified in Equation (1.63).

The static-stiffness matrix can also be written as a function of the strain-nodal displacement matrix $[B]$ (Equation (1.70)) as

$$[K] = \int [B]^T G[B] |J| \mathrm{d}\xi \mathrm{d}\eta \tag{1.74}$$

Addressing the right-hand side of Equation (1.53) yields the *nodal load* vector. For vanishing prescribed surface tractions as in the illustrative example

$$\int w^T p \mathrm{d}x\mathrm{d}y \tag{1.75}$$

results. For p = constant and substituting Equations (1.71) and using Equation (1.72) leads to

$$\{w\}^T p \int [N(\xi, \eta)]^T |J| d\xi d\eta = \{w\}^T \{\bar{P}\} \tag{1.76}$$

with the nodal load

$$\{\bar{P}\} = p \int [N(\xi, \eta)]^T |J| d\xi d\eta \tag{1.77}$$

Substituting Equations (1.51) and (1.66) results in

$$\{\bar{P}\} = p\Delta \begin{Bmatrix} 1/3 \\ 1/3 \\ 1/3 \end{Bmatrix} \tag{1.78}$$

Step 4: Assemblage of Finite Elements and Enforcement of Boundary Conditions

The properties of the individual finite elements are assembled and the displacement boundary conditions are enforced, resulting in the global model. This simple assemblage is possible as at a node the displacement and the weighting function are the same for all finite elements sharing this node and the displacement and the weighting function are defined piecewise locally over each individual element, leading to simple addition of the contributions of these finite elements.

As already discussed in Step 2, the global functions (approximate displacement and weighting function) are generated by patching those of the individual finite elements together. This process permits the contribution of each finite element (consisting of the static-stiffness matrix and the nodal load for the illustrative example) to the global weighted-residual statement to be calculated independently of the others.

For the wedge, the assembly proceeds as follow (Figure 1.8). The wedge is partitioned into four triangles a, b, c, d, each with local nodes 1, 2, 3. The coefficients of the static-stiffness matrix K_{ij}^k, i, $j = 1, 2, 3$ and $k = a, b, c, d$ follow from Equation (1.63). A superscript is introduced to indicate the individual finite element. The nodes of the global model are numbered as 0, 1, 2, 3, 4 and the corresponding displacements (degrees of freedom) are denoted as $\{u\} = [u_0, u_1, u_2, u_3, u_4]^T$. The global nodal force-nodal displacement relationship equals

$$\{P\} = [K]\{u\} - \{\bar{P}\} \tag{1.79}$$

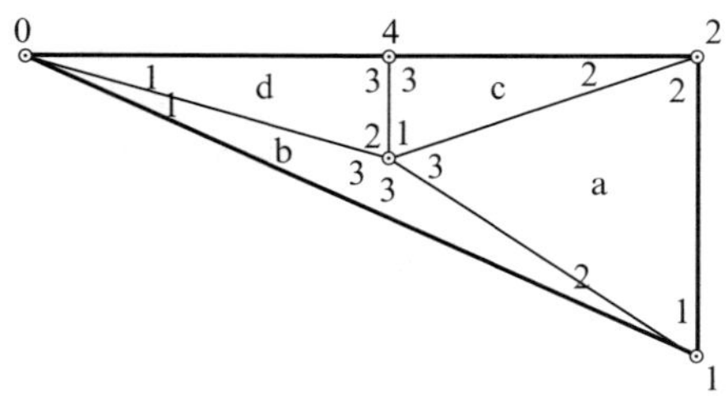

Figure 1.8 Local and global numbering of nodes for assemblage

The global static-stiffness matrix is assembled as

$$
\begin{aligned}
K_{00} &= K_{11}^{b} + K_{11}^{d} \\
K_{01} &= K_{12}^{b} \\
K_{02} &= 0 \\
K_{03} &= K_{13}^{b} + K_{12}^{d} \\
K_{04} &= K_{13}^{d} \\[4pt]
K_{10} &= K_{21}^{b} \\
K_{11} &= K_{11}^{a} + K_{22}^{b} \\
K_{12} &= K_{12}^{a} \\
K_{13} &= K_{13}^{a} + K_{23}^{b} \\
K_{14} &= 0 \\[4pt]
K_{20} &= 0 \\
K_{21} &= K_{21}^{a} \\
K_{22} &= K_{22}^{a} + K_{22}^{c} \\
K_{23} &= K_{23}^{a} + K_{21}^{c} \\
K_{24} &= K_{23}^{c} \\[4pt]
K_{30} &= K_{31}^{b} + K_{21}^{d} \\
K_{31} &= K_{31}^{a} + K_{32}^{b} \\
K_{32} &= K_{32}^{a} + K_{12}^{c} \\
K_{33} &= K_{33}^{a} + K_{33}^{b} + K_{11}^{c} + K_{22}^{d} \\
K_{34} &= K_{13}^{c} + K_{23}^{d} \\[4pt]
K_{40} &= K_{31}^{d} \\
K_{41} &= 0 \\
K_{42} &= K_{32}^{c} \\
K_{43} &= K_{31}^{c} + K_{32}^{d} \\
K_{44} &= K_{33}^{c} + K_{33}^{d}
\end{aligned}
\tag{1.80}
$$

The global nodal load vector is assembled as

$$
\begin{aligned}
\bar{P}_0 &= \bar{P}_1^{b} + \bar{P}_1^{d} \\
\bar{P}_1 &= \bar{P}_1^{a} + \bar{P}_2^{b} \\
\bar{P}_2 &= \bar{P}_2^{a} + \bar{P}_2^{c} \\
\bar{P}_3 &= \bar{P}_3^{a} + \bar{P}_3^{b} + \bar{P}_1^{c} + \bar{P}_2^{d} \\
\bar{P}_4 &= \bar{P}_3^{c} + \bar{P}_3^{d}
\end{aligned}
\tag{1.81}
$$

The contributions of the individual finite elements are added to their appropriate locations in the global static-stiffness matrix. For instance, element c with local nodes 1, 2, 3 leads to contributions at the global nodes 3, 2, 4.

This assemblage process corresponds to formulating *compatibility of the nodal displacements and equilibrium of the nodal forces.* For instance, at global node 4, the nodal forces of element d in local node 3 and of element c in local node 3 are added, and are in equilibrium with the exterior applied nodal load.

The following features apply. First, the assembled static-stiffness matrix $[K]$ is *symmetric*. Second, $[K]$ is *sparse*. The element K_{ij} vanishes, if the global nodes i, j are not directly connected by a finite element. This applies, for instance, for K_{02}. Of course, this is a consequence of selecting piecewise local shape functions for use in the approximate displacement and weighting function. Third, by selecting an appropriate global numbering scheme, $[K]$ will be *banded*, i.e. the zero elements will be placed as far away from the diagonal as possible. For instance, if the nodes 0, 1, 2, 3, 4 shown in Figure 1.8 are placed in the sequence 0, 1, 3, 4, 2 in the columns and rows of the matrix, the element in the upper right corner corresponding to K_{02} in the original numbering scheme will be zero. The nonzero elements in $[K]$ will form a band on both sides of the main diagonal. Efficient algorithms exist to solve a symmetric banded matrix.

As in the weighted-residual formulation of Equation (1.53) the approximate displacement u satisfies the prescribed displacement $\bar{u}$ on S_u, the displacement boundary condition must be enforced. The global nodal force-nodal displacement relationship (Equation (1.79)) is formulated as

$$\begin{bmatrix} K_{00} & K_{01} & K_{02} & K_{03} & K_{04} \\ K_{10} & K_{11} & K_{12} & K_{13} & K_{14} \\ K_{20} & K_{21} & K_{22} & K_{23} & K_{24} \\ K_{30} & K_{31} & K_{32} & K_{33} & K_{34} \\ K_{40} & K_{41} & K_{42} & K_{43} & K_{44} \end{bmatrix} \begin{Bmatrix} u_0 \\ u_1 \\ u_2 \\ u_3 \\ u_4 \end{Bmatrix} = \begin{Bmatrix} \bar{P}_0 \\ \bar{P}_1 \\ \bar{P}_2 \\ \bar{P}_3 \\ \bar{P}_4 \end{Bmatrix} + \begin{Bmatrix} R_0 \\ R_1 \\ R_2 \\ 0 \\ 0 \end{Bmatrix} \tag{1.82}$$

In all global nodes the known nodal loads are applied, and in nodes 0, 1, 2 unknown reaction forces R_0, R_1, R_2 appear. The displacement boundary conditions are equal to

$$\begin{aligned} u_0 &= 0 \\ u_1 &= 0 \\ u_2 &= \bar{u}_2 \end{aligned} \tag{1.83}$$

Substituting Equation (1.83) into Equation (1.82) leads to the final algebraic equations with in this case two unknowns defining the approximate solution of the wedge:

$$\begin{bmatrix} K_{33} & K_{34} \\ K_{43} & K_{44} \end{bmatrix} \begin{Bmatrix} u_3 \\ u_4 \end{Bmatrix} = -\begin{Bmatrix} K_{32} \\ K_{42} \end{Bmatrix} \bar{u}_2 + \begin{Bmatrix} \bar{P}_3 \\ \bar{P}_4 \end{Bmatrix} \tag{1.84}$$

Step 5: Solution of Global Equations

The symmetric linear equations which in general are banded describing the global model can be solved efficiently. This leads to the displacements in all nodes.

From Equation (1.84), u_3, u_4 are calculated. The reaction forces then follow from Equation (1.82) as

$$\begin{aligned} R_0 &= K_{02}\bar{u}_2 + K_{03}u_3 + K_{04}u_4 - \bar{P}_0 \\ R_1 &= K_{12}\bar{u}_2 + K_{13}u_3 + K_{14}u_4 - \bar{P}_1 \\ R_2 &= K_{22}\bar{u}_2 + K_{23}u_3 + K_{24}u_4 - \bar{P}_2 \end{aligned} \tag{1.85}$$

Step 6: Calculation of Additional Results

Starting from the known nodal displacements of the finite elements, additional results such as displacements in the interior of elements and stresses can be straightforwardly calculated. As the calculation of stresses leads to differentiation of the shape functions, the accuracy of the stresses will, in general, be smaller than that of the displacements.

For instance, $u(x, y)$ follows from Equation (1.43) and $\{\tau\}$ from Equation (1.3) with $u_{,x}, u_{,y}$ specified in Equation (1.55).

For the illustrative example, the wedge of Figure 1.3 described in section 1.2, the following results are obtained. The displacement in point P, which is u_3, equals $0.073\bar{u} + 0.015(r_0^2/G)p$ and the (constant) shear stress (surface traction) τ_x on side 1-2 equals $1.353(G/r_0)\bar{u} - 0.031r_0p$.

For the sake of completeness, the extension from statics to dynamics is discussed. In the weighted-residual statement of Equation (1.53), the term $\rho\int_V \ddot{u}(t)w\mathrm{d}V$ is added (Equation (1.38)). Again, the contribution of each individual finite element can be addressed

$$\rho\int w^T\ddot{u}\mathrm{d}x\mathrm{d}y \tag{1.86}$$

The integration is performed over the parent element (Figure 1.7). Substituting Equations (1.52), (1.71) and (1.72) into Equation (1.86) yields

$$\{w\}^T\rho\int[N(\xi,\eta)]^T[N(\xi,\eta)]|J|\mathrm{d}\xi\mathrm{d}\eta\{\ddot{u}\} \tag{1.87}$$

with $|J| = 2\Delta$. Again, $\{w\}^T$ is arbitrary and the remaining vector is equal to nodal forces which are defined as the product of the so-called *mass matrix* $[M]$ and the acceleration vector $\{\ddot{u}\}$. $[M]$ thus equals

$$[M] = \rho\int[N(\xi,\eta)]^T[N(\xi,\eta)]2\Delta\mathrm{d}\xi\mathrm{d}\eta \tag{1.88}$$

$[N(\xi,\eta)]$ are linear functions in ξ, η (Equation (1.51)). Using the integration formula over the right-isosceles triangular parent element

$$\int(1-\xi-\eta)^\alpha\xi^\beta\eta^\gamma\mathrm{d}\xi\mathrm{d}\eta = \frac{\alpha!\beta!\gamma!}{(\alpha+\beta+\gamma+2)!} \tag{1.89}$$

the mass matrix of the triangular element with linear shape functions equals

$$[M] = \Delta\rho \begin{bmatrix} \frac{1}{6} & \frac{1}{12} & \frac{1}{12} \\ \frac{1}{12} & \frac{1}{6} & \frac{1}{12} \\ \frac{1}{12} & \frac{1}{12} & \frac{1}{6} \end{bmatrix} \tag{1.90}$$

Adding the dynamic part to the static part (Equation (1.79)) yields the nodal force-nodal displacement relationship for dynamics

$$\{P\} = [K]\{u\} + [M]\{\ddot{u}\} - \{\bar{P}\} \tag{1.91}$$

As $[M]$ is assembled analogously as $[K]$ is, this equation also applies formally to the global system with all matrices and vectors expanded correspondingly.

The properties of the finite element method can be summarised as follows:

- The *domain is spatially discretised* (Figure 1.9), leading to a large number of nodal unknowns (degrees of freedom).
- The assumption of shape functions for the approximate displacements which are *piecewise local* in a finite element is the central feature of the method.
- As the shape functions are polynomials, *standard integration of regular functions* leads to the static-stiffness matrix and the load vector (and the mass matrix in dynamics).
- Assembling, i.e. *summing* these properties of the individual finite elements enforces equilibrium and compatibility of the global model. The resulting system of equations is *symmetric* (as the same shape functions are used for the approximate displacements and the weighting functions), *sparse* (a few non-zeros per row, as the shape functions are defined locally for each finite element) and *banded* (for an appropriate numbering scheme of the nodes).
- Special algorithms using the banded symmetric structure of the coefficient matrix permit *large systems with many unknowns* to be solved.
- As the *surface tractions* (stresses) follow from differentiation of the shape functions and are not independent unknowns appearing in the system of equations, their *accuracy* will, in general be *smaller* than that of the displacements.

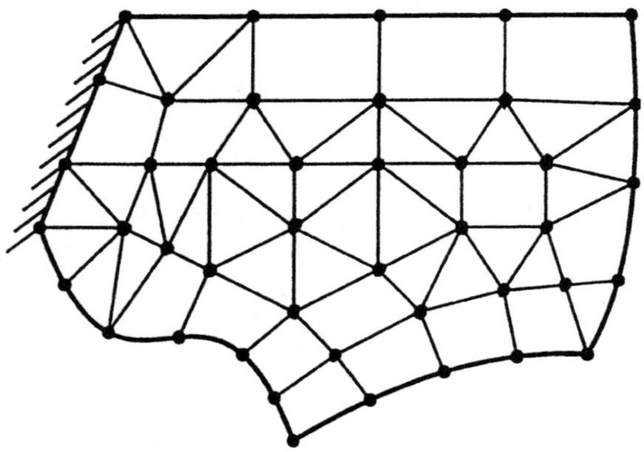

Figure 1.9 Spatial discretisation of finite element method

- The weighted-residual statement can be interpreted as an application of the principle of minimum potential energy. Increasing the number of degrees of freedom by refining the mesh leads to *convergence from the 'stiff' side*, i.e. the strain energy increases.
- The finite element method is well suited for *inhomogeneous* and *anisotropic* material.
- Addressing an unbounded medium such as the soil supporting the dam in Figure 1.1, the standard finite element method *cannot represent accurately the condition to be formulated at infinity*. As it is not possible to model the medium up to infinity with a finite number of elements, the spatial discretisation of finite elements is terminated on a so-called artificial boundary, where the physical behaviour of the truncated domain outside of this boundary is modelled only approximately.

1.5 BOUNDARY ELEMENT METHOD

The essential features of the boundary element method can be summarised as follows. The boundary only is subdivided into non-overlapping subboundaries called boundary elements. As the solution of the problem follows as a combination of exact solutions (called fundamental solutions) in the interior domain corresponding to an assumed variation of the unknown boundary values, no spatial discretisation of the interior is required. This results in a reduction of the dimension of the spatial discretisation by one. For instance, for a two-dimensional problem, only line boundary elements are processed. Based on a boundary-integral equation integrating on each boundary element the product of the assumed variation of the unknowns and the exact solution yields the contribution to the coefficient matrix of the equations describing the global model, which can then be solved.

Before describing the basic steps, the boundary-integral equation is derived. Again, Laplace's equation described in section 1.2 is addressed.

The starting point is the weighted-residual statement of Equation (1.33), which is formulated before any boundary conditions are enforced as ($S = S_u + S_t$)

$$-G\int_V (w_{,xx} + w_{,yy})u\mathrm{d}V - \int_V pw\mathrm{d}V + G\int_S uw_{,n}\mathrm{d}S = \int_S \tau_n w\mathrm{d}S \tag{1.92}$$

To derive an equation where the first term on the left-hand side of Equation (1.92) can be written as a boundary integral, a special weighting function, called the *fundamental solution*, is selected. It is denoted with an asterisk ($w = u^*$, $Gw_{,n} = \tau_n^*$). A concentrated unit load is applied to an infinite two-dimensional plane in point i (Figure 1.10). This so-called Dirac-delta function δ corresponds to a certain distribution of contact pressure on a small area which tends to zero (e.g. constant over a circle of radius ε with $\varepsilon \to 0$) with a resultant equal to 1

$$\int_V \delta\mathrm{d}V = 1 \tag{1.93}$$

The integral of δ applied at point i, denoted as δ_i, multiplied by a displacement u is equal to u evaluated at i, u_i

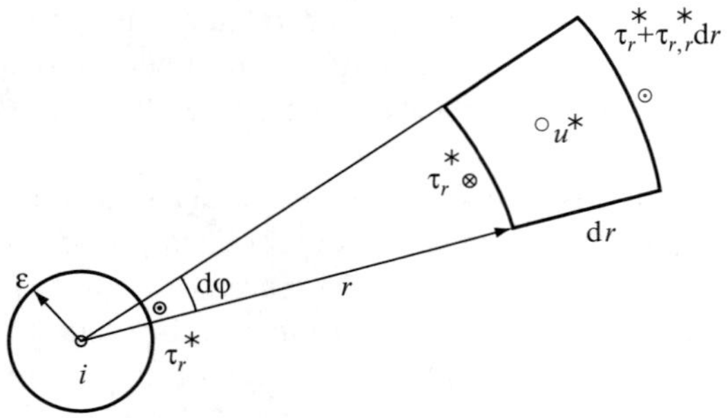

Figure 1.10 Infinitesimal element of fundamental solution

$$\int_V \delta u \mathrm{d}V = u_i \tag{1.94}$$

The governing differential equation (Equation (1.4)) of the fundamental solution with the Dirac delta function at point i equals

$$G(u^*{,_{xx}} + u^*{,_{yy}}) + \delta_i = 0 \tag{1.95}$$

Equation (1.92) is formulated with the fundamental solution equal to the weighting function

$$-G\int_V (u^*{,_{xx}} + u^*{,_{yy}})u\mathrm{d}V - \int_V pu^*\mathrm{d}V + \int_S u\tau_n^*\mathrm{d}S = \int_S \tau_n u^*\mathrm{d}S \tag{1.96}$$

Substituting Equation (1.95) in Equation (1.96) and applying Equation (1.94) yields

$$u_i - \int_V pu^*\mathrm{d}V + \int_S u\tau_n^*\mathrm{d}S = \int_S \tau_n u^*\mathrm{d}S \tag{1.97}$$

In this equation, the concentrated unit load is applied at point i, leading to the corresponding displacement u^* and surface traction τ_n^*, which describe the fundamental solution. Point i lies inside the domain.

To derive the fundamental solution, which, due to symmetry, is a function of the radial co-ordinate r measured from point i (Figure 1.10), formulating equilibrium of an infinitesimal element which is not loaded ($r > \varepsilon$)

$$-\tau_r^* r\mathrm{d}\varphi + (\tau_r^* + \tau_r^*{,_r}\mathrm{d}r)(r + \mathrm{d}r)\mathrm{d}\varphi = 0 \tag{1.98}$$

leads to

$$\frac{1}{r}\tau_r^* + \tau_r^*{,_r} = 0 \tag{1.99}$$

The constitutive law for an isotropic material expressed in the derivative of the displacement (Equation (1.8)) equals

$$\tau_r^* = Gu,^*_r \tag{1.100}$$

Substituting Equation (1.100) into Equation (1.99) results in

$$G\left(u,^*_{rr} + \frac{1}{r}u,^*_r\right) = 0 \tag{1.101}$$

To solve this ordinary differential equation

$$v = u,^*_r \tag{1.102a}$$

is substituted, yielding

$$v,_r + \frac{1}{r}v = 0 \tag{1.102b}$$

or

$$\frac{v,_r}{v} = -\frac{1}{r} \tag{1.102c}$$

which is equal to

$$(\ln v),_r = -\frac{1}{r} \tag{1.102d}$$

Substituting

$$w = \ln v \tag{1.102e}$$

leads to

$$w,_r = -\frac{1}{r} \tag{1.102f}$$

with the solution

$$w = -\ln r = \ln\left(\frac{1}{r}\right) \tag{1.102g}$$

From Equations (1.102e) and (1.102g)

$$v = \frac{1}{r} \tag{1.102h}$$

results, yielding from Equation (1.102a)

$$u_{,r}^* = \frac{1}{r} \tag{1.102i}$$

Its solution equals

$$u^* = \ln r \tag{1.102j}$$

Any constant factor c can be introduced, as

$$u^* = -c \ln r = c \ln \frac{1}{r} \tag{1.102k}$$

also satisfies Equation (1.101). The second solution of this second-order differential equation is a constant, which can be disregarded as it corresponds to a rigid body motion with vanishing shear stress.

To determine the constant c, equilibrium of the loaded circle of infinitesimal radius ε is addressed (Figure 1.10). The shear stress $\tau_r^*(r = \varepsilon)$ multiplied by the circumference $2\pi\varepsilon$ and the concentrated load 1 add up to zero.

$$2\pi\varepsilon\tau_r^*(r = \varepsilon) + 1 = 0 \tag{1.103}$$

τ_r^* is calculated from Equation (1.100) with Equation (1.102k)

$$\tau_r^* = -Gc\frac{1}{r} \tag{1.104}$$

Evaluating Equation (1.104) for $r = \varepsilon$ and substituting into Equation (1.103) leads to

$$c = \frac{1}{2\pi G} \tag{1.105}$$

The fundamental solution thus equals

$$u^* = \frac{1}{2\pi G} \ln \frac{1}{r} \tag{1.106}$$

$$\tau_r^* = -\frac{1}{2\pi}\frac{1}{r} \tag{1.107}$$

with the distance r measured from the source point i (point of application of the Dirac-delta function) to the receiver point where the solution is to be determined. In Equation (1.97) the receiver point varies for any fixed position i of the concentrated load. For $r \to 0$, both u^* and τ_r^* become infinite, the singularity of τ_r^* (involving a derivative) being stronger than that of u^*.

In Equation (1.97), point i lies inside the domain. To derive the boundary-integral equation involving boundary values, point i, where the Dirac-delta function is applied, is moved to the boundary, which is assumed to be smooth locally, i.e. where the concentrated load exists (no corners). An easy procedure to perform this limit is to consider that point i is on the boundary, but to augment the domain by a semi-circle of radius ε with the centre at i (Figure 1.11). Point i can be regarded as an internal point. Equation (1.97) does apply in this case. Taking the limit $\varepsilon \to 0$ then moves point i to the boundary, resulting in the corresponding boundary-integral equation.

The behaviour of the two surface integrals in Equation (1.97) for $\varepsilon \to 0$ is examined. The integral on the right-hand side is formulated after substituting Equation (1.106) with the circumference of the semi-circle $S_c = \pi\varepsilon$

$$\int_{S_c} \tau_n u^* \mathrm{d}S = \frac{1}{2\pi G} \lim_{\varepsilon \to 0} \int_{S_c} \tau_n \ln\frac{1}{\varepsilon}\,\mathrm{d}S = \frac{1}{2\pi G} \lim_{\varepsilon \to 0} \left(\tau_n \ln\frac{1}{\varepsilon}\pi\varepsilon\right) = 0 \tag{1.108}$$

as $\varepsilon \ln\frac{1}{\varepsilon} \to 0$ for $\varepsilon \to 0$. This means that no change occurs on the right-hand side of Equation (1.97) when point i is moved to the boundary. The integral on the left-hand side involving a stronger singularity (Equation (1.107)), however, yields a non-vanishing term

$$\int_{S_c} u\tau_n^* \mathrm{d}S = -\frac{1}{2\pi} \lim_{\varepsilon \to 0} \int_{S_c} u\frac{1}{\varepsilon}\,\mathrm{d}S = -\frac{1}{2\pi} \lim_{\varepsilon \to 0} \left(u\frac{\pi\varepsilon}{\varepsilon}\right) = -\frac{u_i}{2} \tag{1.109}$$

Substituting Equations (1.108) and (1.109) into Equation (1.97) yields the *boundary-integral equation* for a boundary point i:

$$\frac{1}{2}u_i - \int_V pu^* \mathrm{d}V + \int_S u\tau_n^* \mathrm{d}S = \int_S \tau_n u^* \mathrm{d}S \tag{1.110}$$

Equation (1.110) can be formulated for every point i on the boundary, leading to a system of boundary-integral equations.

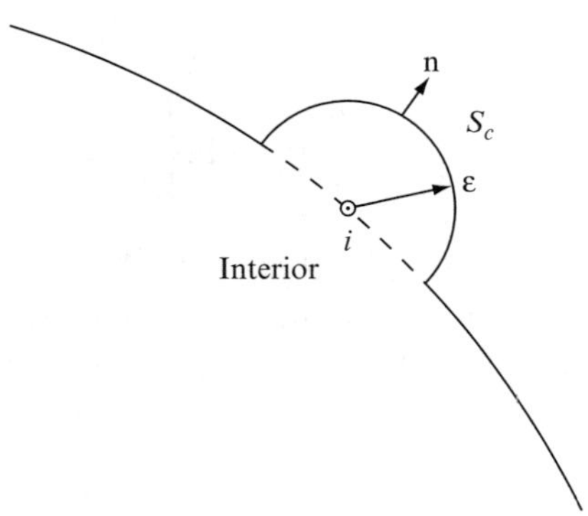

Figure 1.11 Semi-circle with infinitesimal radius around point on boundary

In Equations (1.110) and (1.97), the body load p is represented by an integration over the domain V (second integral on the left-hand side), which can be transformed into an equivalent boundary integral as follows. Subtracting Equation (1.33) from Equation (1.35) yields

$$G\int_V (u_{,xx}+u_{,yy})w\,\mathrm{d}V - G\int_V (w_{,xx}+w_{,yy})u\,\mathrm{d}V = -G\int_S uw_{,n}\mathrm{d}S + G\int_S u_{,n}w\,\mathrm{d}S \quad (1.111)$$

with $\tau_n = Gu_{,n}$ and $S = S_u + S_t$. G can be cancelled. Selecting $u = v^*$, a function which is related to the fundamental solution as defined in the following, and $w = p$, Equation (1.111) is transformed into

$$\int_V (v_{,}{}^*_{xx} + v_{,}{}^*_{yy})p\,\mathrm{d}V - \int_V (p_{,xx}+p_{,yy})v^*\mathrm{d}V = -\int_S v^*p_{,n}\mathrm{d}S + \int_S v_{,}{}^*_{n}p\,\mathrm{d}S \quad (1.112)$$

A limitation for the load p is introduced. p is assumed to be harmonic, i.e.

$$p_{,xx} + p_{,yy} = 0 \quad (1.113)$$

which is satisfied for a constant and a linear variation over the domain. In addition,

$$v_{,}{}^*_{xx} + v_{,}{}^*_{yy} = u^* \quad (1.114)$$

is enforced with u^* denoting the fundamental solution (Equation (1.106)). Equation (1.114) defines v^*. Substituting Equations (1.113) and (1.114) into Equation (1.112) results in

$$\int_V pu^*\mathrm{d}V = -\int_S p_{,n}v^*\mathrm{d}S + \int_S pv_{,n}^*\mathrm{d}S \quad (1.115)$$

The integration over the domain is thus replaced by two integrations over the boundary of the same domain as for the other boundary integrals in the boundary-integral equation. For a constant body load p, the first integral on the right-hand side vanishes ($p_{,n} = 0$).

As v^*, the function

$$v^* = \frac{r^2}{8\pi G}\left(1 + \ln\frac{1}{r}\right) \quad (1.116)$$

is used, which does satisfy Equation (1.114). This is verified as follows. In polar coordinates with no circumferential variation, the Laplace operator (see Equation (1.101)) corresponding to the left-hand side of Equation (1.114) equals

$$v_{,}{}^*_{rr} + \frac{1}{r}v_{,}{}^*_{r} \quad (1.117)$$

The derivatives such as

$$v,^*_r = \frac{r}{8\pi G}\left(1 + 2\ln\frac{1}{r}\right) \tag{1.118}$$

are determined from Equation (1.116). Substituted into Equation (1.117), u^* specified in Equation (1.106) follows.

The basic steps of the boundary element method are explained addressing the illustrative example, the wedge, presented in Figure 1.3 of Laplace's equation with a body load described in section 1.2.

For statics, the following six steps can be identified.

Step 1: Spatial Discretisation of Boundary

The boundary is subdivided into a number of subboundaries, called boundary elements. This procedure is analogous to that used for finite elements (see Step 1 of the Finite Element Method, section 1.4), with the exception that the dimension is reduced by one. For a two-dimensional problem line boundary elements apply.

The boundary of the wedge 0-1-2 (Figure 1.12) is divided into three straight line elements 0-1, 1-2, 2-0.

Step 2: Displacement and Surface Traction within Boundary Element

The approximate solution of the displacement and surface traction is defined *piecewise* over each boundary element. Nodal values of the displacement and surface traction are selected as the unknowns. The displacement and surface traction within the individual boundary element are interpolated in terms of the nodal values by so-called *shape functions* in a local coordinate system. No continuity requirements of the displacements and surface tractions in adjacent boundary elements on the boundary exist.

For each of the three boundary elements, the node is selected in the middle of the element and two nodal values u, τ_n are assigned. Constant elements are chosen, i.e. u and τ_n are assumed to be constant over each boundary element and equal to their nodal values. This means that at a corner node, in general, a discontinuity in displacement occurs. Over each boundary element either the displacement or the surface traction is prescribed as a boundary condition. The known values are indicated in Figure 1.12 with a bar. The nodal values are contained in the vectors $\{u\}, \{\tau_n\}$ with $\{u\} = [u_1, u_2, u_3]^T, \{\tau_n\} = [\tau_{n1}, \tau_{n2}, \tau_{n3}]^T$.

Step 3: Fundamental Solution of Governing Differential Equation

A fundamental solution which satisfies the governing differential equation exactly must be available. The boundary conditions of the problem do not have to be enforced. In general, the exact solution of a concentrated load in an infinite domain represents the fundamental solution.

For the two-dimensional Laplace equation the fundamental solution u^*, τ_r^* is specified in Equations (1.106) and (1.107).

Step 4: Discretised Boundary-Integral Equation and Enforcement of Boundary Conditions

Substituting Equation (1.115) for p = constant in Equation (1.110) leads to the boundary-integral equation

$$\frac{1}{2}u_i + \int_S u\tau_n^* \mathrm{d}S = \int_S \tau_n u^* \mathrm{d}S + \int_S p v,_n^* \mathrm{d}S \tag{1.119}$$

Besides the load term with integration on the boundary, only values of the displacement u and surface traction τ_n on the boundary appear in this equation. Equation (1.119) is discretised into N constant boundary elements to determine the system from which the boundary values can be determined. Point i represents one of the boundary nodes. This leads to

$$\frac{1}{2}u_i + \sum_{j=1}^{N}\left(\int_{S_j} \tau_n^* \mathrm{d}S\right)u_j = \sum_{j=1}^{N}\left(\int_{S_j} u^* \mathrm{d}S\right)\tau_{nj} + \sum_{j=1}^{N}\left(\int_{S_j} v,_n^* \mathrm{d}S\right)p \tag{1.120}$$

The functions u_j and τ_{nj} can be taken outside the integral of the jth boundary element with the boundary S_j, as they are constant. The integrals

$$H_{ij} = \int_{S_j} \tau_n^* \mathrm{d}S \tag{1.121}$$

$$G_{ij} = \int_{S_j} u^* \mathrm{d}S \tag{1.122}$$

$$\bar{P}_{ij} = p\int_{S_j} v,_n^* \mathrm{d}S \tag{1.123}$$

relate the node i, where the concentrated load is applied, to the jth node, involving integrals of the corresponding fundamental solution u^*, τ_n^* or the associated v^* over the jth boundary element. For node i, Equation (1.120) is written as

$$\frac{1}{2}u_i + \sum_{j=1}^{N} H_{ij}u_j = \sum_{j=1}^{N} G_{ij}\tau_{nj} + \sum_{j=1}^{N} \bar{P}_{ij} \tag{1.124}$$

Equation (1.124) can be formulated successively for each node, i.e. for $i = 1, 2, \ldots N$. This results in a system of N discretised boundary-integral equations, the global equations

$$\left(\frac{1}{2}[I] + [H]\right)\{u\} = [G]\{\tau_n\} + \{\bar{P}\} \tag{1.125}$$

with the unit-matrix $[I]$. Of the $2N$ values of $\{u\}$ and $\{\tau_n\}$, N boundary values are known, as they represent the boundary conditions, resulting in N unknowns on the boundary, which are in general both nodal displacements and nodal surface tractions. $\{\bar{P}\}$ consisting of the elements $P_i, i = 1, \ldots, N$ is known. The known values are moved to the right-hand side and the unknowns to the left-hand side, which can be determined by solving the system of equations. Thus, the boundary element method determines all boundary values of the displacements and surface tractions. Note that the coefficient matrix is fully populated and non-symmetric.

For the wedge with the discretisation shown in Figure 1.12, $N = 3$. The discretised boundary-integral equation Equation (1.125) equals

$$\left(\frac{1}{2}\begin{bmatrix} 1 & & \\ & 1 & \\ & & 1 \end{bmatrix} + \begin{bmatrix} H_{11} & H_{12} & H_{13} \\ H_{21} & H_{22} & H_{23} \\ H_{31} & H_{32} & H_{33} \end{bmatrix}\right)\begin{Bmatrix} u_1 \\ u_2 \\ u_3 \end{Bmatrix} = \begin{bmatrix} G_{11} & G_{12} & G_{13} \\ G_{21} & G_{22} & G_{23} \\ G_{31} & G_{32} & G_{33} \end{bmatrix}\begin{Bmatrix} \tau_{n1} \\ \tau_{n2} \\ \tau_{n3} \end{Bmatrix} + \begin{Bmatrix} \bar{P}_1 \\ \bar{P}_2 \\ \bar{P}_3 \end{Bmatrix} \tag{1.126}$$

Enforcing the boundary conditions (see section 1.2)

$$\begin{aligned} u_1 &= \bar{u}_1 = 0 \\ u_2 &= \bar{u}_2 = 0.5\bar{u} \\ \tau_{n3} &= \bar{\tau}_{n3} = 0 \end{aligned} \tag{1.127}$$

and moving all known values and unknowns to the right-hand and left-hand sides, respectively, yields

$$\begin{bmatrix} -G_{11} & -G_{12} & H_{13} \\ -G_{21} & -G_{22} & H_{23} \\ -G_{31} & -G_{32} & \frac{1}{2} + H_{33} \end{bmatrix}\begin{Bmatrix} \tau_{n1} \\ \tau_{n2} \\ u_3 \end{Bmatrix} = -0.5\begin{Bmatrix} H_{12} \\ \frac{1}{2} + H_{22} \\ H_{32} \end{Bmatrix}\bar{u} + \begin{Bmatrix} \bar{P}_1 \\ \bar{P}_2 \\ \bar{P}_3 \end{Bmatrix} \tag{1.128}$$

Note that those coefficients in Equation (1.126) corresponding to homogeneous boundary conditions do not have to be evaluated ($H_{11}, H_{21}, H_{31}, G_{13}, G_{23}, G_{33}$).

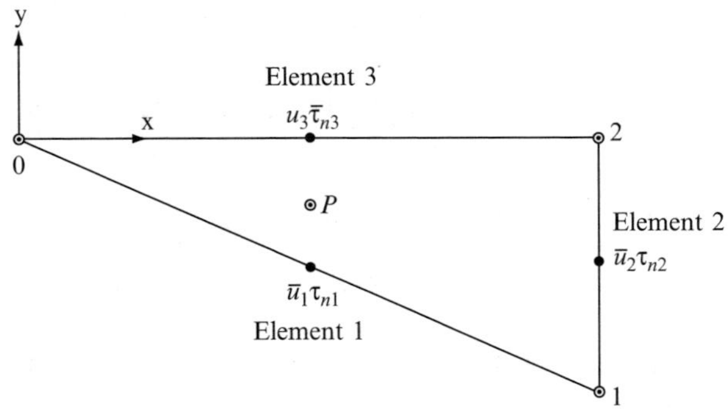

Figure 1.12 Division of boundary of wedge into three boundary elements with constant displacements and surface tractions

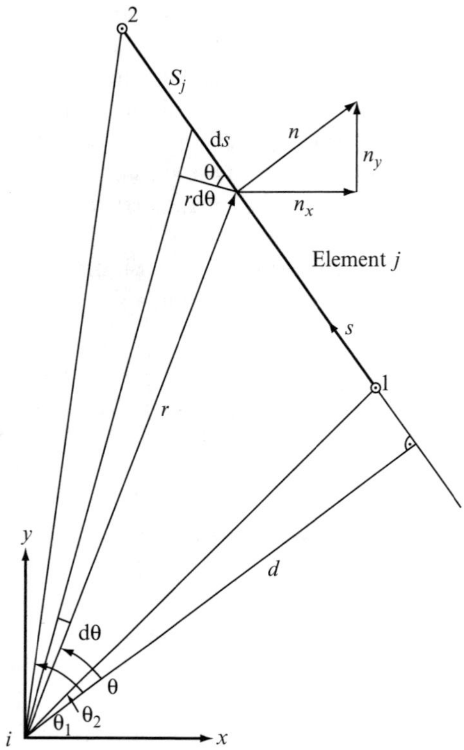

Figure 1.13 Geometry to calculate coefficients of discretised boundary-integral equation

For the constant boundary elements the integrals H_{ij} and G_{ij} (as well as $\bar{P}_{ij}$) can be calculated analytically. For the case where the point i (application of concentrated load) does not lie on the boundary element j (integrations to be performed), the geometrical configuration is shown in Figure 1.13. The integration on S_j from the local node 1 to 2 is performed with the domain lying on the left. A polar coordinate system with its origin at i is introduced where θ is measured from the line perpendicular to the jth boundary element.

The fundamental solution for u^* is specified in Equation (1.106) as

$$u^* = \frac{1}{2\pi G} \ln \frac{1}{r} = -\frac{1}{2\pi G} \ln r \tag{1.129}$$

τ_n^* is expressed as follows. With Equation (1.8)

$$\tau_n^* = G u,^*{}_n \tag{1.130}$$

the partial derivative with respect to the outward unit normal $\{n\}$ with components n_x, n_y is formulated as

$$u,^*{}_n = u,^*{}_r \, r,_n = u,^*{}_r (r,_x n_x + r,_y n_y) \tag{1.131}$$

With

$$r = \sqrt{x^2 + y^2} \tag{1.132}$$

and Equation (1.129), the derivatives are evaluated, and substituting into Equation (1.130) yields

$$\tau_n^* = -\frac{1}{2\pi r^2}(xn_x + yn_y) = -\frac{1}{2\pi}\frac{d}{r^2} \tag{1.133}$$

The expression in parentheses is equal to the scalar product of the vectors $\{r\}$ and $\{n\}$, resulting in the distance d measured from point i to the boundary element j (positive for the configuration in Figure 1.13).

The element H_{ij} in Equation (1.121) follows as

$$H_{ij} = \int_{S_j} \tau_n^* \mathrm{d}s \tag{1.134}$$

With

$$\mathrm{d}s = \frac{r}{\cos\theta}\mathrm{d}\theta \tag{1.135a}$$

and

$$r = \frac{d}{\cos\theta} \tag{1.135b}$$

Equation (1.134) leads to

$$H_{ij} = -\frac{1}{2\pi}\int_{\theta_1}^{\theta_2} \mathrm{d}\theta = \frac{1}{2\pi}(\theta_1 - \theta_2) \tag{1.136}$$

The element G_{ij} in Equation (1.122) defined as

$$G_{ij} = \int_{S_j} u^* \mathrm{d}s \tag{1.137}$$

is calculated analogously substituting Equation (1.129)

$$G_{ij} = \frac{1}{2\pi}\frac{d}{G}\int_{\theta_1}^{\theta_2} \ln\frac{\cos\theta}{d}\frac{1}{\cos^2\theta}\mathrm{d}\theta \tag{1.138}$$

which is equal to

$$G_{ij} = \frac{1}{2\pi}\frac{d}{G}\left(\tan\theta_2 - \tan\theta_1 - \theta_2 + \theta_1 + \tan\theta_2 \ln\frac{\cos\theta_2}{d} - \tan\theta_1 \ln\frac{\cos\theta_1}{d}\right) \tag{1.139}$$

(In the argument of the ln function, d is positive.)

For the case where the point i (application of the concentrated load) lies on the boundary element $j = i$ (integrations to be performed), singularities occur. The length of the ith element is denoted as l. Placing the point i in the middle of the boundary element $j = i$, the vector $\{r\}$ in the direction of element 1-2 will be perpendicular to the outward normal $\{n\}$, leading to $d = 0$ in Equation (1.133), and thus to $\tau_n^* = 0$. This is obvious as the normal corresponds to the circumferential direction in the fundamental solution where the circumferential shear stress vanishes (Figure 1.10). From Equation (1.121)

$$H_{ii} = 0 \tag{1.140}$$

results. Thus, for the terms on the diagonal of the coefficient matrix of the displacements the term ½ remains (Equation (1.126)). For the other coefficient, substituting Equation (1.106) into Equation (1.122), and due to the symmetry of u^* with respect to the middle node,

$$G_{ii} = -\frac{2}{2\pi G}\int_0^{0.5l} \ln r dr \tag{1.141}$$

applies. As the integrand is infinite at $r = 0$, the lower integration limit is selected as ε with $\varepsilon \to 0$:

$$G_{ii} = -\frac{1}{\pi G}\lim_{\varepsilon \to 0}\int_\varepsilon^{0.5l} \ln r dr \tag{1.142}$$

Integration by parts yields

$$G_{ii} = -\frac{1}{\pi G}\lim_{\varepsilon \to 0}(r\ln r - r)\Big|_\varepsilon^{0.5l} = \frac{1}{2\pi}\frac{l}{G}\left(\ln\frac{2}{l} + 1\right) \tag{1.143}$$

as $\varepsilon \ln \varepsilon$ tends to zero for $\varepsilon \to 0$.

The coefficient of the load in Equation (1.123) follows as

$$\bar{P}_{ij} = p\int_{S_j} v_{,n}^* ds \tag{1.144}$$

For the case where the point i does not lie on the boundary element j (Figure 1.13), proceeding as with Equations (1.131) and (1.133),

$$v,^*_n = v,^*_r\, r,_n = v,^*_r \frac{d}{r} \tag{1.145}$$

applies. Substituting Equation (1.118) into Equation (1.145) yields

$$v,^*_n = \frac{d}{8\pi G}\left(1 + 2\ln\frac{1}{r}\right) \tag{1.146}$$

Substituting Equation (1.146) into Equation (1.144) leads to

$$\bar{P}_{ij} = \frac{dp}{8\pi G}\int_{S_j} \mathrm{d}s + \frac{dp}{4\pi G}\int_{S_j} \ln\frac{1}{r}\,\mathrm{d}s \tag{1.147}$$

which, after substitution of Equation (1.135), results in

$$\bar{P}_{ij} = \frac{dlp}{8\pi G} + \frac{d^2 p}{4\pi G}\int_{\theta_1}^{\theta_2} \ln\frac{\cos\theta}{d}\frac{1}{\cos^2\theta}\,\mathrm{d}\theta \tag{1.148}$$

The integral in Equation (1.148) is the same as in Equation (1.138). Thus

$$\bar{P}_{ij} = \frac{dlp}{8\pi G} + \frac{dp}{2}G_{ij} \tag{1.149}$$

results with G_{ij} specified in Equation (1.139). For the case where the point i lies on the boundary element $j = i$, proceeding as with Equation (1.141) yields (Equation (1.118))

$$\bar{P}_{ii} = 2p\int_0^{0.5l} v,^*_r \mathrm{d}r = \frac{p}{4\pi G}\int_0^{0.5l} r\mathrm{d}r - \frac{p}{2\pi G}\int_0^{0.5l} r\ln r\mathrm{d}r \tag{1.150}$$

with

$$\int_0^{0.5l} r\ln r\mathrm{d}r = \lim_{\varepsilon\to 0}\int_\varepsilon^{0.5l} r\ln r\mathrm{d}r = \lim_{\varepsilon\to 0}\left(\frac{r^2}{2}\ln r\,\Big|_\varepsilon^{0.5l} - \frac{1}{2}\int_\varepsilon^{0.5l} r\mathrm{d}r\right) = \frac{l^2}{8}\ln\frac{l}{2} - \frac{l^2}{16} \tag{1.151}$$

Equation (1.150) is formulated as

$$\bar{P}_{ii} = \frac{l^2 p}{16\pi G}(\ln\frac{2}{l} + 1) \tag{1.152}$$

or using Equation (1.143)

$$\bar{P}_{ii} = \frac{lp}{8}G_{ii} \tag{1.153}$$

applies.

Step 5: Solution of Global Equations

The, in general, fully populated non-symmetric equations (Equation (1.125) or (1.128)) describing the global model with enforced boundary conditions are solved. This leads to the N nodal displacements and nodal surface tractions. All N nodal displacements $\{u\}$ and N nodal surface tractions $\{\tau_n\}$ on the boundary are then known.

Step 6: Calculation of Additional Results

With all known N displacements $\{u\}$ and N surface tractions $\{\tau_n\}$ on the boundary, displacements and stresses in the interior can be calculated. For instance, the displacement in point i inside the domain follows from Equation (1.97). For constant p, substituting Equation (1.115) and discretising into N constant boundary elements, this equation yields

$$u_i + \sum_{j=1}^{N} \left(\int_{S_j} \tau_n^* \mathrm{d}S \right) u_j = \sum_{j=1}^{N} \left(\int_{S_j} u^* \mathrm{d}S \right) \tau_{nj} + \sum_{j=1}^{N} \left(\int_{S_j} v,^*{}_n \mathrm{d}S \right) p \qquad (1.154)$$

Substituting Equations (1.121), (1.122) and (1.123) results in

$$u_i = -\sum_{j=1}^{N} H_{ij} u_j + \sum_{j=1}^{N} G_{ij} \tau_{nj} + \sum_{j=1}^{N} \bar{P}_{ij} \qquad (1.155)$$

The fundamental solution u^* and τ_n^* corresponds to the concentrated load applied at the internal point i. $\{u\}$ and $\{\tau_n\}$ are known.

For each different internal point i, u^*, τ_n^* and v^* have to be re-calculated, and the H_{ij} and G_{ij} coefficients as well as the $\bar{P}_{ij}$ re-established. $\{u\}$ and $\{\tau_n\}$ do not change.

For the illustrative example, the wedge, of Figure 1.3 described in section 1.2, the following results are calculated. The (constant) shear stress (surface traction) on side 1-2 $\tau_n = \tau_x$, which is τ_{n2}, follows from solving Equation (1.128), and is equal to $1.807(G/r_0)\bar{u} - 0.134 r_0 p$. The displacement in point P calculated using Equation (1.155) equals $0.026\,\bar{u} + 0.008(r_0^2/G)p$.

The properties of the boundary element method can be summarised as follows:

- Only the *boundary* is *spatially discretised* (Figure 1.14), leading to a reduction of the spatial dimension by one, and thus to a smaller number of degrees of freedom than in the finite element method, resulting in a diminished computational effort. The work of data preparation is also diminished. For certain body loads, no domain discretisation is necessary.
- A *fundamental solution* satisfying exactly the governing differential equations in the domain is required, which differs from problem to problem. This analytical solution is often very *complicated, exhibiting singularities.*
- In each boundary element, shape functions in form of polynomials interpolate the displacements and surface tractions.

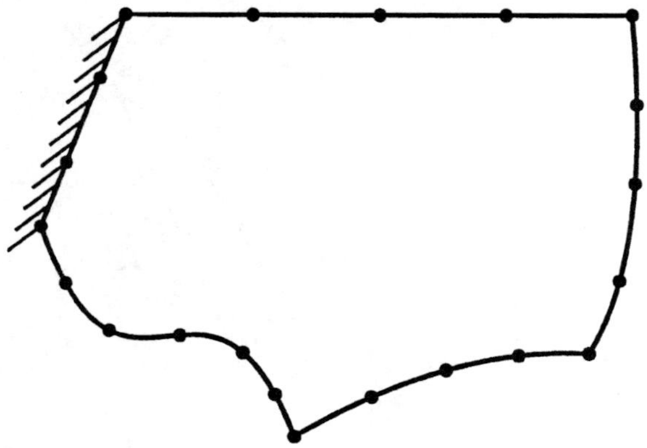

Figure 1.14 Spatial discretisation of boundary element method

- Special *integration* of the polynomials with the fundamental solution involving *singularities* is required.
- The resulting *equations* of the global system are *fully populated* and *non-symmetric.*
- Both *displacements* and *surface tractions* on the boundary appear independently as unknowns in the discretised boundary-integral equations of the global system, and are thus calculated with *similar accuracy*.
- Errors are restricted to the approximation of the boundary values.
- Further results in the interior of the domain are determined selectively.
- The boundary element method is *not well suited for inhomogeneous and anisotropic* material.
- Addressing an unbounded medium such as the soil supporting the dam in Figure 1.1, the boundary element method models *rigorously* the *condition at infinity*, as this condition is satisfied in the fundamental solution.

The finite element and boundary element methods converge, in general, to the exact solution for decreasing element size. The convergence is slow close to a point of stress singularity as occurring around the crack tip in fracture mechanics (e.g. see the crack tip K in Figure 1.1). This is caused by the use of polynomials to interpolate the displacements in both methods. Special techniques and a large number of elements are necessary to achieve high accuracy.

1.6 GLIMPSE AT SCALED BOUNDARY FINITE ELEMENT METHOD

After having solved the wedge problem (Figure 1.3) with the finite element method (section 1.4, Figure 1.5) and the boundary element method (section 1.5, Figure 1.12), it is appropriate to apply the so-called scaled boundary finite element method to the same task. No thorough understanding is envisaged at this stage in this incomplete treatment of the subject. It just serves to whet the appetite!

This boundary element method discretises the boundary. A typical linear (line) element between nodes 1 and 2 is shown for the two-dimensional case in Figure 1.15. To be able to formulate the boundary conditions in a straightforward way, a new

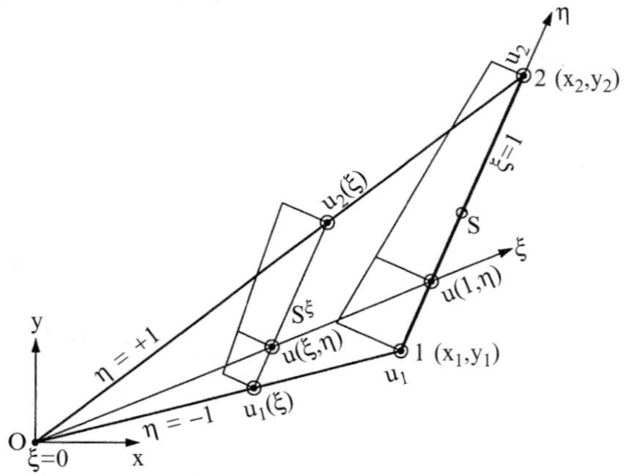

Figure 1.15 Scaled boundary finite element with displacement (Reprinted from *Computers and Structures*, Vol. 78, Wolf and Song, 'The scaled boundary finite element method – a primer: derivations', pp. 191–210, © (2000), with permission from Elsevier Science)

coordinate system ξ, η, called the *scaled boundary coordinates*, is introduced. Its origin O, the *scaling centre*, will for a bounded medium be, in general, inside the domain, and coincides with that of the Cartesian coordinate system x, y. The dimensionless radial coordinate ξ runs from the origin with $\xi = 0$ to the boundary where it is normalised to $\xi = 1$. The bounded domain is thus covered by $0 \leq \xi \leq 1$. The η-coordinate coincides with the circumferential direction on the boundary ($\eta = -1$ at node 1, $\eta = +1$ at node 2). With the nodal coordinates

$$\{x\} = \begin{Bmatrix} x_1 \\ x_2 \end{Bmatrix} \qquad \{y\} = \begin{Bmatrix} y_1 \\ y_2 \end{Bmatrix} \tag{1.156}$$

and the mapping shape functions

$$[N(\eta)] = [N_1(\eta) N_2(\eta)] = \left[\frac{1}{2}(1-\eta)\frac{1}{2}(1+\eta)\right] \tag{1.157}$$

the x- and y-coordinates on the line element on the boundary equal $[N(\eta)]\{x\} = \bar{x} + \frac{1}{2}\Delta_x\eta$ and $[N(\eta)]\{y\} = \bar{y} + \frac{1}{2}\Delta_y\eta$ with the abbreviations

$$\bar{x} = \frac{1}{2}(x_1 + x_2) \qquad \bar{y} = \frac{1}{2}(y_1 + y_2) \tag{1.158}$$

$$\Delta_x = x_2 - x_1 \qquad \Delta_y = y_2 - y_1 \tag{1.159}$$

The scaled boundary transformation relates any point in the domain with the coordinates denoted as x, y to the corresponding point on the line element as

$$\begin{aligned} x &= \xi[N(\eta)]\{x\} \\ y &= \xi[N(\eta)]\{y\} \end{aligned} \tag{1.160}$$

This equation defines the transformation from the Cartesian coordinates x, y to the scaled boundary coordinates ξ, η. Equation (1.160) describing scaling of the boundary has led to the name of the method.

The spatial derivatives in the two coordinate systems are related as (Appendix A, Equations (A.7, A.8))

$$\left\{\begin{matrix}\dfrac{\partial}{\partial x}\\ \dfrac{\partial}{\partial y}\end{matrix}\right\} = \frac{1}{x_{,\xi}y_{,\eta} - x_{,\eta}y_{,\xi}}\begin{bmatrix} y_{,\eta} & -y_{,\xi} \\ -x_{,\eta} & x_{,\xi}\end{bmatrix}\left\{\begin{matrix}\dfrac{\partial}{\partial \xi}\\ \dfrac{\partial}{\partial \eta}\end{matrix}\right\} \tag{1.161}$$

which, after substitution, is written as

$$\left\{\begin{matrix}\dfrac{\partial}{\partial x}\\ \dfrac{\partial}{\partial y}\end{matrix}\right\} = \{b^1\}\frac{\partial}{\partial \xi} + \{b^2\}\frac{1}{\xi}\frac{\partial}{\partial \eta}, \tag{1.162}$$

where

$$\{b^1\} = \frac{1}{|J|}\left\{\begin{matrix}[N(\eta)]_{,\eta}\{y\}\\ -[N(\eta)]_{,\eta}\{x\}\end{matrix}\right\} = \frac{1}{x_1y_2 - x_2y_1}\left\{\begin{matrix}\Delta_y\\ -\Delta_x\end{matrix}\right\} \tag{1.163a}$$

$$\{b^2\} = \frac{1}{|J|}\left\{\begin{matrix}-[N(\eta)]\{y\}\\ [N(\eta)]\{x\}\end{matrix}\right\} = \frac{1}{x_1y_2 - x_2y_1}\left\{\begin{matrix}-2\bar{y} - \Delta_y\eta\\ 2\bar{x} + \Delta_x\eta\end{matrix}\right\} \tag{1.163b}$$

with

$$|J| = \tfrac{1}{2}(x_1y_2 - x_2y_1) \tag{1.164}$$

The infinitesimal area $\mathrm{d}V$ of the domain is calculated as (see explanations in Appendix A, in connection with Equation (A.9) with the different definition of $|J|$ in Equation (A.8), or from geometrical considerations)

$$\mathrm{d}V = \xi|J|\mathrm{d}\xi\mathrm{d}\eta \tag{1.165}$$

and the infinitesimal length $\mathrm{d}S$ on the line element of $\xi = 1$

$$\mathrm{d}S = \sqrt{([N(\eta)]_{,\eta}\{x\})^2 + ([N(\eta)]_{,\eta}\{y\})^2}\mathrm{d}\eta = \Delta_s\mathrm{d}\eta = \frac{1}{2}\sqrt{\Delta_x^2 + \Delta_y^2}\mathrm{d}\eta \tag{1.166}$$

The displacement of the line element on the boundary $\xi = 1$ is interpolated using the mapping functions (Equation (1.157)) as the shape functions $[N(\eta)]$. The discretisation is thus restricted to the boundary. It is postulated that the same shape functions apply for all lines with a constant ξ

$$u(\xi, \eta) = [N(\eta)]\{u(\xi)\} \tag{1.167}$$

with the displacements on the lines passing through O and the nodes 1 and 2, respectively (Figure 1.15)

$$\{u(\xi)\} = \begin{Bmatrix} u_1(\xi) \\ u_2(\xi) \end{Bmatrix} \tag{1.168}$$

Equation (1.167), formulated in the scaled boundary coordinates ξ, η, is the key relationship. As will become apparent, the function $\{u(\xi)\}$ of the variable ξ is determined analytically. Using the same shape functions $[N(\eta)]$ as for the boundary, the displacement $u(\xi, \eta)$ for any ξ is calculated. This equation can be interpreted as a generalisation of the method of separation of variables interpolating in the circumferential direction η the values of $\{u(\xi)\}$. Note that the shape functions $[N(\eta)]$ are a function of one variable η only, in contrast to the procedure used in finite elements (Equation (1.52))

$$\{u(\xi, \eta)\} = [N(\xi, \eta)]\{u\} \tag{1.169}$$

where $\{u\}$ are the displacements in distinct nodes in the domain.

Equation (1.167), together with the definition of the scaled boundary transformation (Equation (1.160)), form the basis of the scaled boundary finite element method.

The shear stresses $\{\tau\}$ are specified in Equation (1.3). Expressing the derivatives with respect to x, y as in Equation (1.161) and substituting Equation (1.167) yields

$$\{\tau\} = G\left([B^1]\{u(\xi)\}_{,\xi} + \frac{1}{\xi}[B^2]\{u(\xi)\}\right) \tag{1.170}$$

with (Equations (1.162) and (1.157))

$$[B^1] = \{b^1\}[N] = \frac{1}{2(x_1y_2 - x_2y_1)}\begin{bmatrix} \Delta_y(1-\eta) & \Delta_y(1+\eta) \\ -\Delta_x(1-\eta) & -\Delta_x(1+\eta) \end{bmatrix} \tag{1.171a}$$

$$[B^2] = \{b^2\}[N]_{,\eta} = \frac{1}{2(x_1y_2 - x_2y_1)}\begin{bmatrix} 2\bar{y} + \Delta_y\eta & -2\bar{y} - \Delta_y\eta \\ -2\bar{x} - \Delta_x\eta & 2\bar{x} + \Delta_x\eta \end{bmatrix} \tag{1.171b}$$

$[B^1]$ and $[B^2]$ are independent of ξ. The right-hand side of Equation (1.170) without the factor G expresses the shear strains.

As the starting point of the derivation of the scaled boundary finite element method, the same virtual work statement as in finite elements (Equation (1.53)) can be selected. The virtual work statement with the virtual displacement w yields

$$\int_V [w_{,x} w_{,y}]^T \{\tau\} \mathrm{d}V = \int_V w^T p \mathrm{d}V + \int_S w^T \tau_n \mathrm{d}S \tag{1.172}$$

with the virtual strains $w_{,x}, w_{,y}$. The left-hand side represents the internal virtual work (elastic strain energy), and the right-hand side the external work of the body load and of the surface traction τ_n on the boundary S.

The integrations of Equation (1.53) apply to the total domain, with S denoting its boundary. It can, however, be decomposed into the subdomains (triangular wedges with apexes in O, 1 and 2, line elements with boundaries 1-2) corresponding to the individual line elements, which are then summed. To simplify the nomenclature, the same symbols are used for quantities associated with the individual elements and with the total domain consisting of the assemblage of all elements. In the following, an individual line element (Figure 1.15) is addressed.

The virtual displacement is discretised in the same way as the displacement (Equation (1.167)) is

$$w(\xi, \eta) = [N(\eta)]\{w(\xi)\} \tag{1.173}$$

with the virtual displacements on the lines passing through O and the nodes 1, 2

$$\{w(\xi)\} = \begin{Bmatrix} w_1(\xi) \\ w_2(\xi) \end{Bmatrix} \tag{1.174}$$

Analogous to Equation (1.170), the virtual strains are equal to

$$\begin{Bmatrix} w_{,x} \\ w_{,y} \end{Bmatrix} = [B^1]\{w(\xi)\}_{,\xi} + \frac{1}{\xi}[B^2]\{w(\xi)\} \tag{1.175}$$

Substituting Equations (1.175), (1.170) and (1.165) in the left-hand side of Equation (1.172) yields

$$\begin{aligned}
&\int_V [w_{,x} w_{,y}]^T \{\tau\} \mathrm{d}V = \\
&\int_0^1 \{w(\xi)\}^T{}_{,\xi} \int_{-1}^{+1} [B^1]^T G [B^1] |J| \mathrm{d}\eta \, \xi \{u(\xi)\}_{,\xi} \mathrm{d}\xi \\
&+ \int_0^1 \{w(\xi)\}^T{}_{,\xi} \int_{-1}^{+1} [B^1]^T G [B^2] |J| \mathrm{d}\eta \{u(\xi)\} \mathrm{d}\xi \\
&+ \int_0^1 \{w(\xi)\}^T \int_{-1}^{+1} [B^2]^T G [B^1] |J| \mathrm{d}\eta \{u(\xi)\}_{,\xi} \mathrm{d}\xi \\
&+ \int_0^1 \{w(\xi)\}^T \int_{-1}^{+1} [B^2]^T G [B^2] |J| \mathrm{d}\eta \tfrac{1}{\xi} \{u(\xi)\} \mathrm{d}\xi
\end{aligned} \tag{1.176}$$

The following coefficient matrices involving integrations on the line element 1-2 on the boundary only are defined:

$$[E^0] = \int_{-1}^{+1} [B^1]^T G [B^1] |J| \mathrm{d}\eta = G \frac{\Delta_x^2 + \Delta_y^2}{6(x_1 y_2 - x_2 y_1)} \begin{bmatrix} 2 & 1 \\ 1 & 2 \end{bmatrix} \tag{1.177a}$$

$$[E^1] = \int_{-1}^{+1} [B^2]^T G[B^1]|J|\mathrm{d}\eta = G\frac{\Delta_x^2 + \Delta_y^2}{12(x_1y_2 - x_2y_1)}\begin{bmatrix} -1 & 1 \\ 1 & -1 \end{bmatrix} - G\frac{\bar{x}\Delta_x + \bar{y}\Delta_y}{2(x_1y_2 - x_2y_1)}\begin{bmatrix} -1 & -1 \\ 1 & 1 \end{bmatrix} \tag{1.177b}$$

$$[E^2] = \int_{-1}^{+1} [B^2]^T G[B^2]|J|\mathrm{d}\eta = G\frac{\Delta_x^2 + \Delta_y^2 + 12(\bar{x}^2 + \bar{y}^2)}{12(x_1y_2 - x_2y_1)}\begin{bmatrix} 1 & -1 \\ -1 & 1 \end{bmatrix} \tag{1.177c}$$

To eliminate the derivative $\{w(\xi)\}_{,\xi}$ integration by parts with respect to the variable ξ is applied

$$\int_0^1 \{w(\xi)\}^T{}_{,\xi}\{f(\xi)\}\mathrm{d}\xi = \{w(\xi=1)\}^T\{f(\xi=1)\} - \int_0^1 \{w(\xi)\}^T\{f(\xi)\}_{,\xi}\mathrm{d}\xi \tag{1.178}$$

Applying Equation (1.178) to the first and second terms of the right-hand sides of Equation (1.176) leads for this equation to

$$\begin{aligned} &\{w(\xi=1)\}^T[E^0]\{u(\xi=1)\}_{,\xi} - \int_0^1 \{w(\xi)\}^T[E^0]\xi\{u(\xi)\}_{,\xi\xi}\mathrm{d}\xi \\ &- \int_0^1 \{w(\xi)\}^T[E^0]\{u(\xi)\}_{,\xi}\mathrm{d}\xi + \{w(\xi=1)\}^T[E^1]^T\{u(\xi=1)\} \\ &- \int_0^1 \{w(\xi)\}^T[E^1]^T\{u(\xi)\}_{,\xi}\mathrm{d}\xi + \int_0^1 \{w(\xi)\}^T[E^1]\{u(\xi)\}_{,\xi}\mathrm{d}\xi \\ &+ \int_0^1 \{w(\xi)\}^T[E^2]\frac{1}{\xi}\{u(\xi)\}\mathrm{d}\xi \end{aligned} \tag{1.179}$$

The right-hand side of Equation (1.172), after substitution of Equations (1.173), (1.165) and (1.166) yields

$$\begin{aligned} &\int_V w^T p \mathrm{d}V + \int_S w^T \tau_n \mathrm{d}S \\ &= \int_0^1 \{w(\xi)\}^T \xi \int_{-1}^{+1} [N(\eta)]^T p|J|\mathrm{d}\eta\mathrm{d}\xi + \{w(\xi=1)\}^T \int_{-1}^{+1} [N(\eta)]^T \tau_n \Delta_s \mathrm{d}\eta \end{aligned} \tag{1.180}$$

Introducing the contributions of the body load and surface traction on the boundary

$$\{F^{\mathrm{b}}\} = \int_{-1}^{+1} [N(\eta)]^T p|J|\mathrm{d}\eta \tag{1.181a}$$

$$\{Q\} = \int_{-1}^{+1} [N(\eta)]^T \tau_n \Delta_s \mathrm{d}\eta \tag{1.181b}$$

leads, in Equation (1.180), to

$$\int_0^1 \{w(\xi)\}^T \xi\{F^{\mathrm{b}}\}\mathrm{d}\xi + \{w(\xi = 1)\}^T\{Q\} \tag{1.182}$$

Equating the left-hand side (Equation (1.179)) and the right-hand side (Equation (1.182)), and multiplying by ξ, yields

$$\begin{aligned}&\{w(\xi = 1)\}^T \xi([E^0]\{u(\xi = 1)\},_\xi + [E^1]^T\{u(\xi = 1)\} - \{Q\})\\ &+ \{w(\xi)\}^T \int_0^1 (-[E^0]\xi^2\{u(\xi)\},_{\xi\xi} - ([E^0] - [E^1] + [E^1]^T)\xi\{u(\xi)\},_\xi\\ &+ [E^2]\{u(\xi)\} - \xi^2\{F^{\mathrm{b}}\})\mathrm{d}\xi = 0\end{aligned} \tag{1.183}$$

As the virtual displacements $\{w(\xi = 1)\}$ are arbitrary, the coefficient matrix of the first term must vanish, leading to

$$[E^0]\{u(\xi = 1)\},_\xi + [E^1]^T\{u(\xi = 1)\} - \{Q\} = 0 \tag{1.184}$$

which defines the internal nodal forces of the boundary ($\xi = 1$) as a function of $\{u(\xi = 1)\},_\xi$ and $\{u(\xi = 1)\}$ (first two terms in Equation (1.184)) and expresses equilibrium with the nodal load due to the surface traction $\{Q\}$. As also the virtual displacements $\{w(\xi)\}$ are arbitrary, the coefficient matrix consisting of an integral along ξ must vanish. Setting the integrand equal to zero yields

$$\begin{aligned}&[E^0]\xi^2\{u(\xi)\},_{\xi\xi} + ([E^0] - [E^1] + [E^1]^T)\xi\{u(\xi))\},_\xi\\ &- [E^2]\{u(\xi)\} + \xi^2\{F^{\mathrm{b}}\} = 0\end{aligned} \tag{1.185}$$

Note that the coefficient matrices $[E^0], [E^1], [E^2]$ are independent of ξ. Integrations over the line element ($\xi = 1$) only are involved. The expressions for $[E^0], [E^1], [E^2]$ (Equations (1.177)) are similar to that for the static-stiffness matrix of a standard line finite element. The same analogy exists for the body loads $\{F^{\mathrm{b}}\}$ (Equation (1.181a)).

Equation (1.185) applies to the domain of the individual line element, a triangular wedge (Figure 1.15). To model the total domain, an assemblage as in the conventional finite element method is performed (Step 4 of section 1.4). As already mentioned, the same symbols are used for the assembled coefficient matrices, the assembled displacements and the assembled nodal loads. For instance, $\{u(\xi)\}$ describes the displacements along all lines passing through the scaling centre and the nodes on the boundary. Formally, Equation (1.185) also represents the total assembled system.

Equation (1.185) represents the *scaled boundary finite element equation in displacement. It is a system of linear non-homogeneous second-order ordinary differential equations for the displacements* $\{u(\xi)\}$ *with the (dimensionless) radial coordinate* ξ *as the independent variable*. The coefficients are not constant, but are polynomials with the same power as the order of the derivative ($\xi^2\{u(\xi)\},_{\xi\xi}, \xi\{u(\xi)\},_{\xi}, \{u(\xi)\}$). The non-homogeneous body load term leads to a particular solution, which can be added to that of the homogeneous case for this linear system. These so-called Euler–Cauchy differential equations can easily be solved analytically. The scaled boundary finite element method is thus a numerical finite element-based procedure in the circumferential direction η and an analytical procedure in the radial direction ξ.

The wedge problem defined in section 1.2 is solved as an illustrative example (top of Figure 1.16). The side 1-2 is discretised with a single line element with two nodes. The scaling centre is selected at O, the intersection point of the other two sides. As will be explained in Chapter 3, straight lines of the boundary passing through the scaling centre do not have to be discretised. The dimensionless radial coordinates $\xi = 0$ in O and $\xi = 1$ on the side 1-2 apply. The prescribed linear displacement on side 1-2 is compatible with the shape function, leading to $u_1 = u(\xi = 1, \eta = -1) = 0, u_2 = u(\xi = 1, \eta = +1) = \bar{u}$.

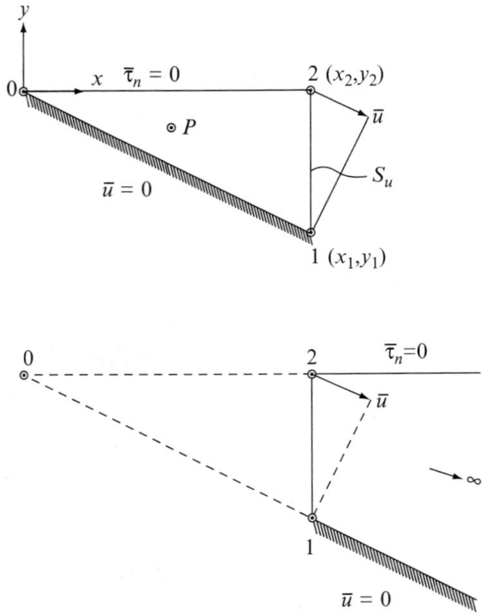

Figure 1.16 Triangular wedge and associated semi-infinite 'trapezoidal' wedge

After enforcing the displacement boundary condition $u_1(\xi) = 0$, the scaled boundary finite element equation in displacement (Equation (1.185)) with the variable $u_2(\xi)$ equals

$$E^0\xi^2 u_2(\xi),_{\xi\xi} + E^0\xi u_2(\xi),_{\xi} - E^2 u_2(\xi) + \xi^2 F_2^{\mathrm{b}} = 0 \tag{1.186}$$

The coefficients equal (Equations (1.177a) and (1.177c))

$$E^0 = G\frac{\Delta_x^2 + \Delta_y^2}{3(x_1 y_2 - x_2 y_1)} \tag{1.187a}$$

$$E^2 = G\frac{\Delta_x^2 + \Delta_y^2 + 12(\bar{x}^2 + \bar{y}^2)}{12(x_1 y_2 - x_2 y_1)} \tag{1.187b}$$

The constant body load p leads to (Equation (1.181a))

$$F_2^{\mathrm{b}} = \tfrac{1}{2}(x_1 y_2 - x_2 y_1)p \tag{1.188}$$

As only one degree of freedom on the boundary remains, Equation (1.186) is a scalar differential equation instead of a system of differential equations. Note that the coefficient E^1 does not appear.

Equation (1.186) is an Euler–Cauchy differential equation which can easily be solved analytically. First, the homogeneous equation, i.e. for vanishing body load, is addressed. Substituting

$$u_2(\xi) = \xi^{\alpha} \tag{1.189}$$

with the unknown coefficient α in Equation (1.186) with $F_2^{\mathrm{b}} = 0$ leads to

$$E^0\alpha^2 - E^2 = 0 \tag{1.190}$$

i.e.

$$\alpha_{\substack{1\\2}} = \pm\sqrt{\frac{E^2}{E^0}} = \pm\frac{1}{2}\sqrt{1 + \frac{12(\bar{x}^2 + \bar{y}^2)}{\Delta_x^2 + \Delta_y^2}} \tag{1.191}$$

The general solution of the homogeneous form of Equation (1.186) is

$$u_2(\xi) = c_1\xi^{\alpha_1} + c_2\xi^{\alpha_2} \tag{1.192}$$

with the integration constants c_1 and c_2.

Secondly, when body loads are present, a particular solution of the non-homogeneous equation is present, which is chosen as $c_3\xi^{\beta}$. Substituting in Equation (1.186) results in $\beta = 2$ and

$$c_3 = -\frac{x_1y_2 - x_2y_1}{2(4E^0 - E^2)}p \tag{1.193}$$

The general solution, the superposition of the homogeneous and particular solutions, equals

$$u_2(\xi) = c_1\xi^{\alpha_1} + c_2\xi^{\alpha_2} - \frac{x_1y_2 - x_2y_1}{2(4E^0 - E^2)}p\xi^2 \tag{1.194}$$

For the wedge, the displacement at $O(\xi = 0)$ must remain finite. As $\alpha_2 < 0$ (Equation (1.191)), c_2 must vanish, resulting in

$$u_2(\xi) = c\xi^{\alpha_1} - \frac{x_1y_2 - x_2y_1}{2(4E^0 - E^2)}p\xi^2 \tag{1.195}$$

The boundary condition $u_2(\xi = 1) = \bar{u}$ leads to

$$c = \bar{u} + \frac{x_1y_2 - x_2y_1}{2(4E^0 - E^2)}p \tag{1.196}$$

yielding

$$u_2(\xi) = \bar{u}\xi^{\alpha_1} + \frac{x_1y_2 - x_2y_1}{2(4E^0 - E^2)}p(\xi^{\alpha_1} - \xi^2) \tag{1.197}$$

The displacement in the wedge follows with $u_1(\xi) = 0$ from Equation (1.167) with Equation (1.157) as

$$u(\xi, \eta) = N_2(\eta)u_2(\xi) = \frac{1}{2}(1 + \eta)u_2(\xi) \tag{1.198}$$

The shear stress τ_x is calculated from Equation (1.170) as

$$\tau_x(\xi, \eta) = G\left(B^1u_2(\xi)_{,\xi} + \frac{1}{\xi}B^2u_2(\xi)\right) \tag{1.199}$$

with (Equation (1.171))

$$B_1 = \frac{\Delta_y}{2(x_1y_2 - x_2y_1)}(1 + \eta) \tag{1.200a}$$

$$B_2 = \frac{1}{2(x_1y_2 - x_2y_1)}(-2\bar{y} - \Delta_y\eta) \tag{1.200b}$$

For the values of the triangular wedge described in section 1.2, $\alpha_1 = 4$. The displacement in point P with $\xi = 0.5, \eta = 0$ is equal to $0.031\bar{u} + 0.012(r_0^2/G)p$.

In principle, the scaled boundary finite element equation in displacement (Equation (1.186)) also applies to the associated semi-infinite 'trapezoidal' wedge (bottom of

Figure 1.16). The sides $1-\infty$ and $2-\infty$ are fixed and free, respectively. Again, a linear variation of the displacement is prescribed on side $1-2$. For the unbounded medium $1 \leq \xi < \infty$ applies. For a vanishing body load, the general solution for the displacement is specified by Equation (1.192). The boundary condition at infinity enforcing a finite displacement leads to $c_1 = 0$. After formulating $u_2(\xi = 1) = \bar{u}$, the displacement of the unbounded domain equals

$$u_2(\xi) = \bar{u}\xi^{\alpha_2} \tag{1.201}$$

For the specific example, $\alpha_2 = -4$ applies.

Although the derivation of the scaled boundary finite element method in this section is very sketchy, some characteristics of the novel procedure can be mentioned:

- The procedure is *semi-analytical, transforming* the *partial differential equation* using a virtual work statement as in finite elements in the circumferential direction *to* an *ordinary differential equation*, with the radial coordinate as the independent variable, which can then be solved analytically.
- Only the *boundary is spatially discretised*, resulting in a reduction of the spatial discretisation by one, but in contrast to the standard boundary element method, *no fundamental solution* is required and *no singular integrals* occur.
- The analytical solution in the radial direction permits the *boundary condition at infinity* to be satisfied *rigorously*.

The scaled boundary finite element method is a boundary element method, but based on finite elements, tending to combine the advantages of the boundary and finite element methods. It also has appealing features of its own, for instance, that certain straight boundaries passing through the scaling centre are not discretised. The novel procedure also combines the advantages of the numerical and analytical procedures.

2 Novel Computational Procedure

The most widely used computational procedures in solid mechanics are the finite element method and the boundary element method, with their own specific features, advantages and disadvantages, described in sections 1.4 and 1.5. For the sake of illustration, the linear (three-dimensional) analysis in elastodynamics of a bounded medium and of an unbounded (infinite) medium is addressed (Figure 2.1). In the domain V the equations of motion apply, which are partial differential equations in displacements u. Body loads p are present. On the boundary S either the displacements u on S_u or the surface tractions t on S_t are prescribed. For an unbounded medium (Figure 2.1b), an additional boundary condition at infinity ('radiation condition') must be satisfied. As initial conditions, the displacements and velocities vanish in the domain.

The key *advantages* of the *finite element method* (see also the end of section 1.4) are listed as follows:

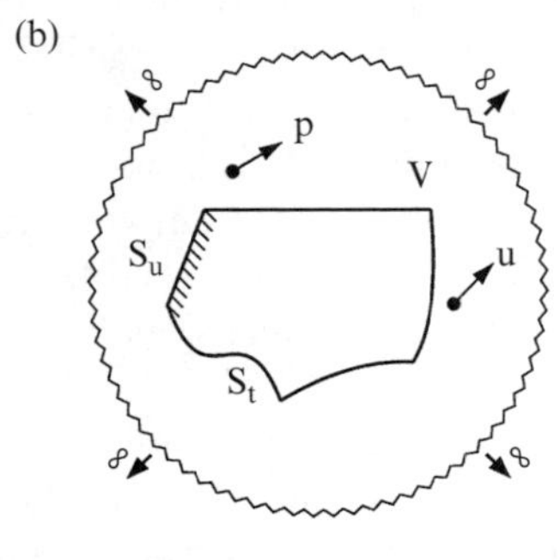

Figure 2.1 Problem definition in elastodynamics (section): (a) bounded medium, (b) unbounded medium (Reprinted from *Computers and Structures*, Vol. 78, Wolf and Song, 'The scaled boundary finite element method – a primer: derivations', pp. 191–210, © (2000), with permission from Elsevier Science)

- No fundamental solution, which is an analytical solution of the governing partial differential equations, and thus problem-dependent, is required, expanding the scope of application to, for example, anisotropic material without any increase in computational effort and avoiding singular integrals.
- The assumption for the approximate displacements and weighting functions of piecewise local shape functions in a finite element in the form of polynomials leads to standard integration of regular functions, and for all problems to a unified procedure yielding a local simple solution reflecting the dynamic behaviour.
- After assemblage of these contributions of the individual finite elements, which corresponds to summing, the resulting system of equations is symmetric, sparse and banded. Symmetry, if present in the exact solution, is also achieved in the numerical procedure.
- For statics, increasing the number of degrees of freedom by refining the mesh leads to convergence from the 'stiff' side.
- Inhomogeneous material can be analysed without any increase in computational effort.

The key *advantages* of the *boundary element method* (see also the end of section 1.5) are listed as follows:

- The spatial dimension of the problem is reduced by one, as only the boundary is discretised with surface boundary elements, reducing the computational and data preparation efforts.
- The boundary condition at infinity ('radiation condition') is satisfied rigorously.

The novel computational procedure called the *scaled boundary finite element method*, is a boundary element method based on finite element concepts. It *combines* the *advantages of the finite element and boundary element methods* described above, but avoids the corresponding disadvantages. In other words, the scaled boundary finite element method stems from addressing the disadvantages of the finite element and boundary element methods, much as an oyster responds to a grain of sand. In addition, the novel procedure presents appealing features of its own. These will just be mentioned in the following. A thorough explanation is not possible at this stage, and a comprehensive understanding is not a requirement to study the chapters of the Model Problem in Part I.

For most practical cases, the complicated geometry, the shape of the boundary, the variations of the material properties and of the prescribed loads in the domain and the boundary conditions preclude an analytical solution of the governing partial differential equations. However, if the physical problem is governed by ordinary differential equations, classical mathematical techniques can, in important cases, lead to an exact analytical solution in the single independent variable. For instance, in certain cases when symmetry exists, leading to a one-dimensional problem, the governing ordinary differential equations in, as an example the radial coordinate, can thus be solved exactly, while the problem in three dimensions cannot be addressed analytically. To make use of this advantage also for the general case without any symmetry, a coordinate system consisting of the radial direction and two local circumferential directions (parallel to the boundary) is introduced. The governing partial differential equations are transformed into this coordinate system. In the circumferential directions the boundary is discretised with surface finite elements, reducing the governing partial differential equations to ordinary differential equations in the radial coordinate. The coefficients of the ordinary differential equations are determined by the numerical finite element approximation in the circumferential directions. These ordinary differential equations are then solved analytically in the radial direction. Effectively, the governing partial differential equations are weakened in the circumferential directions in a finite element manner, but remain strong in the radial direction. Thus, the novel computational procedure *combines the advantages of the analytical and numerical approaches*. The method is thus a *semi-analytical* procedure for solving (linear) partial differential equations. In the circumferential directions (parallel to the boundary), where the behaviour is, in general, smooth, the weighted-residual approximation of finite elements applies, leading to convergence in the finite element sense. The material properties can vary in the circumferential directions. For a bounded domain, the radial coordinate points from the boundary towards the interior, where in the presence of singularities the analytical solution permits the stress-intensity factors to be calculated directly based on their definition. For an unbounded domain the radial coordinate points from the boundary towards infinity, where the boundary conditions at infinity (radiation condition) can be incorporated exactly in the analytical solution. The analytical solution in the radial direction permits, for instance, a displacement at a specific location to be calculated directly without processing all displacements for increasing radial coordinate, starting from the boundary up to the location as in a numerical approach. As a consequence of the fact that the solution is exact in the radial direction, part of the boundary A_u (subscript u for prescribed displacements) and A_t (subscript t for prescribed surface tractions) does not have to be discretised, as shown in Figure 2.2a for a bounded medium, and in Figure 2.2b for an unbounded medium. As will be demonstrated, this feature allows the crack faces in a bounded medium and the free surface in an unbounded (semi-infinite) medium such as a half-space to be represented without discretisation. These boundary conditions on A_u, A_t are satisfied exactly. The same advantage also applies to certain interfaces between different materials (in the standard boundary element method, the

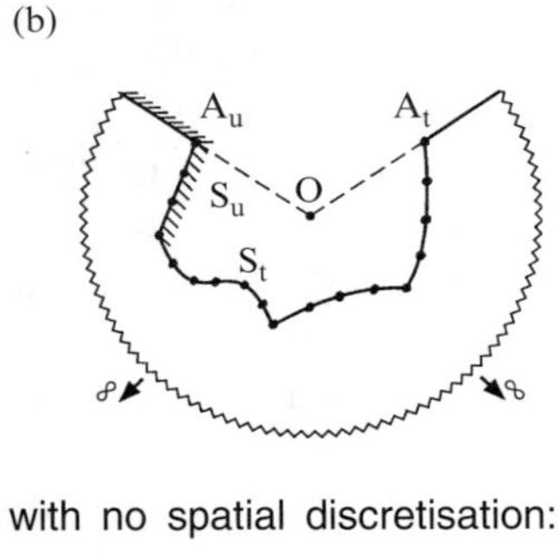

Figure 2.2 Part of boundary with no spatial discretisation: (a) bounded medium, (b) unbounded medium

surfaces A_u, A_t would have to be discretised when the fundamental solution of the full-space is used).

In a nutshell, the scaled boundary finite element method is a *semi-analytical fundamental-solution-less boundary element method based on finite elements*. Paraphrasing the title of the famous paper by Zienkiewicz, the best of both worlds (marriage *à la mode*) is achieved in two ways: with respect to the analytical and numerical methods; and with respect to the finite element and boundary element methods within the numerical procedures.

The scaled boundary finite element method has been developed for unbounded and bounded two- and three-dimensional media during the past few years, addressing elastodynamics and diffusion as examples of linear governing differential equations. Both the frequency and time domain solutions, as well as statics, were examined. Two derivations were presented. The mechanically-based derivation developed originally applies finite element assemblage and similarity, and lead to the name consistent infinitesimal finite element cell method for the novel approach (References [S3], [W10] for the scalar wave equation, [W7], [S5], [W11] for the vector wave equation and [S6] for the incompressible vector wave equation; [W8] for statics and [S4] for diffusion). The so-called scaled boundary transformation-based derivation, developed later in an effort to streamline the procedure, starts from the governing partial differential equations. The weighted-residual technique is then applied, which is mathematically more appealing and is consistent with today's finite element technology (References [S7], [S8], [S9], [S10], [W12], [W14], [S13]). Early results were also described in part of the book [W9] with software, where further references addressing the historical development are specified. This free computer program SIMILAR, including its source code, can be downloaded from the Internet (addresses see under Reference [W9]).

To describe the basic concepts on which the scaled boundary finite element method is based, a primer has been written, expanding two conference papers (References [W13], [S11]). A *model problem* is defined, which is easy to follow. The *simplest wave propagation problem* is discussed, which still *contains all essential features*: the scalar wave equation in two dimensions, which plays a dominant role in many fields of engineering and physics. A wedge as a bounded medium and a truncated semi-infinite wedge as an unbounded medium are used to derive the governing equations and solution procedures step-by-step in great detail. The Model Problem is addressed in Chapters 3–14 of Part I, which is the core of the primer. At the end of Part I, three short appendices lead to deeper insight into certain aspects of the Model Problem, and provide the link to the generalisation to two- and three-dimensional statics, elastodynamics and diffusion of Part II. In Chapters 15–22 a detailed derivation from scratch of the fundamental equations and solution procedures is specified for easy reference, which are necessary to implement the method in a specific area. For the sake of illustration, Chapters 23 and 24 describe examples for bounded and unbounded media, emphasising the salient advantageous features. Error estimation and adaptivity is addressed in Chapter 25, which permits comparisons of the computational efficiency with that of the standard finite element method. Chapter 26 contains concluding remarks, addressing restrictive properties of the novel method (where there is light, there is also shadow!), and suggestions for future research.

Part I

Model Problem: Line Element for Scalar Wave Equation

The salient features of the scaled boundary finite element method are listed in Chapter 2. To discuss the derivation of the governing equations and their solution of this novel procedure in great detail, a Model Problem is addressed, which is easy to understand. The scalar wave equation, the Helmholz equation, in two dimensions is discussed. A wedge as a bounded medium and a truncated semi-infinite wedge as an unbounded medium are examined. This Model Problem exhibits all essential features, but is so simple that after appropriate boundary conditions are enforced, the scaled boundary finite element equations for a single boundary element are scalar. Their solution leads to closed-form expressions, which can be derived using elementary mathematics. Solution procedures which are also applicable to the vector and matrix cases are then discussed. The generalisation of the concepts, derivations and solution procedures to other partial differential equations performed in Part II is possible without much additional effort.

The concepts of the so-called scaled boundary transformation of the geometry and of similarity are discussed in Chapter 3, which permits the domain to be described by a boundary discretisation with finite elements and the radial coordinate. The model problem is defined in Chapter 4, including the derivation of the governing differential equation. Two derivations of the scaled boundary finite element equations are presented. In Chapter 5, the so-called scaled boundary transformation and the weighted-residual technique lead to the scaled boundary finite element equations in displacement and in dynamic stiffness. In Chapter 6, the same equations are obtained by the mechanically-based derivation using similarity and finite element assemblage. Both derivations have their own characteristics: the former, starting from the transformed governing differential equation and applying the method of weighted residuals, will appeal to today's numerical analyst; the latter, starting from the dynamic stiffness familiar to engineers, is physically motivated, providing mechanical insight. As already mentioned, for a single boundary element and appropriate boundary conditions, the scaled boundary finite element equations are scalar, which are solved analytically in this special case in Chapter 7. Mostly analytical solution procedures that also apply to the vector and matrix cases are then discussed. In Chapter 8, the displacements and stiffnesses in statics for bounded and unbounded media are addressed. In Chapter 9, the mass of the bounded medium is examined. In Chapter 10, the high-frequency asymptotic expansion for dynamic stiffness of the unbounded medium satisfying the radiation condition is calculated. In Chapter 11, the dynamic stiffness, the unit-impulse response and the displacement of the unbounded medium are determined numerically. In Chapter 12, an analytical solution in the frequency domain for the displacements and dynamic stiffnesses of the bounded and unbounded media is derived. In Chapter 13, implementation issues are discussed, which also apply to the general matrix equations. Conclusions resulting from addressing the Model Problem are stated in Chapter 14. Three appendices then follow, which provide further insight into key aspects. Appendix A deals with solid modelling; Appendix B discusses the analysis in the frequency domain; also providing the link to the time domain, and Appendix C establishes the equations of motion of a dynamic unbounded medium-structure interaction problem using the properties calculated in the Model Problem (static stiffness and mass of bounded wedge, dynamic stiffness and unit-impulse response of unbounded truncated semi-infinite wedge). Finally, Appendix D describes the early historical development leading up to the scaled boundary finite element method.

Chapter 3–14 of Part I are based on two tutorial papers [W13], [S11] with significant extensions.

3 Concepts of Scaled Boundary Transformation of Geometry and Similarity

To explain the concept of the scaled boundary finite element method in general terms, a three-dimensional linear elastic bounded medium with a section shown in Figure 2.1a is addressed. The so-called *scaling centre O* is chosen in a zone from which the total boundary must be visible (Figure 3.1a). This can always be achieved by subdividing the total domain in subdomains, which can be regarded as superelements. As an example, the domain in Figure 3.2 is subdivided into two superelements, each with its own scaling centre. The superelements do not have to be convex. This creates an additional boundary common to both superelements that is discretised analogously as the rest of the boundary. For a bounded medium the scaling centre will, in general, be located inside the domain. Only the boundary is discretised with doubly-curved surface finite-elements with any arrangement of nodes (Figure 3.1a).

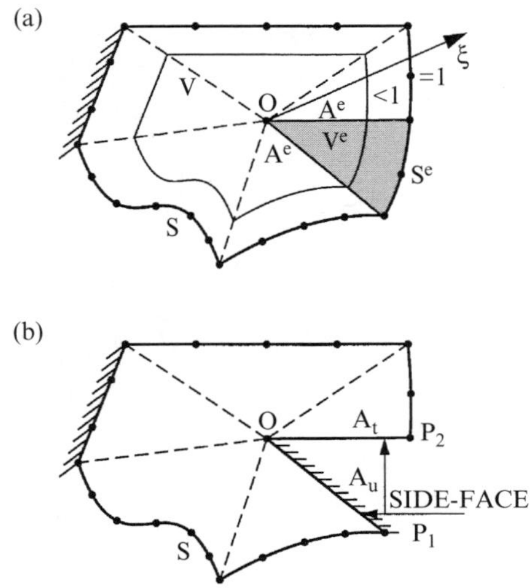

Figure 3.1 Modelling of bounded medium with surface finite elements (section) and choice of scaling centre: (a) inside domain, (b) on boundary (Reprinted from *Computers and Structures*, Vol. 78, Wolf and Song, 'The scaled boundary finite element method – a primer: derivations', pp. 191–210, © (2000), with permission from Elsevier Science)

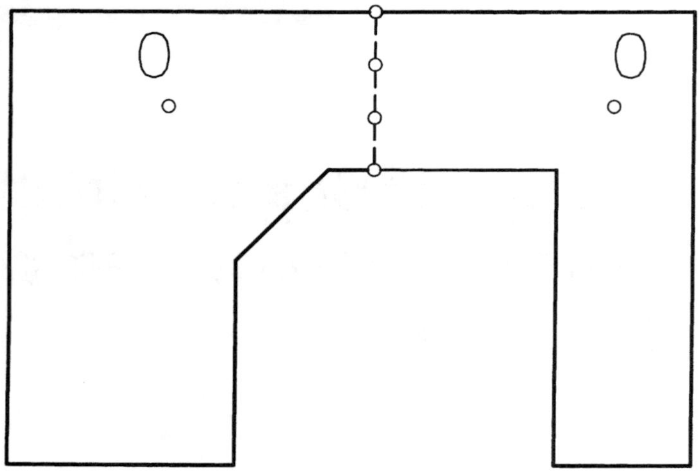

Figure 3.2 Subdivision of total domain into two superelements (subdomains), each with its own scaling centre

This surface finite element could also be called a scaled boundary finite element or scaled boundary element. A typical finite element with the surface S^e (superscript e for element) is shown in Figure 3.3. Two local curvilinear coordinates η, ζ in the circumferential directions of the surface ($-1 \leq \eta \leq +1, -1 \leq \zeta \leq +1$) are defined. This is similar to the procedure used in an isoparametric surface finite element (see Appendix A, Figure A.1). The dimensionless radial coordinate ξ points from the scaling centre O to a point on the boundary (Figures 3.1a, 3.3). Through scaling, $\xi = 0$ in O and $\xi = 1$ on the boundary are chosen. Connecting the edge of the surface finite element by straight lines to the scaling centre O forms the side-face A^e, which corresponds to either $\eta = \pm 1$ or $\zeta = \pm 1$. The side-face A^e and the surface S^e define a pyramid with apex O and volume V^e. The geometry of this pyramid is conveniently defined by

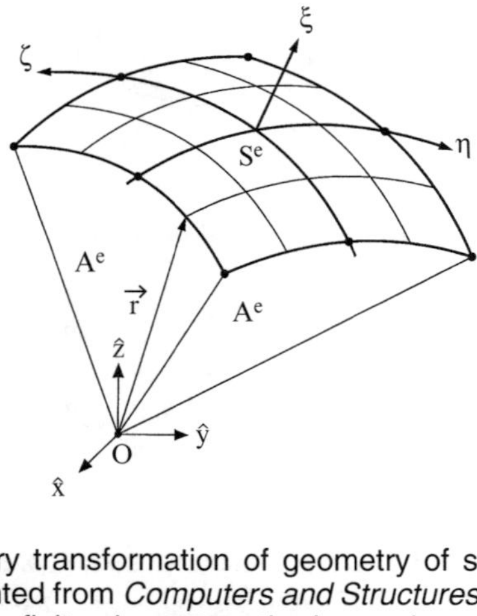

Figure 3.3 Scaled boundary transformation of geometry of surface finite element forming pyramid (Reprinted from *Computers and Structures*, Vol. 78, Wolf and Song, 'The scaled boundary finite element method – a primer: derivations', pp. 191–210, © (2000), with permission from Elsevier Science)

the *scaled boundary coordinates* ξ, η, ζ, with $0 \leq \xi \leq 1$. A transformation of the geometry from the Cartesian coordinates $\hat{x}, \hat{y}, \hat{z}$ to the dimensionless coordinates ξ, η, ζ is performed. The transformation is unique due to the choice of the scaling centre. This transformation is similar to the representation of a spherical surface introducing spherical coordinates where, on the surface, the radial coordinate is constant, but more general.

The dimensionless radial coordinate ξ can be interpreted as a scaling factor. As already mentioned, $\xi = 1$ corresponds to the surface finite element on the boundary. A surface corresponding to a constant ξ smaller than 1 is similar to the surface finite element on the boundary. O is denoted as the scaling centre, as this point is obtained for $\xi = 0$. The point O can also be called the similarity centre, and ξ represents a (dimensionless) characteristic length.

This transformation, defined by the scaling factor and the local coordinates of the surface finite elements, is called the *scaled boundary transformation*. Assembling all the pyramids by connecting their side-faces, which corresponds to enforcing compatibility and equilibrium, results in the total medium with volume V and the closed boundary S. No side-faces A^e passing through the scaling centre remain.

Summarising, *scaling of the boundary discretised with finite elements* is applied, which explains the name scaled boundary finite element method.

When during the assemblage process some side-faces of the pyramids are not connected or some pyramids are missing, an additional boundary A passing through the scaling centre O is created (see also Figure 2.2a). The case of a missing pyramid is illustrated in Figure 3.1b, with the displacements and surface tractions prescribed on A_u and A_t, respectively. As the side-faces are obtained by scaling the edges P_1, P_2 of the surface finite elements on the boundary, the boundary A_u, A_t is modelled without spatial discretisation. Or, in other words, the boundary conditions on A_u and A_t are satisfied exactly by formulating them only in P_1 and P_2, respectively. The spatial dimension of the side-face A is reduced by one when viewed from O. In this case, the total boundary is decomposed into two parts: that part of the boundary passing through the scaling centre denoted as the side-face A; and the remaining part called the interface S (Figure 3.1b). The term 'interface' is used only when a distinction with the word boundary is necessary. Only the interface is discretised with (doubly-curved) surface finite elements. The scaling centre O is thus chosen on the (total) boundary. Applications of this powerful extension exist in fracture mechanics, where the scaling centre O is chosen at the crack tip. The adjacent sides of pyramids forming the crack faces O-P_1, O-P_2 are not connected. Note that the crack faces are not discretised.

The same concept also applies to an unbounded medium shown in Figure 2.1b. The scaling centre O is chosen outside the unbounded medium, from which the total boundary is visible (Figure 3.4a), with a typical finite element shown in Figure 3.3. The curvilinear coordinates η, ζ on the surface of the boundary remain unchanged, which still corresponds to $\xi = 1$. A truncated semi-infinite pyramid with its apex at the scaling centre O is defined by a scaling factor larger than one $1 \leq \xi < \infty$ (Figure 3.4a). Assembling all the truncated semi-infinite pyramids corresponding to the surface finite elements yields the unbounded medium. When during the assemblage process some side-faces are not connected or some truncated semi-infinite pyramids $(O-)P_1 - \infty, (O-)P_2 - \infty$ are missing, the remaining side-faces of the truncated semi-infinite pyramids form an additional boundary (A_u and A_t) extending to infinity

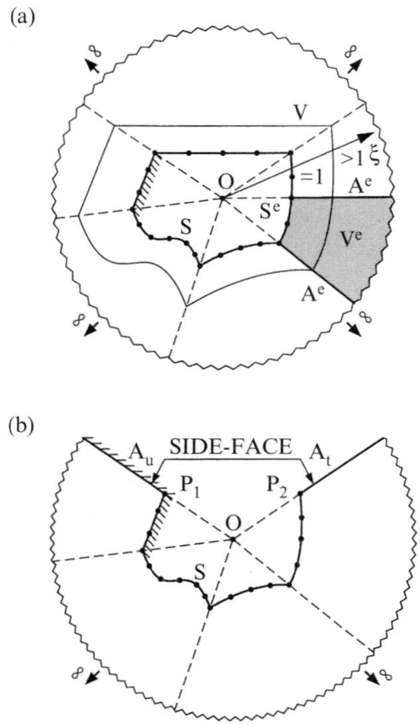

Figure 3.4 Modelling of unbounded medium with surface finite elements (section) and choice of scaling centre: (a) outside domain, (b) on extension of boundary (Reprinted from *Computers and Structures*, Vol. 78, Wolf and Song, 'The scaled boundary finite element method – a primer: derivations', pp. 191–210, © (2000), with permission from Elsevier Science)

(Figure 3.4b). The extension of these side-faces A_u, A_t of the unbounded medium pass through the scaling centre O. Again, the side-face is not discretised. This feature allows the free surface of a half-space to be modelled without discretisation, which can be applied when representing a foundation embedded in a half-space.

It is acknowledged that the discussion of the scaled boundary transformation of geometry discussed in this chapter is quite general. It is more extensive than would be necessary just to understand the procedure applied to the Model Problem. It is hoped that this will permit the reader to gain insight into further aspects while studying the Model Problem.

4 Wedge and Truncated Semi-infinite Wedge of Shear Plate

While the concepts of the scaled boundary transformation of geometry and of similarity are discussed for the general case in Chapter 3, for the derivations and solution procedures a specific simple example that still exhibits all essential features is chosen. The extension to more complicated cases of practical significance is reasonably straightforward.

This Model Problem addresses the two-dimensional out-of-plane (anti-plane) motion of elastodynamics, which is already used for the illustrative example in section 1.2. The detailed derivation of the governing differential equations will not be repeated (see Figure 1.2 and the accompanying text). Only the key relations are summarised for easy reference.

The coordinates of a point in the domain are denoted as $\hat{x}$, $\hat{y}$ as x, y are reserved for the boundary. The out-of-plane motion with the displacement u perpendicular to the plane is discussed. Formulating equilibrium in the Cartesian coordinate system yields (Equation (1.9))

$$\tau_{x,\hat{x}} + \tau_{y,\hat{y}} - \rho\ddot{u} + p = 0 \tag{4.1}$$

with the body load p per unit volume acting perpendicular to the plane (dimension force per length^3) and the mass density ρ (dimension mass per length^3). To simplify the nomenclature, the circumflex ^ is omitted when used in a subscript to indicate a direction. The stresses acting perpendicular to the plane are equal to (Equation (1.3))

$$\{\tau\} = \begin{Bmatrix} \tau_x \\ \tau_y \end{Bmatrix} = G\begin{Bmatrix} u,_{\hat{x}} \\ u,_{\hat{y}} \end{Bmatrix} \tag{4.2}$$

with the shear modulus G and $u,_{\hat{x}}, u,_{\hat{y}}$ denoting the shear strains. Equation (4.1) is reformulated for later use as

$$\begin{Bmatrix} \dfrac{\partial}{\partial\hat{x}} \\ \dfrac{\partial}{\partial\hat{y}} \end{Bmatrix}^T \{\tau\} - \rho\ddot{u} + p = 0 \tag{4.3}$$

Substituting Equation (4.2) into Equation (4.1) leads to the governing partial differential equation, the scalar wave equation (Equation (1.11)) in the domain V, also called the Helmholz equation

$$u_{,\hat{x}\hat{x}} + u_{,\hat{y}\hat{y}} - \frac{1}{c_s^2}\ddot{u} + \frac{p}{G} = 0 \tag{4.4}$$

with the shear-wave velocity (Equation (1.12))

$$c_s = \sqrt{\frac{G}{\rho}} \tag{4.5}$$

The surface traction representing the shear stress τ_n with the outward normal $\{n\}$ with the components $(n_x,\ n_y)$ (subscript n for normal) equals (Equations (1.7) and (1.8))

$$\tau_n = \{n\}^T\{\tau\} = Gu_{,n} \tag{4.6}$$

The boundary value problem consists (with vanishing initial displacements and velocities) of solving the wave equation (Equation (4.4)) subject to the boundary conditions on S_u (prescribed displacement $\bar{u}$) and on S_t (prescribed surface traction $\bar{\tau}_n$)

$$u = \bar{u} \text{ on } S_u \tag{4.7a}$$

$$\tau_n = \bar{\tau}_n \text{ on } S_t \tag{4.7b}$$

Both a bounded medium and an unbounded medium are addressed. For the bounded medium, a wedge in the shape of a triangle with body load p is addressed (Figure 4.1a). The three vertices are the origin O and Points 1 and 2 with the indicated coordinates. On the sides $O-1$ and $O-2$ the displacement $u = \bar{u} = 0$ and the surface traction $\tau_n = \bar{\tau}_n$, respectively, are prescribed. On the side $1-2$, the prescribed displacement varies linearly from 0 at Point 1 to the value $\bar{u}$ at 2. The medium is initially at rest.

For the corresponding unbounded medium, a truncated semi-infinite wedge with body load p is examined (Figure 4.1b). The same displacement boundary condition as for the bounded medium is prescribed on the side $1-2$. The same boundary conditions apply to the sides extending from Points 1 and 2 to infinity as from $O-1$ and $O-2$, respectively.

For dynamics in the frequency domain, constant hysteretic material damping is also addressed. The same equations as for the elastic case apply, but replacing the shear modulus G by

$$G^* = G(1 + 2i\zeta) \tag{4.8}$$

with the material damping ratio ζ (asterisk for material damping). Note that the shear-wave velocity (Equation (4.5)) is also modified

$$c_s{}^* = c_s\sqrt{1 + 2i\zeta} \tag{4.9}$$

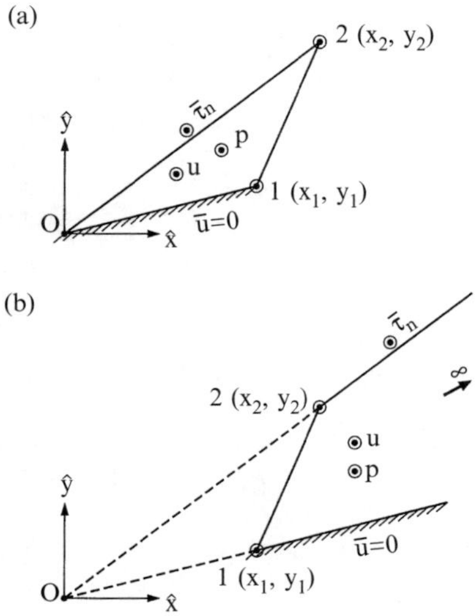

Figure 4.1 Out-of-plane motion of: (a) wedge, (b) truncated semi-infinite wedge (Reprinted from *Computers and Structures*, Vol. 78, Wolf and Song, 'The scaled boundary finite element method – a primer: derivations', pp. 191–210, © (2000), with permission from Elsevier Science)

As already mentioned in section 1.2, the reader who is not familiar with structural mechanics should 'translate' the Model Problem defined above to an example governed by the Helmholz equation in his or her field of expertise.

5 Scaled Boundary Transformation-based Derivation

The first derivation presented in this chapter follows the state-of-the-art discussion of numerical analysis. It will become clear that in the radial direction the governing differential equations are satisfied 'exactly', while in the circumferential directions a weighted residual formulation as in finite elements introduces an approximation. In the following, the ''signs used with exactly will be deleted when appropriate.

As demonstrated in section 1.4, in the finite element method the equations of motion of a single element are derived first, which leads to its static-stiffness and mass matrices. Assemblage of the finite elements enforcing equilibrium and compatibility then yields the equations of motion of the global system with its corresponding property matrices. The derivation of the scaled boundary finite element equations is analogous.

5.1 SCALED BOUNDARY TRANSFORMATION OF GEOMETRY

An individual line finite element is addressed (Figure 5.1a). Interpreting Figure 3.1a for the two-dimensional case, deleting the midnode, and for a straight line S^e with local nodes 1 and 2, this line finite element is shown shaded. The scaling centre O coincides with the origin of the Cartesian coordinate system. Moving from 1 to 2, O must be on the left. (As in Chapter 7, the Model Problem is represented by a single line finite element, on the side faces $O-1$ and $O-2$ the displacement and the surface traction must be able to be prescribed, respectively.) The coordinates of a point on the straight line element are denoted as x, y. Two nodes with the coordinates $(x_1, \ y_1)$ and $(x_2, \ y_2)$ are arranged as

$$\{x\} = \begin{Bmatrix} x_1 \\ x_2 \end{Bmatrix}$$
$$\{y\} = \begin{Bmatrix} y_1 \\ y_2 \end{Bmatrix} \tag{5.1}$$

Using the solid modelling concepts described in Appendix A, the line element is generated from its parent element (Figure 5.1b) with the mapping functions

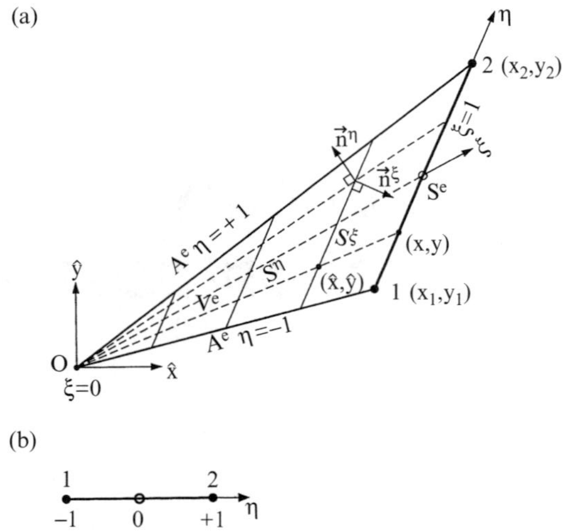

Figure 5.1 Two-node line finite element: (a) scaled boundary transformation, (b) parent element (Reprinted from *Computers and Structures*, Vol. 78, Wolf and Song, 'The scaled boundary finite element method – a primer: derivations', pp. 191–210, © (2000), with permission from Elsevier Science)

$$[N] = [N(\eta)] = \left[\frac{1}{2}(1-\eta)\ \ \frac{1}{2}(1+\eta)\right] \tag{5.2}$$

as

$$x = [N]\{x\} = \bar{x} + \frac{1}{2}\Delta_x \eta$$
$$y = [N]\{y\} = \bar{y} + \frac{1}{2}\Delta_y \eta \tag{5.3}$$

with the abbreviations

$$\Delta_x = x_2 - x_1$$
$$\Delta_y = y_2 - y_1 \tag{5.4}$$

$$\bar{x} = \frac{1}{2}(x_1 + x_2)$$
$$\bar{y} = \frac{1}{2}(y_1 + y_2) \tag{5.5}$$

The dimensionless radial coordinate ξ points from O to a point on the boundary (with $\xi = 1$). The scaled boundary transformation relates any point in the domain with the coordinates denoted as $\hat{x}, \hat{y}$ to the corresponding point on the line element as (Figure 5.1a)

$$\hat{x} = \xi x = \xi\left(\bar{x} + \frac{1}{2}\Delta_x \eta\right)$$
$$\hat{y} = \xi y = \xi\left(\bar{y} + \frac{1}{2}\Delta_y \eta\right) \tag{5.6}$$

This represents the mapping transformation (Equation (A.4)), called the *scaled boundary transformation*. The equation defines the transformation from the Cartesian coordinates $\hat{x}, \hat{y}$ to the scaled boundary coordinates ξ, η, and vice versa, which forms a right-handed system. Lines S^{ξ} with a constant ξ are parallel to the line finite element S^e (Figure 5.1a). Lines S^{η} with a constant η pass through the scaling centre O and the point on the line finite element with the same η.

The spatial derivatives in the two coordinate systems are related as (see Equation (A.6))

$$\begin{Bmatrix} \dfrac{\partial}{\partial \xi} \\ \dfrac{\partial}{\partial \eta} \end{Bmatrix} = [\hat{J}(\xi, \eta)] \begin{Bmatrix} \dfrac{\partial}{\partial \hat{x}} \\ \dfrac{\partial}{\partial \hat{y}} \end{Bmatrix} \tag{5.7}$$

with the Jacobian matrix defined as

$$[\hat{J}(\xi, \eta)] = \begin{bmatrix} \hat{x},_{\xi} & \hat{y},_{\xi} \\ \hat{x},_{\eta} & \hat{y},_{\eta} \end{bmatrix} \tag{5.8}$$

The partial derivatives of $\hat{x}, \hat{y}$ with respect to ξ, η are determined using Equation (5.6):

$$\hat{x},_{\xi} = x = \bar{x} + \frac{1}{2}\Delta_x \eta \tag{5.9a}$$

$$\hat{y},_{\xi} = y = \bar{y} + \frac{1}{2}\Delta_y \eta \tag{5.9b}$$

$$\hat{x},_{\eta} = \xi x,_{\eta} = \frac{1}{2}\Delta_x \xi \tag{5.9c}$$

$$\hat{y},_{\eta} = \xi y,_{\eta} = \frac{1}{2}\Delta_y \xi \tag{5.9d}$$

$[\hat{J}(\xi, \eta)]$ can be written as

$$[\hat{J}(\xi, \eta)] = \begin{bmatrix} 1 & \\ & \xi \end{bmatrix} [J(\eta)] \tag{5.10}$$

where

$$[J(\eta)] = \begin{bmatrix} x & y \\ x,_{\eta} & y,_{\eta} \end{bmatrix} \tag{5.11}$$

and with (Equation 5.3)

$$x,_{\eta} = [N],_{\eta}\{x\} = \frac{1}{2}\Delta_x$$
$$y,_{\eta} = [N],_{\eta}\{y\} = \frac{1}{2}\Delta_y \tag{5.12}$$

$[J(\eta)]$ is a function of the geometry of the line element only. The argument η is omitted for conciseness. The derivatives with respect to $\hat{x}, \hat{y}$ follow from Equation (5.7):

$$\left\{\begin{matrix} \frac{\partial}{\partial \hat{x}} \\ \frac{\partial}{\partial \hat{y}} \end{matrix}\right\} = [\hat{J}(\xi, \eta)]^{-1} \left\{\begin{matrix} \frac{\partial}{\partial \xi} \\ \frac{\partial}{\partial \eta} \end{matrix}\right\} \tag{5.13}$$

Substituting Equations (5.10) and (5.11) into Equation (5.13) yields

$$\left\{\begin{matrix} \frac{\partial}{\partial \hat{x}} \\ \frac{\partial}{\partial \hat{y}} \end{matrix}\right\} = \{b^1\}\frac{\partial}{\partial \xi} + \{b^2\}\frac{1}{\xi}\frac{\partial}{\partial \eta} = \left[\{b^1\}\{b^2\}\frac{1}{\xi}\right] \left\{\begin{matrix} \frac{\partial}{\partial \xi} \\ \frac{\partial}{\partial \eta} \end{matrix}\right\} \tag{5.14}$$

where

$$\{b^1\} = \frac{1}{|J|}\left\{\begin{matrix} y,_{\eta} \\ -x,_{\eta} \end{matrix}\right\} = \frac{1}{x_1 y_2 - x_2 y_1}\left\{\begin{matrix} \Delta_y \\ -\Delta_x \end{matrix}\right\} \tag{5.15a}$$

$$\{b^2\} = \frac{1}{|J|}\left\{\begin{matrix} -y \\ x \end{matrix}\right\} = \frac{1}{x_1 y_2 - x_2 y_1}\left\{\begin{matrix} -2\bar{y} - \Delta_y \eta \\ 2\bar{x} + \Delta_x \eta \end{matrix}\right\} \tag{5.15b}$$

with the determinant (Equation (5.11))

$$|J| = \frac{1}{2}(x_1 y_2 - x_2 y_1) \tag{5.16}$$

For later use, note that

$$(\{b^2\}|J|),_{\eta} = -\{b^1\}|J| \tag{5.17}$$

applies.

The infinitesimal area dV^e of the domain (see Equation (A.9)) is calculated as (Equation (5.10))

$$dV^e = |\hat{J}(\xi,\eta)| d\xi d\eta = \xi |J| d\xi d\eta \tag{5.18}$$

The infinitesimal length of a line S^{ξ} with constant ξ (Figure 5.1a) equals (Equations (5.9c) and (5.9d))

$$\mathrm{d}S^{\xi} = \sqrt{(\hat{x}_{,\eta})^2 + (\hat{y}_{,\eta})^2}\mathrm{d}\eta = \frac{1}{2}\xi\sqrt{\Delta_x^2 + \Delta_y^2}\mathrm{d}\eta \tag{5.19}$$

To be able to calculate the surface traction, the unit normal vectors $\{n^{\xi}\}$ and $\{n^{\eta}\}$ of the lines S^{ξ} and S^{η} with constant ξ and η, respectively, are addressed (Figure 5.1a):

$$\{n^{\xi}\} = \frac{1}{\sqrt{\Delta_x^2 + \Delta_y^2}}\begin{Bmatrix} \Delta_y \\ -\Delta_x \end{Bmatrix} \tag{5.20a}$$

$$\{n^{\eta}\} = \frac{1}{\sqrt{x^2 + y^2}}\begin{Bmatrix} -y \\ x \end{Bmatrix} \tag{5.20b}$$

Note that $\{n^{\xi}\}$ and $\{n^{\eta}\}$ are proportional to $\{b^1\}$ and $\{b^2\}$, respectively (Equations (5.15) and (5.20)):

$$\{b^1\} = \frac{\sqrt{\Delta_x^2 + \Delta_y^2}}{x_1 y_2 - x_2 y_1}\{n^{\xi}\} = \frac{\sqrt{\Delta_x^2 + \Delta_y^2}}{2|J|}\{n^{\xi}\} \tag{5.21a}$$

$$\{b^2\} = \frac{2\sqrt{x^2 + y^2}}{x_1 y_2 - x_2 y_1}\{n^{\eta}\} = \frac{\sqrt{x^2 + y^2}}{|J|}\{n^{\eta}\} \tag{5.21b}$$

5.2 GOVERNING EQUATIONS IN SCALED BOUNDARY COORDINATES

The scaled boundary transformation is applied to the geometry of the domain only. The shear stresses τ_x, τ_y are still defined in the original Cartesian coordinate system $\hat{x}, \hat{y}$. The equilibrium equation (Equation (4.3)) is reformulated in the scaled boundary coordinates as follows. Substituting Equation (5.14) yields

$$\{b^1\}^T\{\tau\}_{,\xi} + \frac{1}{\xi}\{b^2\}^T\{\tau\}_{,\eta} - \rho\ddot{u} + p = 0 \tag{5.22}$$

The stress-displacement relationship (Equation (4.2)) is transformed, substituting Equation (5.14), into

$$\{\tau\} = G\left(\{b^1\}u_{,\xi} + \frac{1}{\xi}\{b^2\}u_{,\eta}\right) \tag{5.23}$$

The boundary conditions are straightforwardly formulated in the scaled boundary coordinate system, as each side (line element and two side-faces) of the triangle corresponds to either $\xi = 1$ or $\eta = \pm 1$ (Figure 5.2). The surface traction is calculated as (Equation (4.6))

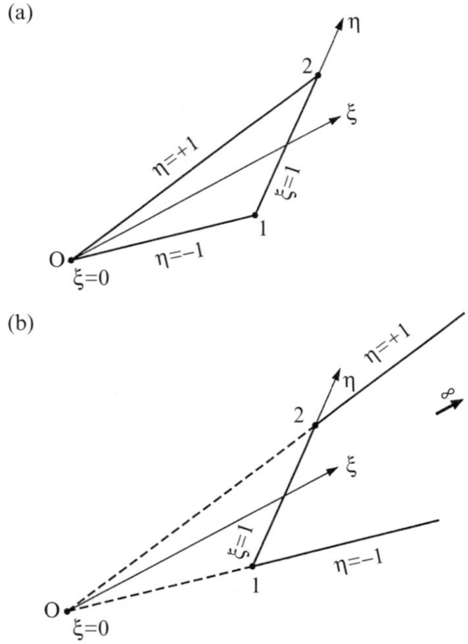

Figure 5.2 (a) Bounded medium with positive side $1-2(\xi=1)$, positive side-face $O-2(\eta=+1)$, and negative side-face $O-1(\eta=-1)$; (b) unbounded medium with negative side $1-2(\xi=1)$, positive side-face $2\text{-}\infty(\eta=+1)$ and negative side-face $1\text{-}\infty(\eta=-1)$ (Reprinted from *Computers and Structures*, Vol. 78, Wolf and Song, 'The scaled boundary finite element method – a primer: derivations', pp. 191–210, © (2000), with permission from Elsevier Science)

$$\tau_n = \{n\}^T\{\tau\} \tag{5.24}$$

with $\{n\}$ denoting the unit outward normal of the side. A side is positive or negative when the outward normal forms an angle in absolute value smaller than 90° with the positive or negative direction of the corresponding coordinate axis, respectively. For the bounded and unbounded cases, the sides 1–2 corresponding to the line elements are positive and negative, respectively. The sides $\eta=+1$ and $=-1$ are positive and negative, respectively, for both the bounded and unbounded cases. The normals calculated in Equation (5.20) correspond to the outward normals of the positive sides (Figure 5.1). For the outward normals of the negative sides sign changes occur.

5.3 METHOD OF WEIGHTED RESIDUALS

The triangle defined by scaling a single line element on the boundary is addressed (Figure 5.1a). The superscript e is dropped for conciseness. To derive a finite element approximation, the weighted-residual technique is applied to the equilibrium equation in the scaled boundary coordinates without any boundary conditions (Equation (5.22), strong form). After performing integration by parts with respect to the ξ-direction, the boundary conditions are introduced. This derivation is the same as in section 1.3, where starting from the weighted-residual statement of the governing differential equation only (Equation (1.21)) and after performing integrations by parts the boundary conditions are enforced (Equation (1.33)). Multiplication by a weighting function $w = w(\xi, \eta)(= w^T)$ and integration over the domain of the triangle V yields

$$\int_V w^T \{b^1\}^T \{\tau\},_\xi \mathrm{d}V + \int_V w^T \frac{1}{\xi} \{b^2\}^T \{\tau\},_\eta \mathrm{d}V - \int_V w^T \rho \ddot{u} \mathrm{d}V + \int_V w^T p \mathrm{d}V = 0 \quad (5.25)$$

with $\mathrm{d}V$ specified in Equation (5.18). The second term of Equation (5.25) denoted as I is examined. After substituting Equation (5.18) and introducing the limits of integration

$$I = \int_0^1 \int_{-1}^{+1} w^T \{b^2\}^T \{\tau\},_\eta |J| \mathrm{d}\eta \mathrm{d}\xi \quad (5.26)$$

holds (Figure 5.1a). $\{b^2\}$ is a function of η (Equation (5.15b)). Applying integration by parts in the η-direction leads to

$$I = \int_0^1 \left((w^T \{b^2\}^T |J| \{\tau\}) \Big|_{\eta=-1}^{\eta=+1} - \int_{-1}^{+1} (w^T \{b^2\}^T |J|),_\eta \{\tau\} \mathrm{d}\eta \right) \mathrm{d}\xi \quad (5.27)$$

Substituting Equation (5.21b) and (5.17) results in

$$I = \int_0^1 \left(\left(w^T \sqrt{x^2 + y^2} \{n^\eta\}^T \{\tau\} \right)\Big|_{\eta=+1} - \left(w^T \sqrt{x^2 + y^2} \{n^\eta\}^T \{\tau\} \right)\Big|_{\eta=-1} \right.$$
$$\left. - \int_{-1}^{+1} (-w^T \{b^1\}^T + w,_\eta^T \{b^2\}^T) \{\tau\} |J| \mathrm{d}\eta \right) \mathrm{d}\xi \quad (5.28)$$

On the negative side $\eta = -1, \tau_n = -\{n^\eta\}^T \{\tau\}$ as $\{n^\eta\}$ corresponds to the outward normal of the positive side (Equation (5.21b)). Substituted into Equation (5.28) yields

$$I = \int_0^1 \left(\sqrt{x_2^2 + y_2^2} (w^T \tau_n)\Big|_{\eta=+1} + \sqrt{x_1^2 + y_1^2} (w^T \tau_n)\Big|_{\eta=-1} \right.$$
$$\left. - \int_{-1}^{+1} (-w^T \{b^1\}^T + w,_\eta^T \{b^2\}^T) \{\tau\} |J| \mathrm{d}\eta \right) \mathrm{d}\xi \quad (5.29)$$

Substituting Equations (5.29) and (5.18) into Equation (5.25) leads to

$$\int_0^1 \left(\xi \int_{-1}^{+1} w^T \{b^1\}^T \{\tau\}_{,\xi} |J| \mathrm{d}\eta + \sqrt{x_2^2 + y_2^2}(w^T \tau_n)\big|_{\eta=+1} + \sqrt{x_1^2 + y_1^2}(w^T \tau_n)\big|_{\eta=-1} \right.$$
$$\left. - \int_{-1}^{+1} (-w^T \{b^1\}^T + w_{,\eta}^T \{b^2\}^T)\{\tau\}|J|\mathrm{d}\eta - \xi \int_{-1}^{+1} w^T \rho \ddot{u} |J| \mathrm{d}\eta + \xi \int_{-1}^{+1} w^T p |J| \mathrm{d}\eta \right) \mathrm{d}\xi = 0 \tag{5.30}$$

The displacement of the line element on the boundary $\xi = 1$ is interpolated using the mapping functions (Equation (5.2)) as the shape functions $[N(\eta)]$. It is postulated that the *same shape functions apply for all lines with a constant* ξ

$$u = u(\xi, \eta) = [N(\eta)]\{u(\xi)\} \tag{5.31}$$

with the displacement (Figure 5.3)

$$\{u(\xi)\} = \begin{Bmatrix} u_1(\xi) \\ u_2(\xi) \end{Bmatrix} \tag{5.32}$$

The shear stresses are calculated substituting Equation (5.31) into Equation (5.23):

$$\{\tau\} = G\left([B^1]\{u(\xi)\}_{,\xi} + \frac{1}{\xi}[B^2]\{u(\xi)\} \right) \tag{5.33}$$

with (Equations (5.15) and (5.2))

$$[B^1] = \{b^1\}[N] = \frac{1}{2(x_1 y_2 - x_2 y_1)} \begin{bmatrix} \Delta_y(1-\eta) & \Delta_y(1+\eta) \\ -\Delta_x(1-\eta) & -\Delta_x(1+\eta) \end{bmatrix} \tag{5.34a}$$

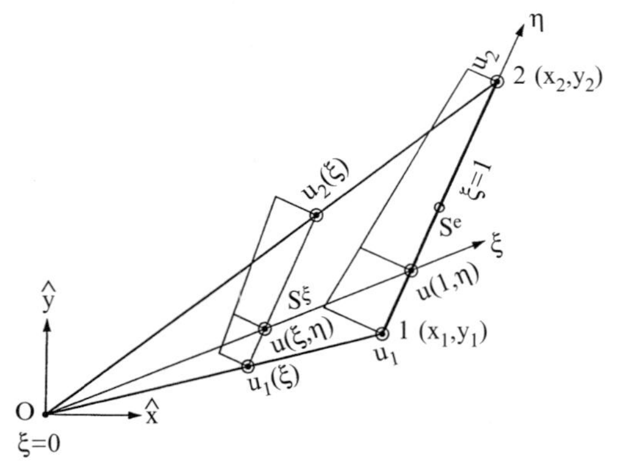

Figure 5.3 Displacement in domain (Reprinted from *Computers and Structures*, Vol. 78, Wolf and Song, 'The scaled boundary finite element method – a primer: derivations', pp. 191–210, © (2000), with permission from Elsevier Science)

$$[B^2] = \{b^2\}[N],_{\eta} = \frac{1}{2(x_1y_2 - x_2y_1)} \begin{bmatrix} 2\bar{y} + \Delta_y\eta & -2\bar{y} - \Delta_y\eta \\ -2\bar{x} - \Delta_x\eta & 2\bar{x} + \Delta_x\eta \end{bmatrix} \tag{5.34b}$$

$[B^1]$ and $[B^2]$ are independent of ξ. As in the standard finite element method (section 1.4), the weighting function is discretised in the same way as the displacement:

$$w = w(\xi, \eta) = [N(\eta)]\{w(\xi)\} \tag{5.35}$$

with

$$\{w(\xi)\} = \begin{Bmatrix} w_1(\xi) \\ w_2(\xi) \end{Bmatrix} \tag{5.36}$$

Substituting Equation (5.35) into Equation (5.30) and using Equation (5.34) yields

$$\int_0^1 \{w(\xi)\}^T \left(\xi \int_{-1}^{+1} [B^1]^T \{\tau\},_{\xi} |J| \mathrm{d}\eta - \int_{-1}^{+1} (-[B^1]^T + [B^2]^T)\{\tau\}|J|\mathrm{d}\eta \right.$$
$$\left. -\xi \int_{-1}^{+1} [N]^T \rho\ddot{u}|J|\mathrm{d}\eta + \xi\{F^{\mathrm{b}}\} + \{F^{\mathrm{t}}\} \right) \mathrm{d}\xi = 0 \tag{5.37}$$

with

$$\{F^{\mathrm{b}}\} = \int_{-1}^{+1} [N]^T p|J|\mathrm{d}\eta \tag{5.38}$$

$$\{F^{\mathrm{t}}\} = \begin{Bmatrix} \sqrt{x_1^2 + y_1^2}\,\tau_n|_{\eta=-1} \\ \sqrt{x_2^2 + y_2^2}\,\tau_n|_{\eta=+1} \end{Bmatrix} \tag{5.39}$$

$\{F^{\mathrm{b}}\}$, which can be a function of ξ, represents the nodal loads resulting from the body load. As the surface traction τ_n acting on the side-faces $O-1$ and $O-2$ can be a function of ξ, the corresponding nodal forces $\{F^{\mathrm{t}}\}$ can also be a function of ξ.

Equation (5.37) is a weighted-residual statement over the domain of an element $O-1-2$ with integration in the ξ- and η-directions of the equilibrium equation (Equation (5.22)). For an arbitrary $\{w(\xi)\}$, its coefficient in Equation (5.37) will vanish for all ξ:

$$\xi \int_{-1}^{+1} [B^1]^T \{\tau\},_{\xi} |J| \mathrm{d}\eta - \int_{-1}^{+1} (-[B^1]^T + [B^2]^T)\{\tau\}|J|\mathrm{d}\eta - \xi \int_{-1}^{+1} [N]^T \rho\ddot{u}|J|\mathrm{d}\eta$$
$$+ \xi\{F^{\mathrm{b}}\} + \{F^{\mathrm{t}}\} = 0 \tag{5.40}$$

As $\{w(\xi)\}$ is a continuous function in the ξ-direction, Equation (5.40) expresses that the *equilibrium equation* (Equation (5.22)) *is enforced 'exactly' in the* ξ*-direction.* Thus, no error exists in the radial direction. Note that Equation (5.40) is a weighted-residual statement in the η-direction (circumferential) with integrals in this direction only. Substituting the shear stresses (Equation (5.33)) and the displacement (Equation (5.31)) leads to the differential equation expressed in the displacement $\{u(\xi)\}$:

$$\xi \int_{-1}^{+1} [B^1]^T G\left([B^1]\{u(\xi)\},_{\xi\xi} + \frac{1}{\xi}[B^2]\{u(\xi)\},_{\xi} - \frac{1}{\xi^2}[B^2]\{u(\xi)\}\right)|J|\mathrm{d}\eta$$
$$- \int_{-1}^{+1} \left(-[B^1]^T + [B^2]^T\right) G\left([B^1]\{u(\xi)\},_{\xi} + \frac{1}{\xi}[B^2]\{u(\xi)\}\right)|J|\mathrm{d}\eta$$
$$- \xi \int_{-1}^{+1} [N]^T \rho [N]\{\ddot{u}(\xi)\}|J|\mathrm{d}\eta + \xi\{F^{\mathrm{b}}\} + \{F^{\mathrm{t}}\} = 0 \tag{5.41}$$

Multiplying Equation (5.41) by ξ and introducing the coefficient matrices

$$[E^0] = \int_{-1}^{+1} [B^1]^T G [B^1]|J|\mathrm{d}\eta = G\frac{\Delta_x^2 + \Delta_y^2}{6(x_1 y_2 - x_2 y_1)}\begin{bmatrix} 2 & 1 \\ 1 & 2 \end{bmatrix} \tag{5.42a}$$

$$[E^1] = \int_{-1}^{+1} [B^2]^T G [B^1]|J|\mathrm{d}\eta = G\frac{\Delta_x^2 + \Delta_y^2}{12(x_1 y_2 - x_2 y_1)}\begin{bmatrix} -1 & 1 \\ 1 & -1 \end{bmatrix}$$
$$- G\frac{\bar{x}\Delta_x + \bar{y}\Delta_y}{2(x_1 y_2 - x_2 y_1)}\begin{bmatrix} -1 & -1 \\ 1 & 1 \end{bmatrix} \tag{5.42b}$$

$$[E^2] = \int_{-1}^{+1} [B^2]^T G [B^2]|J|\mathrm{d}\eta = G\frac{\Delta_x^2 + \Delta_y^2 + 12(\bar{x}^2 + \bar{y}^2)}{12(x_1 y_2 - x_2 y_1)}\begin{bmatrix} 1 & -1 \\ -1 & 1 \end{bmatrix} \tag{5.42c}$$

and

$$[M^0] = \int_{-1}^{+1} [N]^T \rho [N]|J|\mathrm{d}\eta = \rho\frac{x_1 y_2 - x_2 y_1}{6}\begin{bmatrix} 2 & 1 \\ 1 & 2 \end{bmatrix} \tag{5.43}$$

yields

$$[E^0]\xi^2\{u(\xi)\},_{\xi\xi} + ([E^0] - [E^1] + [E^1]^T)\xi\{u(\xi)\},_{\xi} - [E^2]\{u(\xi)\}$$
$$- [M^0]\xi^2\{\ddot{u}(\xi)\} + \{F(\xi)\} = 0 \tag{5.44}$$

with the contribution of the body load and the surface traction

$$\{F(\xi)\} = \xi^2\{F^{\mathrm{b}}\} + \xi\{F^{\mathrm{t}}\} \tag{5.45}$$

Note that the coefficient matrices $[E^0]$, $[E^1]$, $[E^2]$, $[M^0]$ are independent of ξ. Integrations over the line element (at $\xi = 1$) only are involved. The expressions for $[E^0]$, $[E^1]$, $[E^2]$ are similar to that for the static-stiffness matrix of a standard line finite element, and $[M^0]$ resembles the mass matrix. Integrations of polynomials only in η occur. $[E^0]$ and $[M^0]$ are positive definite, $[E^2]$ is semi-positive definite, and $[E^1]$ is non-symmetric.

The derivation of Equation (5.44) is performed for the bounded medium (Figure 5.2a). Equation (5.44) also applies to the unbounded medium (Figure 5.2b) with the same definition of the coefficient matrices $[E^0]$, $[E^1]$, $[E^2]$, $[M^0]$ (Equations (5.42) and (5.43)). The only modification in the derivation consists of replacing the domain of the triangle $V(0 \leq \xi \leq 1)$ in Equation (5.25) by that of corresponding 'trapezoid' $(1 \leq \xi < \infty)$.

Equation (5.44) applies to the domain of the triangular (or trapezoid) corresponding to one line finite element on the boundary. To model the total domain, an *assemblage* as in the conventional finite element method is performed, and the *displacement boundary conditions* on the side-face A_u, if present, *are enforced* (Figures 3.1b and 3.4b). To simplify the nomenclature, the same symbols are used for the assembled coefficient matrices, the assembled displacements and the assembled nodal loads in the following. In the assemblage process, the nodal forces $\{F^{\mathrm{t}}\}$ in Equation (5.45) will cancel, with the exception of those on the side-face A_t of the total boundary with prescribed surface tractions (Figures 3.1b and 3.4b). The assemblage process yields

$$[E^0]\xi^2\{u(\xi)\}_{,\xi\xi} + ([E^0] - [E^1] + [E^1]^T)\xi\{u(\xi)\}_{,\xi} - [E^2]\{u(\xi)\}$$
$$- [M^0]\xi^2\{\ddot{u}(\xi)\} + \{F(\xi)\} = 0 \tag{5.46}$$

Equation (5.46) represents the *scaled boundary finite element equation in displacement* formulated in the time domain. The corresponding equation in the frequency domain (see Appendix B with Equation (B.16)) equals

$$[E^0]\xi^2\{u(\xi)\}_{,\xi\xi} + ([E^0] - [E^1] + [E^1]^T)\xi\{u(\xi)\}_{,\xi} - [E^2]\{u(\xi)\}$$
$$+ \omega^2[M^0]\xi^2\{u(\xi)\} + \{F(\xi)\} = 0 \tag{5.47}$$

with the amplitudes of the displacements $\{u(\xi)\}$ and of the nodal loads $\{F(\xi)\}$ for a specific frequency ω. Equation (5.47) is a system of *linear non-homogeneous second-order ordinary differential equations for the displacement amplitudes* $\{u(\xi)\}$ *with the dimensionless radial coordinate* ξ *as the independent variable*. As already pointed out, the coefficient matrices $[E^0]$, $[E^1]$, $[E^2]$, $[M^0]$ are independent of ξ. The required boundary conditions are addressed as part of the solution procedures in Chapter 11.

5.4 DYNAMIC-STIFFNESS MATRIX

To determine the nodal force-(nodal) displacement relationship as in standard finite elements, the amplitudes of the internal nodal forces $\{Q(\xi)\}$, which are equal to the resultants of the shear stresses, on a line with a constant ξ are addressed. Applying the principle of virtual work yields

$$\{w(\xi)\}^T\{Q(\xi)\} = \int_{S^\xi} w^T\{n^\xi\}^T\{\tau\}\mathrm{d}S^\xi \tag{5.48}$$

Substituting Equations (5.19), (5.21a) and (5.35) leads for an arbitrary $\{w(\xi)\}$ to

$$\{Q(\xi)\} = \xi\int_{-1}^{+1}[N]^T\{b^1\}^T\{\tau\}|J|\mathrm{d}\eta \tag{5.49}$$

Substituting Equations (5.33) and (5.34a) results in

$$\{Q(\xi)\} = \xi\int_{-1}^{+1}[B^1]^T G\left([B^1]\{u(\xi)\}_{,\xi} + \frac{1}{\xi}[B^2]\{u(\xi)\}\right)|J|\mathrm{d}\eta \tag{5.50}$$

Comparing with Equations (5.42a) and (5.42b) yields

$$\{Q(\xi)\} = [E^0]\xi\{u(\xi)\}_{,\xi} + [E^1]^T\{u(\xi)\} \tag{5.51}$$

For a bounded medium (Figure 5.2a), the side $1-2$ is positive, i.e. $\{n^\xi\}$ is the unit outward normal. The amplitudes of the nodal forces $\{R(\xi)\}$ are equal to the amplitudes of the internal nodal forces $\{Q(\xi)\}$. For an unbounded medium (Figure 5.2b), the opposite applies. The two cases can be unified as

$$\{R(\xi)\} = \pm\{Q(\xi)\} \tag{5.52}$$

with the upper and lower signs corresponding to the bounded and unbounded media, respectively. The same Equations (5.51) and (5.52) apply to the assembled system. In the frequency domain, the amplitudes of the displacements $\{u(\xi)\}$ are related to those of the nodal forces $\{R(\xi)\}$ as

$$\{R(\xi)\} = [S(\omega,\xi)]\{u(\xi)\} - \{R^F(\xi)\} \tag{5.53}$$

$[S(\omega,\xi)]$ denotes the dynamic-stiffness matrix on a line with a constant ξ, and $\{R^F(\xi)\}$ represents the amplitudes of the nodal loads due to the body load and the surface traction. Equation (5.53) can be illustrated by the weighted-residual statement of Equation (1.38). The left-hand side transformed into the frequency domain corresponds to $[S(\omega,\xi)]\{u(\xi)\}$, and the right-hand side to $\{R^F(\xi)\}$. Substituting Equations (5.51) and (5.53) into Equation (5.52) yields

$$\pm[S(\omega,\xi)]\{u(\xi)\} \mp \{R^F(\xi)\} = [E^0]\xi\{u(\xi)\},_\xi + [E^1]^T\{u(\xi)\} \tag{5.54}$$

Differentiating Equation (5.54) with respect to ξ

$$\begin{aligned}\pm[S(\omega,\xi)],_\xi\{u(\xi)\} &\pm [S(\omega,\xi)]\{u(\xi)\},_\xi \mp \{R^F(\xi)\},_\xi \\ &- [E^0]\xi\{u(\xi)\},_{\xi\xi} - ([E^0]+[E^1]^T)\{u(\xi)\},_\xi = 0\end{aligned} \tag{5.55}$$

is obtained. Adding Equations (5.47) and (5.55) multiplied by ξ results in

$$\begin{aligned}\pm\xi[S(\omega,\xi)],_\xi\{u(\xi)\} &+ (\pm[S(\omega,\xi)] - [E^1])\xi\{u(\xi)\},_\xi \mp \xi\{R^F(\xi)\},_\xi \\ &- [E^2]\{u(\xi)\} + \omega^2[M^0]\xi^2\{u(\xi)\} + \{F(\xi)\} = 0\end{aligned} \tag{5.56}$$

Solving Equation (5.54) for $\xi\{u(\xi)\},_\xi$ and substituting in Equation (5.56) yields

$$\begin{aligned}((\pm[S(\omega,\xi)] - [E^1])[E^0]^{-1}(\pm[S(\omega,\xi)] - [E^1]^T) \pm \xi[S(\omega,\xi)],_\xi - [E^2] + \omega^2\xi^2[M^0])\{u(\xi)\} \\ \mp \xi\{R^F(\xi)\},_\xi \mp (\pm[S(\omega,\xi)] - [E^1])[E^0]^{-1}\{R^F(\xi)\} + \{F(\xi)\} = 0\end{aligned} \tag{5.57}$$

For arbitrary $\{u(\xi)\}$, its coefficient matrix must vanish leading to

$$(\pm[S(\omega,\xi)] - [E^1])[E^0]^{-1}(\pm[S(\omega,\xi)] - [E^1]^T) \pm \xi[S(\omega,\xi)],_\xi - [E^2] + \omega^2\xi^2[M^0] = 0 \tag{5.58}$$

which will permit $[S(\omega,\xi)]$ to be determined. In addition, the remaining part of Equation (5.57) leads to

$$\xi\{R^F(\xi)\},_\xi + (\pm[S(\omega,\xi)] - [E^1])[E^0]^{-1}\{R^F(\xi)\} = \pm\{F(\xi)\} \tag{5.59}$$

with $[S(\omega,\xi)]$ determined from Equation (5.58). This is a system of linear ordinary differential equations of the first order for the amplitudes of the nodal load $\{R^F(\xi)\}$. The boundary condition is $\{R^F(\xi=0)\} = 0$ for a bounded medium, and $\{R^F(\xi\to\infty)\} = 0$ for an unbounded medium. As $\xi[S(\omega,\xi)],_\xi = \omega\xi[S(\omega,\xi)],_{\omega\xi}$ applies and the matrices $[E^0]$, $[E^1]$, $[E^2]$, $[M^0]$ are independent of ω and ξ, it follows from Equation (5.58) that $[S(\omega,\xi)]$ is actually a function of the product $\omega\xi$ only. This confirms the well-known fact that the dynamic-stiffness matrix $[S(\omega,\xi)]$ *is a function of the dimensionless frequency* defined for the line with a constant ξ as

$$a = \frac{\omega\xi r_0}{c_s} \tag{5.60}$$

with the shear-wave velocity $c_s = \sqrt{G/\rho}$ and the characteristic length r_0 of the boundary ($\xi = 1$). As r_0 any length associated with the boundary can be selected, e.g. the distance $O - 1$ or $O - 2$ (Figure 4.1). An independent proof based on dimensional analysis that the dynamic stiffness is a function of the dimensionless frequency is specified in section 6.1. The derivative with respect to a can be interpreted either for varying ξ with fixed ω or for varying ω with fixed ξ. Thus, for the two-dimensional case $[S(\omega, \xi)] = [S(a)]$ satisfies

$$a[S(a)]_{,a} = \xi[S(\omega, \xi)]_{,\xi} = \omega[S(\omega, \xi)]_{,\omega} \tag{5.61}$$

This permits the spatial derivative to be replaced by the frequency derivative in Equation (5.58)

$$(\pm[S(\omega, \xi)] - [E^1])[E^0]^{-1}(\pm[S(\omega, \xi)] - [E^1]^T) \pm \omega[S(\omega, \xi)]_{,\omega} - [E^2] + \omega^2\xi^2[M^0] = 0 \tag{5.62}$$

On the boundary ($\xi = 1$) the dynamic-stiffness matrix for the bounded medium $[S^{\mathrm{b}}(\omega)]$ (superscript b for bounded) is expressed as

$$([S^{\mathrm{b}}(\omega)] - [E^1])[E^0]^{-1}([S^{\mathrm{b}}(\omega)] - [E^1]^T) + \omega[S^{\mathrm{b}}(\omega)]_{,\omega} - [E^2] + \omega^2[M^0] = 0 \tag{5.63}$$

and for the unbounded medium $[S^{\infty}(\omega)]$ (superscript ∞ for unbounded) as

$$([S^{\infty}(\omega)] + [E^1])[E^0]^{-1}([S^{\infty}(\omega)] + [E^1]^T) - \omega[S^{\infty}(\omega)]_{,\omega} - [E^2] + \omega^2[M^0] = 0 \tag{5.64}$$

These represent the *scaled boundary finite element equations in dynamic stiffness* formulated in the frequency domain for two-dimensional bounded and unbounded media. Each of them is a system of *nonlinear first-order ordinary differential equations with the frequency ω as the independent variable*. For an unbounded medium, the required boundary condition is formulated for $\omega \to \infty$ (or equivalently for $\xi \to \infty$ at infinity). This topic is discussed as part of the solution procedures in Chapter 10.

For the bounded medium $[S^{\mathrm{b}}(\omega)]$ is usually not calculated. The static-stiffness matrix $[K^{\mathrm{b}}]$ and the mass matrix $[M^{\mathrm{b}}]$ are used, which follow from the low-frequency expansion of $[S^{\mathrm{b}}(\omega)]$

$$[S^{\mathrm{b}}(\omega)] = [K^{\mathrm{b}}] - \omega^2[M^{\mathrm{b}}] + O(\omega^4) \tag{5.65}$$

The first two terms on the right-hand side are the same as those in Equation (C.7) of Appendix C.

6 Mechanically-based Derivation

As an alternative to the scaled boundary transformation-based derivation discussed in Chapter 5, the scaled boundary finite element equations in dynamic stiffness and displacement can also be established using a mechanically-based derivation, which is more familiar to engineers. The piecewise definition of the approximate solution in each finite element permits the properties such as the static and dynamic stiffnesses and the loads to be determined on an element basis, which can then be assembled. This assemblage leading to the global model is an appealing central feature, analogous to the well known analysis of a frame structure consisting of individual beam members with a finite number of interconnection points.

The dynamic characteristics of a bounded or an unbounded medium can be described by the force-displacement relationship with respect to the degrees of freedom of the nodes on the boundary (interface). In the frequency domain, the amplitudes of the nodal displacements $\{u\}$ are related to those of the nodal forces $\{R\}$ (for vanishing body loads) as

$$\{R\} = [S(\omega)]\{u\} \tag{6.1}$$

with the dynamic-stiffness matrix $[S(\omega)]$.

The concept of the mechanically-based derivation based on assemblage is explained using the bounded case of the Model Problem (Figure 4.1a), the wedge. The side $1-2$ represents the boundary (interface) which is discretised (Figure 6.1a). O denotes the similarity centre (scaling centre), and $1-2$ corresponds to $\xi = 1$ with the characteristic length r_e (e for exterior). A similar fictitious boundary with the scaling factor $\xi_i < 1$ and the characteristic length r_i ($r_i = \xi_i r_e$) is constructed (i for interior). The discretisation of the similar fictitious boundary is selected to be similar to that of the boundary. The wedge (referred to as r_e) is divided by the fictitious boundary in two parts: the wedge (Figure 6.1c) with a similar boundary (referred to as r_i), and a trapezoid with its exterior boundary coinciding with the boundary and its interior boundary with the fictitious boundary (Figure 6.1b). The discretisations of the exterior and interior boundaries are identical to those of the boundary and the fictitious boundary of the wedge. The finite element discretisation of the trapezoid is defined by connecting the

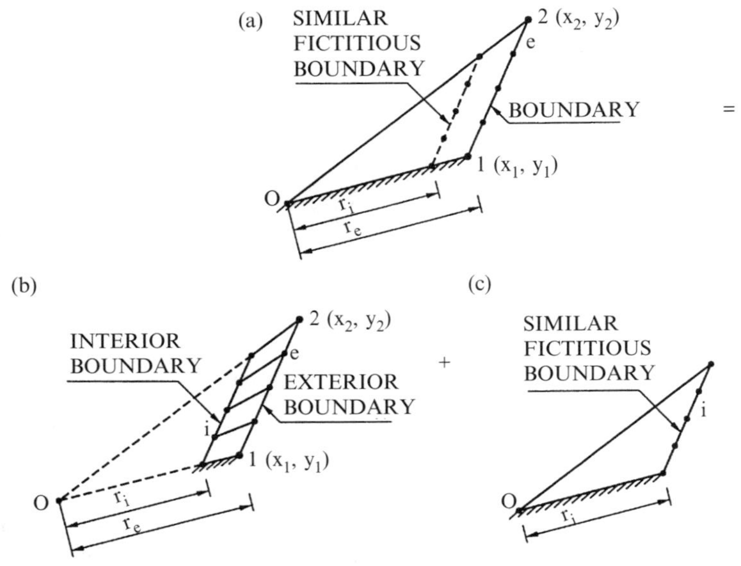

Figure 6.1 Concept of mechanically-based derivation for bounded medium (Reprinted from *Computers and Structures*, Vol. 78, Wolf and Song, 'The scaled boundary finite element method – a primer: derivations', pp. 191–210, © (2000), with permission from Elsevier Science)

corresponding nodes on its exterior and interior boundaries. The discretised trapezoid is called the finite element cell. Adding the finite element cell (Figure 6.1b) to the wedge referred to as r_i (Figure 6.1c) results in the wedge referred to as r_e (Figure 6.1a). The same applies to their dynamic-stiffness matrices. This assemblage enforcing compatibility and equilibrium leads to a relationship linking the dynamic-stiffness matrices at the two similar boundaries of the wedge via that of the finite element cell. Another relationship of the dynamic-stiffness matrices at the two boundaries of the wedge follows from similarity. Thus the dynamic-stiffness matrices at the two boundaries of the wedge can be expressed as a function of that of the finite-element cell. After performing the limit of the cell width $r_e - r_i \rightarrow 0$ analytically, an equation for the dynamic-stiffness matrix of the wedge at the boundary is obtained.

The same concept also applies to the truncated semi-infinite wedge (Figure 6.2). Note that the boundary and the similar fictitious boundary correspond to r_i and r_e, respectively. The scaling factor ξ_e is larger than 1. Again, the dynamic-stiffness matrices at the two boundaries of the truncated semi-infinite wedge can be expressed as a function of that of the finite-element cell, which, being a bounded domain, is straightforward to construct. Performing the limit of the cell width $r_e - r_i \rightarrow 0$ yields an equation for the dynamic-stiffness matrix of the truncated semi-infinite wedge at the boundary.

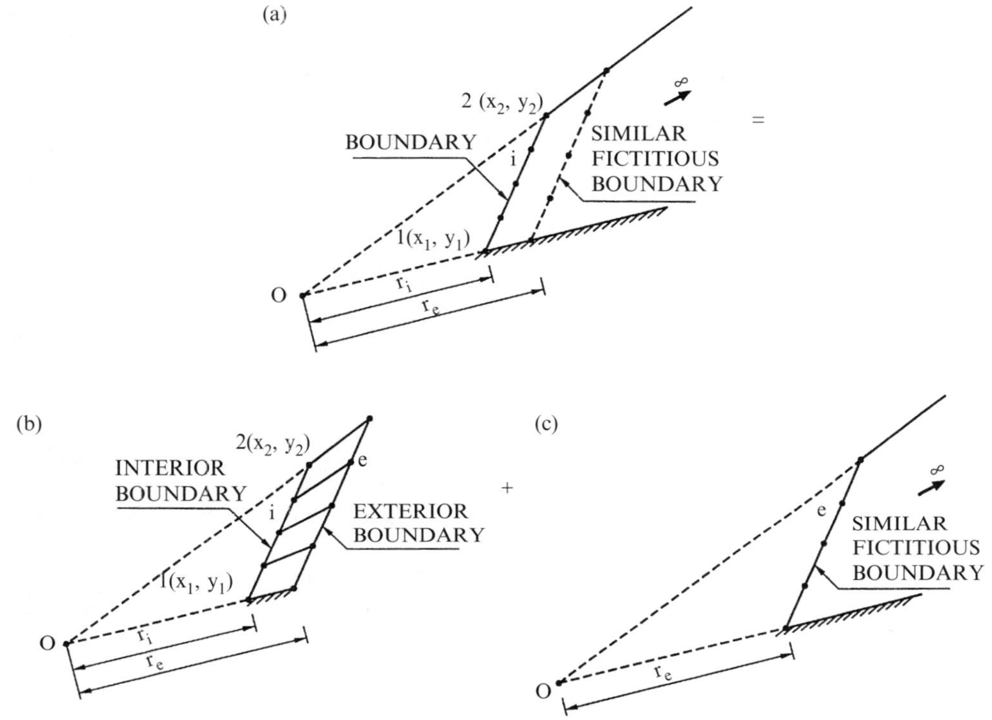

Figure 6.2 Concept of mechanically-based derivation for unbounded medium (Reprinted from *Computers and Structures*, Vol. 78, Wolf and Song, 'The scaled boundary finite element method – a primer: derivations', pp. 191–210, © (2000), with permission from Elsevier Science)

6.1 DYNAMIC-STIFFNESS MATRICES AT SIMILAR BOUNDARIES OF A MEDIUM

The dynamic-stiffness matrices $[S(\omega, r)]$ of similar boundaries with the characteristic length r are addressed.

A dimensional analysis is performed to identify the independent dimensionless variables of which $[S(\omega, r)]$ is a function. The characteristic length r, the shear modulus G, the mass density ρ and the frequency ω are sufficient to determine $[S(\omega, r)]$ for the out-of-plane motion in two dimensions. A bracket denotes the dimension; $[L]$, $[M]$ and $[T]$ are the dimensions of length, mass and time. The dimensions of the variables are

$$[[S]] = [L]^{-1}[M][T]^{-2}$$
$$[r] = [L]$$
$$[G] = [L]^{-1}[M][T]^{-2} \tag{6.2}$$

$$[\rho] = [L]^{-3}[M]$$

$$[\omega] = [T]^{-1}$$

The product of all these variables raised to unknown powers n_i $(i = 1, 2, \ldots, 5)$ must be dimensionless

$$[[S]]^{n_1}[r]^{n_2}[G]^{n_3}[\rho]^{n_4}[\omega]^{n_5} = [L]^{-n_1+n_2-n_3-3n_4}[M]^{n_1+n_3+n_4}[T]^{-2n_1-2n_3-n_5} \tag{6.3}$$

This results in

$$\begin{aligned} -n_1 + n_2 - n_3 - 3n_4 &= 0 \\ n_1 + n_3 + n_4 &= 0 \\ -2n_1 - 2n_3 - n_5 &= 0 \end{aligned} \tag{6.4}$$

The rank of the coefficient matrix of the system of three equations (6.4) with five unknowns equals 3, which permits the two unknowns n_1, n_5 to be chosen arbitrarily. For $n_1 = 1$, $n_5 = 0$, the other unknowns are $n_2 = 0$, $n_3 = -1$, $n_4 = 0$, yielding the first dimensionless variable $[S]G^{-1}$. For $n_1 = 0$, $n_5 = 1$, the other unknowns are $n_2 = 1$, $n_3 = -0.5$, $n_4 = 0.5$, resulting in the second dimensionless variable $rG^{-0.5}\rho^{0.5}\omega$, which is the dimensionless frequency

$$a_0 = \frac{\omega r}{c_s} \tag{6.5}$$

with the shear-wave velocity $c_s = \sqrt{G/\rho}$. Note that a_0 varies with r. The first dimensionless variable $[S^{-1}]$ will be a function $[\bar{S}]$ of the second dimensionless variable a_0, yielding

$$[S(\omega, r)] = G[\bar{S}(a_0)] \tag{6.6}$$

ω and r do not appear explicitly in $[\bar{S}(a_0)]$, but only as a product in a_0 (Equation (6.5)).

The variation of the dynamic-stiffness matrix as a function of the characteristic length r and of the excitation frequency ω is studied. Equation (6.6) is addressed. $[\bar{S}(a_0)]$ is a function of one independent variable, i.e. a_0. For constant ω, $[\bar{S}(a_0)]$ is a function of r, or just as valid, for constant r, $[\bar{S}(a_0)]$ is a function of ω. The same change in a_0 can be achieved by varying the values of either r or ω, with the other fixed. The derivative thus follows for a constant ω varying r as

$$[\bar{S}(a_0)]_{,a_0} = \frac{c_s}{\omega}[\bar{S}(a_0)]_{,r} \tag{6.7a}$$

and the same result is calculated for a constant r, but varying ω, as

$$[\bar{S}(a_0)]_{,a_o} = \frac{c_s}{r}[\bar{S}(a_0)]_{,\omega} \tag{6.7b}$$

Equating the two right-hand sides of Equation (6.7) yields

$$r[\bar{S}(a_0)]_{,r} = \omega[\bar{S}(a_0)]_{,\omega} \tag{6.8}$$

The partial derivative of $[\bar{S}(a_0)]$ with respect to r can thus be replaced by that with respect to ω for similar boundaries. Equations (6.6) and (6.8) lead to

$$r[S(r,\omega)]_{,r} = \omega[S(\omega, r)]_{,\omega} \tag{6.9}$$

which is applicable to the two-dimensional case.

6.2 COEFFICIENT MATRICES

The static-stiffness and mass matrices and the nodal load vector of a single finite element of the cell are calculated, which after assemblage determine the coefficient matrices of the scaled boundary finite-element equations.

A single two-node line finite element with the local nodes 1 and 2 on the interior boundary i is addressed (Figure 6.3a). (As in Chapter 7 the Model Problem is solved with a single line finite element only, on the side-faces $O-1$ and $O-2$ the displacement and the surface traction must be able to be prescribed.) Moving from 1 to 2, O must be on the left. Starting from this line element, a two-dimensional four-node finite element of the cell with its exterior boundary e obtained by scaling the interior boundary with a factor larger than 1 is constructed. The coordinates of the two nodes of the line element are arranged as

$$\{x\} = \begin{Bmatrix} x_1 \\ x_2 \end{Bmatrix} \qquad \{y\} = \begin{Bmatrix} y_1 \\ y_2 \end{Bmatrix} \tag{6.10}$$

On the interior boundary, $[N] = [N(\eta)]$ denotes the mapping functions of the line parent element shown in Figure 6.3c (see solid modelling described in Appendix A):

$$[N] = \left[\frac{1}{2}(1-\eta)\ \frac{1}{2}(1+\eta)\right] \tag{6.11}$$

The coordinates of a point on the line element are generated as

$$x = [N]\{x\} = \bar{x} + \frac{1}{2}\Delta_x\eta \qquad y = [N]\{y\} = \bar{y} + \frac{1}{2}\Delta_y\eta \tag{6.12}$$

with the abbreviations

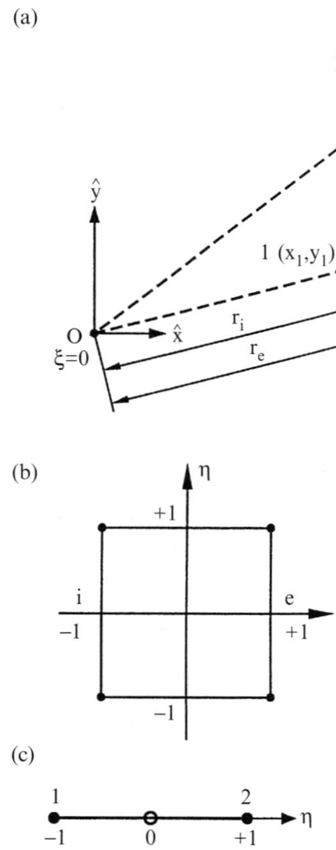

Figure 6.3 (a) Line finite element 1 – 2 with adjacent two-dimensional finite element of cell; (b) two-dimensional parent element of finite element of cell; (c) line parent element (Reprinted from *Computers and Structures*, Vol. 78, Wolf and Song, 'The scaled boundary finite element method – a primer: derivations', pp. 191–210, © (2000), with permission from Elsevier Science)

$$\Delta_x = x_2 - x_1$$
$$\Delta_y = y_2 - y_1 \tag{6.13}$$

$$\bar{x} = \frac{1}{2}(x_1 + x_2)$$
$$\bar{y} = \frac{1}{2}(y_1 + y_2) \tag{6.14}$$

The coordinates of the nodes on the exterior boundary are expressed as

$$\{x_e\} = (1 + w)\{x\}$$
$$\{y_e\} = (1 + w)\{y\} \tag{6.15}$$

with the dimensionless cell width

$$w = \frac{r_e - r_i}{r_i} \tag{6.16}$$

The mapping functions of the two-dimensional parent element $[\hat{N}] = [\hat{N}(\hat{\xi}, \eta)]$ (Figure 6.3b) are generated from those of the line element (Equation (6.11)) with the local coordinates $\hat{\xi}, \eta$. The symbol circumflex denotes the two-dimensional finite element. The mapping functions are decomposed with respect to the interior and exterior boundaries:

$$[\hat{N}] = [[\hat{N}_i]\,[\hat{N}_e]] \tag{6.17}$$

where

$$[\hat{N}_j] = \frac{1}{2}(1 + \hat{\xi}_j\hat{\xi})[N] \; (j = i,e) \tag{6.18}$$

with $\hat{\xi}_j = -1$ for $j = i$ and $\hat{\xi}_j = +1$ for $j = e$. The coordinates $\hat{x}, \hat{y}$ of a point in the two-dimensional finite element are expressed as

$$\hat{x} = [\hat{N}_i]\{x\} + [\hat{N}_e]\{x_e\} = \left(1 + \frac{w}{2}(1 + \hat{\xi})\right)x$$
$$\hat{y} = [\hat{N}_i]\{y\} + [\hat{N}_e]\{y_e\} = \left(1 + \frac{w}{2}(1 + \hat{\xi})\right)y \tag{6.19}$$

The Jacobian matrix of the two-dimensional finite element equals (Equation (A.6))

$$[\hat{J}(\hat{\xi}, \eta)] = \begin{bmatrix} \hat{x},_{\hat{\xi}} & \hat{y},_{\hat{\xi}} \\ \hat{x},_\eta & \hat{y},_\eta \end{bmatrix} = \begin{bmatrix} \frac{w}{2} & \\ & 1 + \frac{w}{2}(1 + \hat{\xi}) \end{bmatrix} [J(\eta)] \tag{6.20}$$

where

$$[J(\eta)] = \begin{bmatrix} x & y \\ x,_\eta & y,_\eta \end{bmatrix} \tag{6.21}$$

and with

$$x,_\eta = [N],_\eta\{x\} = \frac{1}{2}\Delta_x$$
$$y,_\eta = [N],_\eta\{y\} = \frac{1}{2}\Delta_y \tag{6.22}$$

$[J(\eta)]$ is a function of the geometry of the line element only. The argument η is omitted for conciseness. The determinant of the Jacobian matrix (Equation (6.20)) equals

$$|\hat{J}(\hat{\xi},\eta)| = \frac{w}{2}\left(1+\frac{w}{2}(1+\hat{\xi})\right)|J| \tag{6.23}$$

with (Equation (6.21))

$$|J| = xy,_{\eta} - x,_{\eta}\, y = \frac{1}{2}(\bar{x}\Delta_y - \Delta_x\bar{y}) = \frac{1}{2}(x_1y_2 - x_2y_1) \tag{6.24}$$

The inverse of the Jacobian matrix equals

$$[\hat{J}(\hat{\xi},\eta)]^{-1} = [J]^{-1}\begin{bmatrix} \frac{2}{w} & \\ & \dfrac{1}{1+\frac{w}{2}(1+\hat{\xi})} \end{bmatrix} \tag{6.25}$$

with

$$[J]^{-1} = \frac{1}{|J|}\begin{bmatrix} y,_{\eta} & -y \\ -x,_{\eta} & x \end{bmatrix} \tag{6.26}$$

The derivatives with respect to $\hat{x}, \hat{y}$ follow from (see Equation (A.6))

$$\begin{Bmatrix} \dfrac{\partial}{\partial\hat{\xi}} \\ \dfrac{\partial}{\partial\eta} \end{Bmatrix} = [\hat{J}(\hat{\xi},\eta)]\begin{Bmatrix} \dfrac{\partial}{\partial\hat{x}} \\ \dfrac{\partial}{\partial\hat{y}} \end{Bmatrix} \tag{6.27}$$

as

$$\begin{Bmatrix} \dfrac{\partial}{\partial\hat{x}} \\ \dfrac{\partial}{\partial\hat{y}} \end{Bmatrix} = [\hat{J}(\hat{\xi},\eta)]^{-1}\begin{Bmatrix} \dfrac{\partial}{\partial\hat{\xi}} \\ \dfrac{\partial}{\partial\eta} \end{Bmatrix} = \{b^1\}\frac{\partial}{\partial\hat{\xi}} + \{b^2\}\frac{1}{1+\frac{w}{2}(1+\hat{\xi})}\frac{\partial}{\partial\eta} \tag{6.28}$$

where

$$\{b^1\} = \frac{1}{|J|}\begin{Bmatrix} y,_{\eta} \\ -x,_{\eta} \end{Bmatrix} = \frac{1}{x_1y_2 - x_2y_1}\begin{Bmatrix} \Delta_y \\ -\Delta_x \end{Bmatrix} \tag{6.29a}$$

$$\{b^2\} = \frac{1}{|J|}\begin{Bmatrix} -y \\ x \end{Bmatrix} = \frac{1}{x_1y_2 - x_2y_1}\begin{Bmatrix} -2\bar{y} - \Delta_y\eta \\ 2\bar{x} + \Delta_x\eta \end{Bmatrix} \tag{6.29b}$$

The infinitesimal area dV of the finite element (see Equation (A.9)) is calculated as

$$dV = |\hat{J}(\hat{\xi},\eta)|\,d\hat{\xi}d\eta = \frac{w}{2}\left(1+\frac{w}{2}(1+\hat{\xi})\right)|J|\,d\hat{\xi}d\eta \tag{6.30}$$

The shape functions of the finite element are chosen to be identical to the mapping functions (Equation (6.17)). Applying Equation (6.28) to the shape functions (Equation (6.18)) leads to the strain-nodal displacement matrix

$$[B_j] = \frac{\hat{\xi}_j}{w}[B^1] + \frac{1+\hat{\xi}_j\hat{\xi}}{2(1+\frac{w}{2}(1+\hat{\xi}))}[B^2] \qquad (j = i, e) \tag{6.31}$$

with

$$[B^1] = \{b^1\}[N] = \frac{1}{2(x_1y_2 - x_2y_1)}\begin{bmatrix} \Delta_y(1-\eta) & \Delta_y(1+\eta) \\ -\Delta_x(1-\eta) & -\Delta_x(1+\eta) \end{bmatrix} \tag{6.32a}$$

$$[B^2] = \{b^2\}[N]_{,\eta} = \frac{1}{2(x_1y_2 - x_2y_1)}\begin{bmatrix} 2\bar{y}+\Delta_y\eta & -2\bar{y}-\Delta_y\eta \\ -2\bar{x}-\Delta_x\eta & 2\bar{x}+\Delta_x\eta \end{bmatrix} \tag{6.32b}$$

Note that $[B^1]$ and $[B^2]$ defined in Equation (6.32) are the same as specified in Equation (5.34). The static-stiffness matrix $[K]$ is decomposed into submatrices with respect to the interior and exterior boundaries as ($j = i, e$; $l = i, e$), and calculated as (see Equation (1.74))

$$[K_{jl}] = \int_V [B_j]^T G[B_l]\mathrm{d}V \tag{6.33}$$

See the observation in connection with Equation (1.71). Substituting Equations (6.31) and (6.30) into Equation (6.33) yields

$$\begin{aligned}[K_{jl}] = {} & \frac{\hat{\xi}_j\hat{\xi}_l}{2w}[E^0]\int_{-1}^{+1}\left(1+\frac{w}{2}(1+\hat{\xi})\right)\mathrm{d}\hat{\xi} + \frac{\hat{\xi}_l}{4}[E^1]\int_{-1}^{+1}(1+\hat{\xi}_j\hat{\xi})\mathrm{d}\hat{\xi} \\ & + \frac{\hat{\xi}_j}{4}[E^1]^T\int_{-1}^{+1}(1+\hat{\xi}_l\hat{\xi})\mathrm{d}\hat{\xi} + \frac{w}{8}[E^2]\int_{-1}^{+1}\frac{(1+\hat{\xi}_j\hat{\xi})(1+\hat{\xi}_l\hat{\xi})}{1+\frac{w}{2}(1+\hat{\xi})}\mathrm{d}\hat{\xi}\end{aligned} \tag{6.34}$$

with the coefficient matrices

$$[E^0] = \int_{-1}^{+1}[B^1]^T G[B^1]|J|\mathrm{d}\eta = G\frac{\Delta_x^2+\Delta_y^2}{6(x_1y_2 - x_2y_1)}\begin{bmatrix} 2 & 1 \\ 1 & 2 \end{bmatrix} \tag{6.35a}$$

$$\begin{aligned}[E^1] = \int_{-1}^{+1}[B^2]^T G[B^1]|J|\mathrm{d}\eta = {} & G\frac{\Delta_x^2+\Delta_y^2}{12(x_1y_2 - x_2y_1)}\begin{bmatrix} -1 & 1 \\ 1 & -1 \end{bmatrix} \\ & - G\frac{\bar{x}\Delta_x + \bar{y}\Delta_y}{2(x_1y_2 - x_2y_1)}\begin{bmatrix} -1 & -1 \\ 1 & 1 \end{bmatrix}\end{aligned} \tag{6.35b}$$

$$[E^2] = \int_{-1}^{+1}[B^2]^T G[B^2]|J|\mathrm{d}\eta = G\frac{\Delta_x^2+\Delta_y^2+12(\bar{x}^2+\bar{y}^2)}{12(x_1y_2 - x_2y_1)}\begin{bmatrix} 1 & -1 \\ -1 & 1 \end{bmatrix} \tag{6.35c}$$

Performing the integration over $\hat{\xi}$ analytically, and decomposing with respect to the dimensionless cell width w, yields

$$[K_{jl}] = \frac{1}{w}[K^0_{jl}] + [K^1_{jl}] + w[K^2_{jl}] + O(w^2) \tag{6.36}$$

where

$$[K^0_{jl}] = \hat{\xi}_j\hat{\xi}_l[E^0] \tag{6.37a}$$

$$[K^1_{jl}] = \frac{\hat{\xi}_j\hat{\xi}_l}{2}[E^0] + \frac{\hat{\xi}_l}{2}[E^1] + \frac{\hat{\xi}_j}{2}[E^1]^T \tag{6.37b}$$

$$[K^2_{jl}] = \frac{1}{4}\left(1 + \frac{\hat{\xi}_j\hat{\xi}_l}{3}\right)[E^2] \tag{6.37c}$$

The mass matrix $[M]$ follows from

$$[M_{jl}] = \int_V [\hat{N}_j]^T \rho[\hat{N}_l]\mathrm{d}V \tag{6.38}$$

Substituting Equations (6.18) and (6.30) results in

$$[M_{jl}] = \rho\frac{w}{8}[M^0]\int_{-1}^{+1}(1 + \hat{\xi}_j\hat{\xi})(1 + \hat{\xi}_l\hat{\xi})\left(1 + \frac{w}{2}(1 + \hat{\xi})\right)\mathrm{d}\hat{\xi} \tag{6.39}$$

with the coefficient matrix

$$[M^0] = \int_{-1}^{+1}[N]^T\rho[N]|J|\mathrm{d}\eta = \rho\frac{x_1y_2 - x_2y_1}{6}\begin{bmatrix} 2 & 1 \\ 1 & 2 \end{bmatrix} \tag{6.40}$$

Note that the coefficient matrices $[E^0]$, $[E^1]$, $[E^2]$, and $[M^0]$ defined in Equations (6.35) and (6.40) are the same as specified in Equations (5.42) and (5.43). Performing the integration over $\hat{\xi}$ analytically yields

$$[M_{jl}] = w[M^2_{jl}] + O(w^2) = \frac{w}{4}\left(1 + \frac{\hat{\xi}_j\hat{\xi}_l}{3}\right)[M^0] + O(w^2) \tag{6.41}$$

To derive the scaled boundary finite element equation, the following relationships will be used:

$$[K^0_{ii}] = -[K^0_{ie}] = -[K^0_{ei}] = [K^0_{ee}] = [E^0] \tag{6.42a}$$

$$[K_{ie}^1] + [K_{ee}^1] = -([K_{ii}^1] + [K_{ie}^1])^T = [E^1] \tag{6.42b}$$

$$[K_{ii}^2] + [K_{ie}^2] + [K_{ei}^2] + [K_{ee}^2] = [E^2] \tag{6.42c}$$

$$[M_{ii}^2] + [M_{ie}^2] + [M_{ei}^2] + [M_{ee}^2] = [M^0] \tag{6.43}$$

Equation (6.42) follows from Equation (6.37) with $\hat{\xi}_i = -1$, $\hat{\xi}_e = +1$ and Equation (6.43) from Equation (6.41). The right-hand sides of Equations (6.42) and (6.43) are defined in Equations (6.35) and (6.40). $[E^0]$, $[E^1]$, $[E^2]$ and $[M^0]$ are the coefficient matrices of one line finite element.

The nodal load $\{P^b\}$ of the finite element due to the body load p (superscript b for body load) is examined. Decomposing into subvectors with respect to the interior and exterior boundaries (Figure 6.3a) yields ($j = i,\ e$)

$$\{P_j^b\} = \int_V [\hat{N}_j]^T p \mathrm{d}V \tag{6.44}$$

Substituting Equations (6.18) and (6.30) results in

$$\{P_j^b\} = \frac{w}{4} \int_{-1}^{+1} (1 + \hat{\xi}_j \hat{\xi}) \left(1 + \frac{w}{2}(1 + \hat{\xi})\right) \mathrm{d}\hat{\xi} \{F^b\} \tag{6.45}$$

with

$$\{F^b\} = \int_{-1}^{+1} [N]^T p |J| \mathrm{d}\eta \tag{6.46}$$

Integrating Equation (6.45) over $\hat{\xi}$ leads to

$$\{P_j^b\} = \left(\frac{w}{2} + \frac{w^2}{4}\left(1 + \frac{1}{3}\hat{\xi}_j\right)\right) \{F^b\} \tag{6.47}$$

Analogously, the nodal load $\{P^t\}$ of the finite element due to the prescribed surface traction τ_n (superscript t for traction), if present, on the two sides $S^{\eta=-1}$, $S^{\eta=+1}$ (Figure 6.3a) is calculated. Again, decomposing with respect to the interior and exterior boundaries yields ($j = i,\ e$)

$$\{P_j^t\} = \int_{S^{\eta=-1}} [\hat{N}_j]^T \tau_n \mathrm{d}S^\eta + \int_{S^{\eta=+1}} [\hat{N}_j]^T \tau_n \mathrm{d}S^\eta \tag{6.48}$$

The infinitesimal length for a constant η follows from Equation (6.19) as

$$\mathrm{d}S^\eta = \sqrt{(\mathrm{d}\hat{x})^2 + (\mathrm{d}\hat{y})^2} = \frac{w}{2}\sqrt{x^2 + y^2} \mathrm{d}\hat{\xi} \tag{6.49}$$

Substituting Equations (6.18) and (6.49) into Equation (6.48) results in

$$\{P_j^{\mathrm{t}}\} = \frac{w}{4}\int_{-1}^{+1}(1+\hat{\xi}_j\hat{\xi})d\hat{\xi}\left(([N]^T\tau_n)\Big|_{\eta=-1}\sqrt{x_1^2+y_1^2}+([N]^T\tau_n)\Big|_{\eta=+1}\sqrt{x_2^2+y_2^2}\right) = \frac{w}{2}\{F^{\mathrm{t}}\} \tag{6.50}$$

with

$$\{F^{\mathrm{t}}\} = \begin{Bmatrix} \sqrt{x_1^2+y_1^2}\tau_n\big|_{\eta=-1} \\ \sqrt{x_2^2+y_2^2}\tau_n\big|_{\eta=+1} \end{Bmatrix} \tag{6.51}$$

Note that $\{F^{\mathrm{b}}\}$ and $\{F^{\mathrm{t}}\}$ defined in Equations (6.46) and (6.51) are the same as specified in Equations (5.38) and (5.39).

As for the static-stiffness and mass matrices of the finite-element cell, the coefficient matrices $[E^0], [E^1], [E^2]$ and $[M^0]$ are assembled to form those of the boundary (Figure 6.2) enforcing the displacement boundary condition on A_u (Figure 3.4b). The assembled coefficient matrices will thus be banded. To simplify the nomenclature, the same symbols are used for the assembled coefficient, static-stiffness and mass matrices in the following. For the assembled static-stiffness and mass matrices, Equations (6.42) and (6.43) still hold. The same assemblage also applies to the nodal loads $\{P^{\mathrm{b}}\}$, $\{P^{\mathrm{t}}\}$ (Equations (6.47) and (6.50)), and the corresponding vectors $\{F^{\mathrm{b}}\}$, $\{F^{\mathrm{t}}\}$ (Equations (6.46) and (6.51)).

6.3 ASSEMBLAGE OF FINITE ELEMENT CELL AND MEDIUM

For both the bounded and unbounded cases (Figures 6.1 and 6.2), the assemblage enforcing compatibility and equilibrium formulated in the frequency domain links the dynamic-stiffness matrices of the medium at the boundary and at the similar fictitious boundary. This relationship involves the dynamic-stiffness matrix of the finite element cell, which can be expressed by its static-stiffness and mass matrices. In addition, the nodal load of the medium due to the body load and the surface traction can be expressed by that of the finite element cell.

The force-displacement relationship of the finite element cell located between the interior and exterior boundaries (Figures 6.1b and 6.2b) is written as

$$\{P\} = [S(\omega)]\{u\} - \{\bar{P}\} \tag{6.52}$$

with the amplitudes of the nodal displacements $\{u\}$ and of the nodal forces $\{P\}$. The dynamic-stiffness matrix equals

$$[S(\omega)] = [K] - \omega^2[M] \tag{6.53}$$

with the static-stiffness matrix $[K]$ and the mass matrix $[M]$ of the finite element cell. The amplitudes of the known nodal load of the finite element cell due to the body load and prescribed surface traction are formulated as (Equations (6.47) and (6.50))

$$\{\bar{P}\} = \{P^{\mathrm{b}}\} + \{P^{\mathrm{t}}\} \tag{6.54}$$

Partitioning Equation (6.52) into the nodes lying on the interior and exterior boundaries yields

$$\begin{Bmatrix} \{P_i\} \\ \{P_e\} \end{Bmatrix} = \begin{bmatrix} [S_{ii}(\omega)] & [S_{ie}(\omega)] \\ [S_{ei}(\omega)] & [S_{ee}(\omega)] \end{bmatrix} \begin{Bmatrix} \{u_i\} \\ \{u_e\} \end{Bmatrix} - \begin{Bmatrix} \{\bar{P}_i\} \\ \{\bar{P}_e\} \end{Bmatrix} \tag{6.55}$$

The force-displacement relationship of the medium at the boundaries (interfaces) corresponding to the interior and the exterior boundaries of the finite element cell is formulated as

$$\{R_i\} = [S_i(\omega)]\{u_i\} - \{R_i^{\mathrm{F}}\} \tag{6.56a}$$

$$\{R_e\} = [S_e(\omega)]\{u_e\} - \{R_e^{\mathrm{F}}\} \tag{6.56b}$$

with the amplitudes of the unknown nodal loads of the medium $\{R^{\mathrm{F}}\}$ due to the body load and prescribed surface traction. For the bounded medium, the outward normals of the boundary of the medium (Figure 6.1a) and of the exterior boundary of the finite-element cell (Figure 6.1b) point in the same direction yielding $\{R_e\} = \{P_e\}$, while the opposite applies for the similar fictitious boundary of the medium (Figure 6.1c) and the interior boundary of the finite-element cell (Figure 6.1b), which leads to $\{R_i\} = -\{P_i\}$. Or in other words, assembling the medium and the cell by formulating equilibrium on the similar fictitious boundary i, which coincides with the interior boundary i, yields $\{R_i\} + \{P_i\} = 0$, and at the boundary e the nodal forces of the medium must be the same as those of the cell leading to $\{R_e\} = \{P_e\}$. For the unbounded medium (Figure 6.2) the opposite applies, resulting in $\{R_i\} = \{P_i\}$ and $\{R_e\} = -\{P_e\}$. The two cases can be unified as

$$\{R_i\} = \mp\{P_i\} \tag{6.57a}$$

$$\{R_e\} = \pm\{P_e\} \tag{6.57b}$$

with the upper and lower signs corresponding to the bounded and unbounded media, respectively. Eliminating $\{P\}$ and $\{R\}$ by substituting Equations (6.55) and (6.56) into Equation (6.57) yields

$$\begin{bmatrix} [S_{ii}(\omega)] & [S_{ie}(\omega)] \\ [S_{ei}(\omega)] & [S_{ee}(\omega)] \end{bmatrix} \begin{Bmatrix} \{u_i\} \\ \{u_e\} \end{Bmatrix} - \begin{Bmatrix} \{\bar{P}_i\} \\ \{\bar{P}_e\} \end{Bmatrix} = \begin{bmatrix} \mp[S_i(\omega)] & \\ & \pm[S_e(\omega)] \end{bmatrix} \times \begin{Bmatrix} \{u_i\} \\ \{u_e\} \end{Bmatrix} - \begin{Bmatrix} \mp\{R_i^{\mathrm{F}}\} \\ \pm\{R_e^{\mathrm{F}}\} \end{Bmatrix} \tag{6.58}$$

Solving the second row of Equation (6.58) for $\{u_e\}$ and substituting in the first row results in

$$(\mp[S_i(\omega)] - [S_{ii}(\omega)] + [S_{ie}(\omega)](\mp[S_e(\omega)] + [S_{ee}(\omega)])^{-1}[S_{ei}(\omega)])\{u_i\}$$
$$= \{\bar{P}_i\} \pm \{R_i^{\mathrm{R}}\} - [S_{ie}(\omega)](\mp[S_e(\omega)] + [S_{ee}(\omega)])^{-1}(\{\bar{P}_e\} \mp \{R_e^{\mathrm{F}}\}) \quad (6.59)$$

As $\{u_i\}$ is arbitrary, its coefficient matrix must vanish, leading to

$$\mp[S_i(\omega)] = [S_{ii}(\omega)] - [S_{ie}(\omega)](\mp[S_e(\omega)] + [S_{ee}(\omega)])^{-1}[S_{ei}(\omega)] \quad (6.60)$$

This is the so-called 'dynamic condensation equation', which links the dynamic-stiffness matrices of the medium at the boundary and the similar fictitious boundary. In addition, the right-hand side of Equation (6.59) is equal to zero, yielding a relationship for the amplitudes of the nodal loads at the same two boundaries of the medium

$$\pm\{R_i^{\mathrm{F}}\} = -\{\bar{P}_i\} + [S_{ie}(\omega)](\mp[S_e(\omega)] + [S_{ee}(\omega)])^{-1}(\{\bar{P}_e\} \mp \{R_e^{\mathrm{F}}\}) \quad (6.61)$$

6.4 ANALYTICAL LIMIT OF FINITE ELEMENT CELL WIDTH

Equations (6.60) and (6.61) are valid for an arbitrary cell width. *Performing the analytical limit of the cell width towards zero eliminates the discretisation error in the radial direction.*

The dynamic condensation Equation (6.60) is reformulated as

$$(\mp[S_e(\omega)] + [S_{ee}(\omega)])[S_{ie}(\omega)]^{-1}(\mp[S_i(\omega)] - [S_{ii}(\omega)]) + [S_{ei}(\omega)] = 0 \quad (6.62)$$

The dynamic-stiffness submatrices of the cell are written as follows. For instance,

$$[S_{ii}(\omega)] = [K_{ii}] - \omega^2[M_{ii}] \quad (6.63)$$

applies with $[K_{ii}]$ specified in Equation (6.36) and $[M_{ii}]$ in Equation (6.41). This yields

$$[S_{ii}(\omega)] = \frac{1}{w}[K_{ii}^0] + [K_{ii}^1] + w[K_{ii}^2] - w\omega^2[M_{ii}^2] + O(w^2) \quad (6.64)$$

Using Equation (6.42a) leads to

$$[S_{ii}(\omega)] = \frac{1}{w}[E^0] + [K_{ii}^1] + w([K_{ii}^2] - \omega^2[M_{ii}^2]) + O(w^2) \quad (6.65a)$$

Analogously

$$[S_{ie}(\omega)] = -\frac{1}{w}[E^0] + [K_{ie}^1] + w([K_{ie}^2] - \omega^2[M_{ie}^2]) + O(w^2) \quad (6.65b)$$

$$[S_{ei}(\omega)] = -\frac{1}{w}[E^0] + [K_{ei}^1] + w([K_{ei}^2] - \omega^2[M_{ei}^2]) + O(w^2) \quad (6.65c)$$

$$[S_{ee}(\omega)] = \frac{1}{w}[E^0] + [K^1_{ee}] + w([K^2_{ee}] - \omega^2[M^2_{ee}]) + O(w^2) \tag{6.65d}$$

follow. The inverse $[S_{ie}(\omega)]^{-1}$ appearing in Equation (6.62) is expressed as a polynomial in w with the unknown coefficient matrices $[A]$ and $[B]$

$$[S_{ie}(\omega)]^{-1} = -w[E^0]^{-1} + w^2[A] + w^3[B] + O(w^4) \tag{6.66}$$

The coefficient matrix of w is equal to the inverse of that of $1/w$ in Equation (6.65b). $[A]$ and $[B]$ follow from (Equations (6.65b) and (6.66))

$$\begin{aligned}[I] = [S_{ie}(\omega)]\,[S_{ie}(\omega)]^{-1} &= [I] - w([K^1_{ie}]\,[E^0]^{-1} + [E^0]\,[A]) \\ &- w^2(([K^2_{ie}] - \omega^2[M^2_{ie}])[E_0]^{-1} - [K^1_{ie}]\,[A] + [E^0]\,[B]) + O(w^3)\end{aligned} \tag{6.67}$$

Setting the coefficient matrices of w and w^2 equal to zero yields

$$[A] = -[E^0]^{-1}[K^1_{ie}]\,[E^0]^{-1} \tag{6.68a}$$

$$[B] = -[E^0]^{-1}([K^1_{ie}]\,[E^0]^{-1}[K^1_{ie}] - ([K^2_{ie}] - \omega^2[M^2_{ie}]))\,[E^0]^{-1} \tag{6.68b}$$

Substituting Equations (6.65) and (6.66) with Equation (6.68) into Equation (6.62) leads to

$$\begin{aligned}&\underbrace{[K^1_{ii}] + [K^1_{ie}] + [K^1_{ei}] + [K^1_{ee}]}_{1} \\ &- w(\underbrace{\mp[S_e(\omega)] + [K^1_{ie}] + [K^1_{ee}])[E^0]^{-1}(\mp[S_i(\omega)] - [K^1_{ii}] - [K^1_{ie}])}_{2} \\ &\mp(\underbrace{[S_e(\omega)] - [S_i(\omega)]}_{3}) \qquad\qquad\qquad\qquad\qquad\qquad = O(w^2) \\ &+ w(\underbrace{[K^2_{ii}] + [K^2_{ie}] + [K^2_{ei}] + [K^2_{ee}]}_{4} - \omega^2\underbrace{([M^2_{ii}] + [M^2_{ie}] + [M^2_{ei}] + [M^2_{ee}])}_{5})\end{aligned} \tag{6.69}$$

The sum identified by 1 in Equation (6.69) vanishes according to Equation (6.42b). Equation (6.42b) is substituted in the term identified by 2. The sums identified by 4 and 5 are transformed using Equations (6.42c) and (6.43), respectively. Dividing Equation (6.69) by w results in

$$\begin{aligned}(\pm[S_e(\omega)] - [E^1])[E^0]^{-1}(\pm[S_i(\omega)] - [E^1]^T) \pm \frac{[S_e(\omega)] - [S_i(\omega)]}{w} \\ - [E^2] + \omega^2[M^0] = O(w)\end{aligned} \tag{6.70}$$

The limit of the dimensionless cell width $w \rightarrow 0$ can now be performed. With w defined in Equation (6.16)

$$\lim_{w \rightarrow 0} \frac{[S_e(\omega)] - [S_i(\omega)]}{w} = \lim_{r_e \rightarrow r_i} r_i \frac{[S_e(\omega)] - [S_i(\omega)]}{r_e - r_i} = r[S(\omega)],_r \tag{6.71}$$

results with $[S(\omega)] = [S_i(\omega)]$ and $r = r_i$. With $[S_e(\omega)] = [S(\omega)] + O(w)$, the limit of Equation (6.70) equals

$$(\pm[S(\omega)] - [E^1])[E^0]^{-1}(\pm[S(\omega)] - [E^1]^T) \pm r[S(\omega)],_r - [E^2] + \omega^2[M^0] = 0 \tag{6.72}$$

The derivative with respect to r can be replaced by that with respect to ω using the relationship of the dynamic-stiffness matrices of similar boundaries (Equation (6.9)) for the two-dimensional case). This yields

$$(\mp[S(\omega)] + [E^1])[E^0]^{-1}(\mp[S(\omega)] + [E^1]^T) \pm \omega[S(\omega)],_\omega - [E^2] + \omega^2[M^0] = 0 \tag{6.73}$$

In an actual application, a specific boundary is addressed which fixes r. The dynamic-stiffness matrix thus becomes a function of ω only. The partial derivative $[S(r,\omega)],_\omega$ in Equation (6.9) is replaced by $[S(\omega)],_\omega$.

The dynamic-stiffness matrix for the bounded medium $[S^b(\omega)]$ (superscript b for bounded) is expressed as

$$([S^b(\omega)] - [E^1])[E^0]^{-1}([S^b(\omega)] - [E^1]^T) + \omega[S^b(\omega)],_\omega - [E^2] + \omega^2[M^0] = 0 \tag{6.74}$$

and for the unbounded medium $[S^\infty(\omega)]$ (superscript ∞ for unbounded) as

$$([S^\infty(\omega)] + [E^1])[E^0]^{-1}([S^\infty(\omega)] + [E^1]^T) - \omega[S^\infty(\omega)],_\omega - [E^2] + \omega^2[M^0] = 0 \tag{6.75}$$

This represents the *scaled boundary finite element equation in dynamic stiffness* formulated in the frequency domain. It is a system of *nonlinear ordinary differential equations of first order with the frequency* ω *as the independent variable*. For an unbounded medium, the required boundary condition is formulated for $\omega \rightarrow \infty$. This topic is discussed as part of the solution procedures in Chapter 10.

Turning to the relationship for the nodal loads of the medium, Equation (6.61) is reformulated as

$$\{\bar{P}_e\} \mp \{R_e^F\} = (\mp[S[\omega] + S_{ee}(\omega)])[S_{ie}(\omega)]^{-1}(\{\bar{P}_i\} \pm \{R_i^F\}) \tag{6.76}$$

Substituting Equations (6.65d), (6.66) with Equation (6.68), and using Equation (6.42b), yields

$$(\mp[S(\omega)] + [S_{ee}(\omega)])[S_{ie}(\omega)]^{-1} = -[I] - w(\mp[S_e(\omega)] + [E^1])[E^0]^{-1} + O(w^2) \tag{6.77}$$

Substituting Equation (6.77) into Equation (6.76) results in

$$\frac{\{\bar{P}_i\}+\{\bar{P}_e\}}{w} \mp \frac{\{R_e^{\mathrm{F}}\}-\{R_i^{\mathrm{F}}\}}{w} + (\mp[S_e(\omega)]+[E^1])[E^0]^{-1}(\{\bar{P}_i\} \pm \{R_i^{\mathrm{F}}\}) = O(w) \tag{6.78}$$

From Equations (6.47) and (6.50) and using Equation (6.54),

$$\frac{\{\bar{P}_i\}+\{\bar{P}_e\}}{w} = \{F\} + O(w) \tag{6.79}$$

results, with

$$\{F\} = \{F^{\mathrm{b}}\} + \{F^{\mathrm{t}}\} \tag{6.80}$$

The limit of $w \to 0$ can now be performed. With w as defined in Equation (6.16),

$$\lim_{w\to 0} \frac{\{R_e^{\mathrm{F}}\}-\{R_i^{\mathrm{F}}\}}{w} = \lim_{r_e\to r_i} r_i \frac{\{R_e^{\mathrm{F}}\}-\{R_i^{\mathrm{F}}\}}{r_e - r_i} = r\{R^{\mathrm{F}}\}_{,r} \tag{6.81}$$

results with $\{R^{\mathrm{F}}\} = \{R_i^{\mathrm{F}}\}$ and $r = r_i$. With $[S_e(\omega)] = [S(\omega)] + O(w)$ and $\{\bar{P}_i\} = O(w)$ (Equations (6.54), (6.47) and (6.50)), the limit of Equation (6.78) yields

$$r\{R^{\mathrm{F}}\}_{,r} + (\pm[S(\omega)] - [E^1])[E^0]^{-1}\{R^{\mathrm{F}}\} = \pm\{F\} \tag{6.82}$$

After determining $[S(\omega)]$ from Equation (6.73), this is a system of *linear ordinary differential equations of the first order for the amplitudes of the nodal load of the medium* $\{R^{\mathrm{F}}\}$. The boundary condition is $\{R^{\mathrm{F}}(r=0)\} = 0$ for a bounded medium, and $\{R^{\mathrm{F}}(r\to\infty)\} = 0$ for an unbounded medium.

In a practical application, the dynamic-stiffness matrix $[S(\omega)]$ on the boundary only is required, which can be calculated from Equation (6.73). To derive this equation, it is sufficient to introduce a finite element cell adjacent to the boundary characterised only by r (Figures 6.1b and 6.2b). To be able to determine the response in the interior of the medium, the derivation can be performed for a similar boundary characterised by ξr with its corresponding cell. This corresponds to multiplying the coordinates of the original boundary by the scaling factor ξ. All equations still apply, but formulated at the boundary characterised by ξr. The determinant $|J|$ in Equation (6.24) is multiplied by ξ^2. $[B^1]$ and $[B^2]$ in Equation (6.32) are divided by ξ, and $[M^0]$ (Equation (6.40) is multiplied by ξ^2. Thus, Equation (6.72) for ξr is formulated as

$$(\pm[S(\omega,\xi)] - [E^1])[E^0]^{-1}(\pm[S(\omega,\xi)] - [E^1]^T) \pm \xi[S(\omega,\xi)]_{,\xi} - [E^2] + \omega^2\xi^2[M^0] = 0 \tag{6.83}$$

Addressing the nodal loads, $\{F^{\mathrm{b}}\}$ (Equation (6.46)) is multiplied by ξ^2 and $\{F^{\mathrm{t}}\}$ (Equation (6.51)) by ξ, resulting in (Equation (6.80))

$$\{F(\xi)\} = \xi^2\{F^{\mathrm{b}}\} + \xi\{F^{\mathrm{t}}\} \tag{6.84}$$

Equation (6.82) for ξr is formulated as

$$\xi\{R^{\mathrm{F}}(\xi)\},_{\xi} + (\pm[S(\omega,\xi)] - [E^1])[E^0]^{-1}\{R^{\mathrm{F}}(\xi)\} = \pm\{F(\xi)\} \tag{6.85}$$

The scaled boundary finite element equation in displacement is derived starting from that in dynamic stiffness (Equation (6.83)). The first row of Equation (6.58) formulated at the boundary characterized by ξr is addressed

$$[S_{ii}(\omega,\xi)]\{u_i(\xi)\} + [S_{ie}(\omega,\xi)]\{u_e(\xi)\} - \{\bar{P}_i(\xi)\} = \mp[S_i(\omega,\xi)]\{u_i(\xi)\} \pm \{R_i^{\mathrm{F}}(\xi)\} \tag{6.86}$$

The dimensionless cell width w (Equation (6.16)) is independent of ξ. The same applies for the two-dimensional case to $[K^1]$ in Equation (6.65) as $[E^0]$ is independent of ξ (Equation (6.36)). From Equation (6.65)

$$[S_{ii}(\omega,\xi)] = \frac{1}{w}[E^0] + [K_{ii}^1] + O(w) \tag{6.87a}$$

$$[S_{ie}(\omega,\xi)] = -\frac{1}{w}[E^0] + [K_{ie}^1] + O(w) \tag{6.87b}$$

follows. Substituting Equation (6.87) and

$$\{u_e(\xi)\} = \{u_i(\xi)\} + w\frac{\{u_e(\xi)\} - \{u_i(\xi)\}}{w} \tag{6.88}$$

into the first two terms of Equation (6.86) yields

$$\begin{aligned}[S_{ii}(\omega,\xi)]\{u_i(\xi)\} + [S_{ie}(\omega,\xi)]\{u_e(\xi)\} &= ([K_{ii}^1] + [K_{ie}^1])\{u_i(\xi)\} \\ &+ \left(-\frac{1}{w}[E^0] + [K_{ie}^1]\right) w\frac{\{u_e(\xi)\} - \{u_i(\xi)\}}{w} + O(w)\end{aligned} \tag{6.89}$$

For the limit $w \to 0$

$$\begin{aligned}\lim_{w \to 0} \frac{\{u_e(\xi)\} - \{u_i(\xi)\}}{w} &= \lim_{r_e \to r_i} r_i \frac{\{u_e(\xi)\} - \{u_i(\xi)\}}{r_e - r_i} \\ &= \lim_{r_e \to r_i} \xi r_i \frac{\{u_e(\xi)\} - \{u_i(\xi)\}}{\xi r_e - \xi r_i} = \xi r\{u(\xi)\},_{\xi r} = \xi\{u(\xi)\},_{\xi}\end{aligned} \tag{6.90}$$

applies. Performing the limit of $w \to 0$ of Equation (6.86) with Equations (6.89) and (6.90) using Equation (6.42b) leads to

$$[E^0]\xi\{u(\xi)\},_{\xi} = (\pm[S(\omega,\xi)] - [E^1]^T)\{u(\xi)\} \mp \{R^{\mathrm{F}}(\xi)\} \tag{6.91}$$

Note that $\{\bar{P}_i(\xi)\}$ vanishes, as it is proportional to w (Equations (6.54), (6.47) and (6.50)). $[S(\omega,\xi)] = [S_i(\omega,\xi)]$, $\{u(\xi)\} = \{u_i(\xi)\}$ and $\{R^{\mathrm{F}}(\xi)\} = \{R_i^{\mathrm{F}}(\xi)\}$ apply.

Equation (6.83) is multiplied by $\{u(\xi)\}$ and rearranged

$$((\pm[S(\omega,\xi)]-[E^1])[E^0]^{-1}(\pm[S(\omega,\xi)]-[E^1]^T)$$
$$\pm\xi[S(\omega,\xi)]_{,\xi}-[E^2]+\omega^2\xi^2[M^0])\{u(\xi)\}=0 \tag{6.92}$$

Substituting Equation (6.91) into Equation (6.92) yields

$$(\pm[S(\omega,\xi)]-[E^1])[E^0]^{-1}([E^0]\xi\{u(\xi)\}_{,\xi}\pm\{R^{\mathrm{F}}(\xi)\})$$
$$\pm\xi[S(\omega,\xi)]_{,\xi}\{u(\xi)\}-[E^2]\{u(\xi)\}+\omega^2\xi^2[M^0]\{u(\xi)\}=0 \tag{6.93}$$

Substituting Equation (6.85) leads to

$$-[E^1]\xi\{u(\xi)\}_{,\xi}\mp\xi\{R^{\mathrm{F}}(\xi)\}_{,\xi}+\{F(\xi)\}$$
$$\pm\xi([S(\omega,\xi)]\{u(\xi)\})_{,\xi}-[E^2]\{u(\xi)\}+\omega^2\xi^2[M^0]\{u(\xi)\}=0 \tag{6.94}$$

Differentiating Equation (6.91) with respect to ξ yields

$$\pm([S(\omega,\xi)]\{u(\xi)\})_{,\xi}\mp\{R^{\mathrm{F}}(\xi)\}_{,\xi}=[E^0]\xi\{u(\xi)\}_{,\xi\xi}+([E^0]+[E^1]^T)\{u(\xi)\}_{,\xi} \tag{6.95}$$

Substituting Equation (6.95) into Equation (6.94) results in the *scaled boundary finite element equation in displacement*

$$[E^0]\xi^2\{u(\xi)\}_{,\xi\xi}+([E^0]-[E^1]+[E^1]^T)\xi\{u(\xi)\}_{,\xi}$$
$$-[E^2]\{u(\xi)\}+\omega^2\xi^2[M^0]\{u(\xi)\}+\{F(\xi)\}=0 \tag{6.96}$$

Note that this equation applies to both bounded and unbounded media. Equation (6.96) is a system of *linear non-homogeneous second-order ordinary differential equations for the displacement amplitude* $\{u(\xi)\}$ *with the dimensionless radial coordinate* ξ *as the independent variable*. The required boundary conditions are addressed as part of the solution procedures in Chapter 11.

The two derivations lead to identical fundamental equations: the scaled boundary finite-element equations in displacement (Equations (5.47) and (6.96)) and in dynamic stiffness (Equations (5.62) and (6.73)) and the relationships governing the nodal load (Equations (5.59) and (6.85)).

A final observation compares the mechanically-based derivation to that based on the scaled boundary transformation. The mechanically-based derivation starts with a discrete formulation (Equation (6.55)). This appealing feature for engineers skips the derivation from the continuous formulation in the governing differential equations. Taking the analytical limit, a continuous formulation (e.g. Equation (6.75)) is again established. Thus, this derivation first discretises, and then performs an analytical limit in the radial direction, which represents a detour!

7 Modelisation with Single Line Finite Element

The remaining chapters of Part I discuss the solution procedures of the scaled boundary finite element equations in displacement and in dynamic stiffness for bounded and unbounded media. The wedge, a bounded medium, and the truncated semi-infinite wedge, an unbounded medium, for out-of-plane motion defined in Chapter 4 are discretised with a single line finite element. This finite element can also be called a scaled boundary element. For the Model Problem, after enforcing the displacement boundary condition, the scaled boundary finite element equations are scalar. Working with scalars permits closed-form expressions to be derived using elementary mathematics, which is discussed in this chapter. In general, these analytical solutions for scalars cannot be generalised to the matrix case. To provide insight and to ease the extension, the general procedures of the matrix equations developed in the further chapters are applied to the scalar equations. All procedures are analytical when not indicated otherwise.

The Model Problem described in Chapter 4 addresses the two-dimensional out-of-plane (anti-plane) motion with the displacement u and body load p perpendicular to the plane. A wedge as a bounded medium (Figure 7.1a) and a truncated semi-infinite wedge (Figure 7.1b) are addressed. For the wedge, the side $O-1$ is fixed ($u=0$), and the side $O-2$ is free (surface traction $\tau_n = 0$). For the truncated semi-infinite wedge, the sides $1-\infty$ and $2-\infty$ are fixed and free, respectively.

The scaling centre is selected at O. As the sides $O-1$ and $O-2$ pass through the scaling centre, no discretisation is required on these side-faces of the wedge. As the extensions of the sides $1-\infty$ and $2-\infty$ also pass through O, again, no discretisation is applied on these side-faces extending to infinity of the truncated semi-infinite wedge. This appealing feature of the scaled boundary finite element method is described in connection with Figures 3.1b and 3.4b.

The side $1-2$ is discretised with a single line element with two nodes (Figure 7.1). The dimensionless radial coordinates $\xi = 0$ in O and $\xi = 1$ on the side $1-2$ apply. A body load $p(\xi)$ as a power function in the radial direction ξ is addressed

$$p(\xi) = p_0 \xi^d \tag{7.1}$$

with the power d. After enforcing the displacement boundary condition by setting $u_1(\xi) = 0$, the scaled boundary finite-element equation in the displacement amplitude $u_2(\xi)$ equals (Equations (5.47) and (5.45))

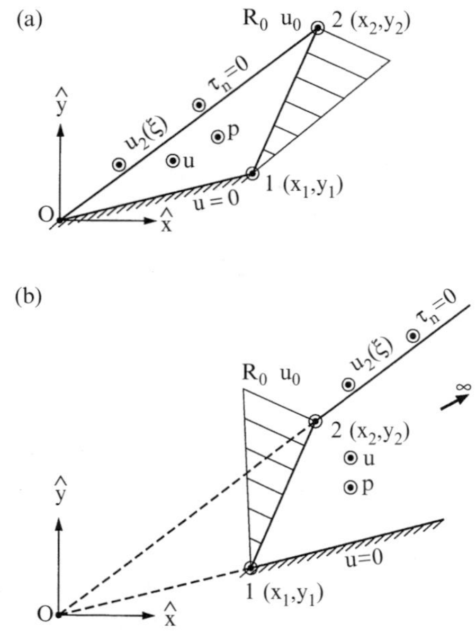

Figure 7.1 Model problem discretised with single line finite element with two nodes: (a) wedge, (b) truncated semi-infinite wedge

$$E^0\xi^2 u_2(\xi)_{,\xi\xi} + E^0\xi u_2(\xi)_{,\xi} - E^2 u_2(\xi) + \omega^2\xi^2 M^0 u_2(\xi) + \xi^2 F_2^{\mathrm{b}}(\xi) = 0 \tag{7.2}$$

The coefficients equal (Equations (5.42a), (5.42c) and (5.43))

$$E^0 = G\frac{\Delta_x^2 + \Delta_y^2}{3(x_1y_2 - x_2y_1)} \tag{7.3a}$$

$$E^2 = G\frac{\Delta_x^2 + \Delta_y^2 + 12(\bar{x}^2 + \bar{y}^2)}{12(x_1y_2 - x_2y_1)} \tag{7.3b}$$

$$M^0 = \rho\frac{x_1y_2 - x_2y_1}{3} \tag{7.4}$$

$F_2^{\mathrm{t}} = 0$ in Equation (5.45), as $\tau_n = 0$ on the side-face $O - 2$ or $2 - \infty$ (Equation (5.39). The body load leads to (Equation (5.38))

$$F_2^{\mathrm{b}}(\xi) = \frac{1}{2}(x_1y_2 - x_2y_1)p_0\xi^d \tag{7.5}$$

As only one degree of freedom on the boundary remains, Equation (7.2) is a scalar instead of a matrix equation. It applies to both the wedge (Figure 7.1a) and the truncated semi-infinite wedge (Figure 7.1b) with the same coefficients.

The scaled boundary finite-element equations in dynamic stiffness for the wedge and the truncated semi-infinite wedge follow from Equations (5.63) and (5.64) as

$$\frac{1}{E^0}(S^{\mathrm{b}}(\omega))^2 - 2\frac{E^1}{E^0}S^{\mathrm{b}}(\omega) + \omega(S^{\mathrm{b}}(\omega))_{,\omega} + \frac{(E^1)^2}{E^0} - E^2 + \omega^2 M^0 = 0 \tag{7.6}$$

$$\frac{1}{E^0}(S^{\infty}(\omega))^2 + 2\frac{E^1}{E^0}S^{\infty}(\omega) - \omega(S^{\infty}(\omega))_{,\omega} + \frac{(E^1)^2}{E^0} - E^2 + \omega^2 M^0 = 0 \tag{7.7}$$

with the coefficient (Equation (5.42b))

$$E^1 = -G\frac{\Delta_x^2 + \Delta_y^2 + 6(\bar{x}\Delta_x + \bar{y}\Delta_y)}{12(x_1 y_2 - x_2 y_1)} \tag{7.8}$$

For the truncated semi-infinite wedge in dynamics without body loads, the nodal force-displacement relationship of Equation (5.54) equals

$$-S^{\infty}(\omega, \xi)u_2(\xi) = E^0\xi u_2(\xi)_{,\xi} + E^1 u_2(\xi) \tag{7.9}$$

Introducing the independent variable

$$\bar{\xi} = \omega\xi \tag{7.10}$$

which is proportional to the dimensionless frequency a, defined in Equation (5.60), Equation (7.9) is formulated as

$$-S^{\infty}(\bar{\xi})u_2(\bar{\xi}) = E^0\bar{\xi}\, u_2(\bar{\xi})_{,\bar{\xi}} + E^1 u_2(\bar{\xi}) \tag{7.11}$$

The analytical solutions of the Model Problem are addressed. In *statics* ($\omega = 0$), for vanishing body loads ($F_2^{\mathrm{b}}(\xi) = 0$) the scaled boundary finite element equations in displacement and stiffness follow from Equations (7.2), (7.6) and (7.7):

$$E^0\xi^2 u_2(\xi)_{,\xi\xi} + E^0\xi u_2(\xi)_{,\xi} - E^2 u_2(\xi) = 0 \tag{7.12}$$

$$\frac{1}{E^0}(K^{\mathrm{b}})^2 - 2\frac{E^1}{E^0}K^{\mathrm{b}} + \frac{(E^1)^2}{E^0} - E^2 = 0 \tag{7.13}$$

$$\frac{1}{E^0}(K^{\infty})^2 + 2\frac{E^1}{E^0}K^{\infty} + \frac{(E^1)^2}{E^0} - E^2 = 0 \tag{7.14}$$

with the static-stiffness coefficients K^{b} and K^{∞} for the wedge and the truncated semi-infinite wedge. Equation (7.12) is an Euler–Cauchy equation, and Equations (7.13) and (7.14) are quadratic equations (Riccati equations). Substituting

$$u_2(\xi) = \xi^{\alpha} \tag{7.15}$$

into Equation (7.12) yields

$$E^0\alpha^2 - E^2 = 0 \tag{7.16}$$

i.e.

$$\alpha_{\frac{1}{2}} = \pm\sqrt{\frac{E^2}{E^0}} = \pm\frac{1}{2}\sqrt{1 + \frac{12(\bar{x}^2 + \bar{y}^2)}{\Delta_x^2 + \Delta_y^2}} \tag{7.17}$$

The general solution of Equation (7.12) is

$$u_2(\xi) = c_1\xi^{\alpha_1} + c_2\xi^{\alpha_2} \tag{7.18}$$

with the integration constants c_1 and c_2. For the bounded case, u_2 must remain finite at $\xi = 0$. As α_2 is negative, $c_2 = 0$ results and with $c = c_1$

$$u_2(\xi) = c\xi^{\alpha_1} \tag{7.19}$$

The static-stiffness coefficient follows from Equation (7.13). The positive root equals

$$K^{\mathrm{b}} = E^1 + \sqrt{E^0E^2} \tag{7.20}$$

For the unbounded case, $u_2 \to 0$ for $\xi \to \infty$. As α_1 is positive, $c_1 = 0$ results and

$$u_2(\xi) = c\xi^{\alpha_2} \tag{7.21}$$

The static-stiffness coefficient is equal to the positive root of Equation (7.14)

$$K^{\infty} = -E^1 + \sqrt{E^0E^2} \tag{7.22}$$

In statics, when body loads are present, Equation (7.2) yields

$$E^0\xi^2u_2(\xi),_{\xi\xi} + E^0\xi u_2(\xi),_{\xi} - E^2u_2(\xi) + \xi^2F_2^{\mathrm{b}}(\xi) = 0 \tag{7.23}$$

with $F_2^{\mathrm{b}}(\xi)$ specified in Equation (7.5). The particular solution of this non-homogeneous equation is chosen as $c_3\xi^{\beta}$. Substituting in Equation (7.23) results in

$$c_3(\beta(\beta-1)\xi^{\beta}E^0 + \beta\xi^{\beta}E^0 - \xi^{\beta}E^2) + \frac{1}{2}(x_1y_2 - x_2y_1)p_0\xi^{d+2} = 0 \tag{7.24}$$

which yields

$$\beta = d + 2 \tag{7.25a}$$

$$c_3 = -\frac{x_1y_2 - x_2y_1}{2((d+2)^2E^0 - E^2)}p_0 \tag{7.25b}$$

The general solution equals

$$u_2(\xi) = c_1\xi^{\alpha_1} + c_2\xi^{\alpha_2} - \frac{x_1y_2 - x_2y_1}{2((d+2)^2E^0 - E^2)}p_0\xi^{d+2} \tag{7.26}$$

For the bounded case, to achieve a finite $u_2(\xi = 0)$ $d > -2$ must be satisfied, which restricts the body load (Equation (7.1)). The same condition leads to $c_2 = 0$ (as for vanishing body loads), resulting in

$$u_2(\xi) = c\xi^{\alpha_1} - \frac{x_1y_2 - x_2y_1}{2((d+2)^2E^0 - E^2)}p_0\xi^{d+2} \tag{7.27}$$

To determine the nodal load due to the body load, the nodal force is calculated (Equations (5.51) and (5.52)):

$$R_2(\xi) = E^0\xi u_2(\xi)_{,\xi} + E^1u_2(\xi) \tag{7.28}$$

Substituting Equation (7.27) results in

$$R_2(\xi) = c(E^1 + \alpha_1E^0)\xi^{\alpha_1} - \frac{(x_1y_2 - x_2y_1)((d+2)E^0 + E^1)}{2((d+2)^2E^0 - E^2)}p_0\xi^{d+2} \tag{7.29}$$

Eliminating the integration constant c from Equation (7.29) using Equation (7.27) yields

$$R_2(\xi) = \left(E^1 + \sqrt{E^0E^2}\right)u_2(\xi) - \frac{(x_1y_2 - x_2y_1)\sqrt{E^0}}{2\left((d+2)\sqrt{E^0} + \sqrt{E^2}\right)}p_0\xi^{d+2} \tag{7.30}$$

The coefficient of $u_2(\xi)$ is equal to K^{b} (Equation (7.20)), and the nodal load due to the body load follows as (Equation (5.53) formulated for statics)

$$R_2^{\mathrm{F}}(\xi) = \frac{(x_1y_2 - x_2y_1)\sqrt{E^0}}{2\left((d+2)\sqrt{E^0} + \sqrt{E^2}\right)}p_0\xi^{d+2} \tag{7.31}$$

Alternatively, the non-homogeneous differential equation for $R_2^F(\xi)$ (Equation (5.59)) formulated for statics) could be solved with K^{b} specified in Equation (7.20).

For the unbounded case with a body load decaying with ξ ($d < -2$ must be satisfied in Equation (7.1)), $u_2 \to 0$ for $\xi \to \infty$ yields $c_1 = 0$ in Equation (7.26) and

$$u_2(\xi) = c\xi^{\alpha_2} - \frac{x_1y_2 - x_2y_1}{2((d+2)^2E^0 - E^2)}p_0\xi^{d+2} \tag{7.32}$$

The nodal force is calculated (Equations (5.51) and (5.52))

$$R_2(\xi) = -E^0\xi u_2(\xi)_{,\xi} - E^1u_2(\xi) \tag{7.33}$$

Substituting Equation (7.32) leads to

$$R_2(\xi) = c\left(-E^1 + \alpha_2 E^0\right)\xi^{\alpha_2} + \frac{(x_1y_2 - x_2y_1)\left((d+2)E^0 + E^1\right)}{2\left((d+2)^2E^0 - E^2\right)}p_0\xi^{d+2} \tag{7.34}$$

Eliminating the integration constant c from Equation (7.34) using Equation (7.32) yields

$$R_2(\xi) = \left(-E^1 + \sqrt{E^0E^2}\right)u_2(\xi) - \frac{(x_1y_2 - x_2y_1)\sqrt{E^0}}{2\left(-(d+2)\sqrt{E^0} + \sqrt{E^2}\right)}p_0\xi^{d+2} \tag{7.35}$$

The coefficient of $u_2(\xi)$ is equal to K^∞ (Equation (7.22)), and the nodal load due to the body load follows as (Equation (5.53))

$$R_2^{\mathrm{F}}(\xi) = \frac{(x_1y_2 - x_2y_1)\sqrt{E^0}}{2\left(-(d+2)\sqrt{E^0} + \sqrt{E^2}\right)}p_0\xi^{d+2} \tag{7.36}$$

$R_2^{\mathrm{F}}(\xi)$ could also be calculated from Equation (5.59) with K^∞ specified in Equation (7.22).

As the last item in the static analysis, the stress singularity in τ_n along the fixed side $O-1$ at the corner point O (with the fixed side $O-1$ and the free side $O-2$) (Figure 7.1a), which can occur for certain configurations, is addressed. An analytical expression can be derived as follows. The displacement of the element $O-1-2$ specified in Equation (5.31) with Equation (5.2) equals

$$u(\xi,\eta) = \frac{1}{2}(1+\eta)u_2(\xi) \tag{7.37}$$

where $u_2(\xi)$ is specified in Equation (7.19) as

$$u_2(\xi) = \bar{u}\xi^{\alpha_1} \tag{7.38}$$

with $\bar{u}$ denoting the prescribed displacement in Point 2. The side $O-1$ corresponds to $\eta = -1$, τ_n on side $O-1$ follows from Equation (5.24) as

$$\tau_n = \{n\}^T\{\tau\} \tag{7.39}$$

with the outward normal equal to (Equation (5.20b))

$$\{n\} = -\frac{1}{\sqrt{x_1^2 + y_1^2}}\begin{Bmatrix} -y_1 \\ x_1 \end{Bmatrix} \tag{7.40}$$

$\{\tau\}$ is specified in Equation (5.23), which is written as

$$\{\tau\} = G\frac{1}{\xi}\{b^2\}u_{,\eta} \tag{7.41}$$

as $u_{,\xi} = 0$ for $\eta = -1$ (Equation (7.37)). $\{b^2\}$ follows for $\eta = -1$ as (Equation (5.15b))

$$\{b^2\} = \frac{1}{x_1 y_2 - x_2 y_1} \left\{ \begin{array}{c} -2\bar{y} + \Delta_y \\ 2\bar{x} - \Delta_x \end{array} \right\} \tag{7.42}$$

Calculating $u,_\eta$ from Equation (7.37) and substituting into Equation (7.41) yields, with Equations (7.42) and (7.40) from Equation (7.39),

$$\tau_n = G \frac{\sqrt{x_1^2 + y_1^2}}{x_2 y_1 - x_1 y_2} \frac{1}{\xi} u_2(\xi) \tag{7.43}$$

or with Equation (7.38),

$$\tau_n = G \frac{\sqrt{x_1^2 + y_1^2}}{x_2 y_1 - x_1 y_2} \bar{u} \xi^{\alpha_1 - 1} \tag{7.44}$$

α_1 is formulated in Equation (7.17). For $\alpha_1 - 1 < 0$ a singularity for τ_n occurs for $\xi \rightarrow 0$ on side $O - 1$ with the power $\alpha_1 - 1$. This condition

$$\frac{1}{2}\sqrt{1 + \frac{12(\bar{x}^2 + \bar{y}^2)}{\Delta_x^2 + \Delta_y^2}} < 1 \tag{7.45}$$

is reformulated as

$$\sqrt{\bar{x}^2 + \bar{y}^2} < \frac{\sqrt{\Delta_x^2 + \Delta_y^2}}{2} \tag{7.46}$$

A stress singularity thus occurs, when the distance from O to the midpoint of side $1 - 2$ is smaller than half the side length $1 - 2$, or in other words, when the angle at O between the sides $O - 1$ and $O - 2$ is larger than 90°.

In *dynamics*, for vanishing body load $\left(F_2^b(\xi) = 0\right)$, the scaled boundary finite element equation in displacement (Equation (7.2)) is a Bessel equation

$$\xi^2 u_2(\xi),_{\xi\xi} + \xi u_2(\xi),_\xi - \alpha_1^2 u_2(\xi) + a^2 u_2(\xi) = 0 \tag{7.47}$$

with the dimensionless frequency as a function of ξ (Equation (5.60))

$$a = \frac{\omega \xi r_0}{c_s} \tag{7.48}$$

where the characteristic length is defined as

$$r_0 = c_s \sqrt{\frac{M^0}{E^0}} = \frac{x_1 y_2 - x_2 y_1}{\sqrt{\Delta_x^2 + \Delta_y^2}} \tag{7.49}$$

and α_1 is specified in Equation (7.17). Alternatively, Equation (7.47) can also be formulated in the independent variable a as

$$a^2 u_2(a)_{,aa} + a u_2(a)_{,a} - \alpha_1^2 u_2(a) + a^2 u_2(a) = 0 \tag{7.50}$$

where no change in nomenclature is introduced for the dependent variable u_2.

For the bounded case, the solution with a finite $u_2(\xi)$ in the domain equals

$$u_2(\xi) = c J_{\alpha_1}(a) \tag{7.51}$$

J_{α_1} is the Bessel function of the first kind of order α_1 and c is the integration constant. The nodal force amplitude equals (Equations (5.51) and (5.52))

$$\begin{aligned} R_2(\xi) &= E^0 \xi u_2(\xi)_{,\xi} + E^1 u_2(\xi) \\ &= c(E^0 a J_{\alpha_1 - 1}(a) + (E^1 - E^0 \alpha_1) J_{\alpha_1}(a)) \end{aligned} \tag{7.52}$$

The dynamic-stiffness coefficient for the bounded case follows as (Equation (5.53))

$$S^{\mathrm{b}}(\omega,\xi) = \frac{R_2(\xi)}{u_2(\xi)} = E^0 a \frac{J_{\alpha_1 - 1}(a)}{J_{\alpha_1}(a)} + E^1 - E^0 \alpha_1 \tag{7.53}$$

This result confirms that $S^{\mathrm{b}}(\omega, \xi)$ is a function of the dimensionless frequency a only (Equation (7.48)). On the boundary 1 – 2 of the wedge, $\xi = 1$ applies.

The low-frequency expansion of the dynamic-stiffness coefficient at $\xi = 1$ (Equation (7.53)) leads to

$$S^{\mathrm{b}}(\omega) = E^1 + \sqrt{E^0 E^2} - \omega^2 \frac{\sqrt{E^0} M^0}{2\left(\sqrt{E^2} + \sqrt{E^0}\right)} + O(\omega^4) \tag{7.54}$$

This defines the static-stiffness coefficient (Equation (5.65))

$$K^{\mathrm{b}} = E^1 + \sqrt{E^0 E^2} \tag{7.55}$$

and the mass coefficient

$$M^{\mathrm{b}} = \frac{\sqrt{E^0} M^0}{2\left(\sqrt{E^2} + \sqrt{E^0}\right)} \tag{7.56}$$

For the unbounded case, it is convenient to formulate the solution of Equation (7.47) or (7.50) using Hankel functions of the first kind $H_{\alpha_1}^{(1)}$ and the second kind $H_{\alpha_1}^{(2)}$ of order α_1:

$$u_2(\xi) = c_1 H_{\alpha_1}^{(1)}(a) + c_2 H_{\alpha_1}^{(2)}(a) \tag{7.57}$$

To introduce the boundary condition at infinity, called the 'radiation condition', the asymptotic behaviour for $\xi \to \infty (a \to \infty)$ is examined

$$H^{(1)}_{\alpha_1}(a) \approx \sqrt{\frac{2}{\pi a}}\, \mathrm{e}^{+\mathrm{i}(a - \frac{\alpha_1 \pi}{2} - \frac{\pi}{4})} \tag{7.58a}$$

$$H^{(2)}_{\alpha_1}(a) \approx \sqrt{\frac{2}{\pi a}}\, \mathrm{e}^{-\mathrm{i}(a - \frac{\alpha_1 \pi}{2} - \frac{\pi}{4})} \tag{7.58b}$$

As the displacement equals the product of the corresponding amplitude and the factor $\mathrm{e}^{+\mathrm{i}\omega t}$, Equations (7.58a) and (7.58b) correspond to waves propagating from infinity and towards infinity, respectively. As only waves propagating towards infinity are permissible, $c_1 = 0$ results, yielding

$$u_2(\xi) = c H^{(2)}_{\alpha_1}(a) \tag{7.59}$$

Note that both Hankel functions tend towards zero for $a \to \infty$. The nodal force amplitude equals (Equations (5.51) and (5.52) with Equation (7.59))

$$\begin{aligned} R_2(\xi) &= -E^0 \xi u_2(\xi),_{\xi} - E^1 u_2(\xi) \\ &= -c(E^0 a H^{(2)}_{\alpha_1 - 1}(a) + (E^1 - E^0 \alpha_1) H^{(2)}_{\alpha_1}(a)) \end{aligned} \tag{7.60}$$

The dynamic-stiffness coefficient for the unbounded case follows as (Equation (5.54))

$$S^{\infty}(\omega, \xi) = \frac{R_2(\xi)}{u_2(\xi)} = -E^0 a \frac{H^{(2)}_{\alpha_1 - 1}(a)}{H^{(2)}_{\alpha_1}(a)} - E^1 + E^0 \alpha_1 \tag{7.61}$$

Again, $S^{\infty}(\omega, \xi)$ is a function of the dimensionless frequency a (Equation (7.48)). On the boundary $1 - 2$ of the truncated semi-infinite wedge, $\xi = 1$ applies

$$S^{\infty}(a_0) = -E^0 a_0 \frac{H^{(2)}_{\alpha_1 - 1}(a_0)}{H^{(2)}_{\alpha_1}(a_0)} - E^1 + E^0 \alpha_1 \tag{7.62}$$

with the dimensionless frequency referred to the boundary

$$a_0 = \frac{\omega r_0}{c_s} \tag{7.63}$$

In the presence of constant hysteretic material damping with damping ratio ζ, the relationships developed above still apply. The shear modulus G is replaced by (Equation (4.8))

$$G^* = G(1 + 2\mathrm{i}\zeta) \tag{7.64}$$

in all equations, which also modifies c_s (Equation (4.9)) and the dimensionless frequencies (Equations (7.48) and (7.63)):

$$a^* = \frac{a}{\sqrt{1 + 2i\zeta}} \tag{7.65}$$

$$a_0^* = \frac{a_0}{\sqrt{1 + 2i\zeta}} \tag{7.66}$$

Thus, in the case of hysteretic material damping E^0, E^1, E^2 (but not M^0), c_s, a and a_0 become complex. The displacement amplitude is still specified in Equation (7.59) with a^* replacing a (Equation (7.65)). The dynamic-stiffness coefficient $S^\infty(a_0^*)$ follows from Equation (7.62), using E^{0*}, E^{1*}, a_0^*, as

$$S^\infty(a_0^*) = -E^{0*} a_0^* \frac{H_{\alpha_1 - 1}^{(2)}(a_0^*)}{H_{\alpha_1}^{(2)}(a_0^*)} - E^{1*} + E^{0*}\alpha_1 \tag{7.67}$$

The static-stiffness coefficient K^∞ of the unbounded case is equal to $S^\infty(a_0 = 0)$, which leads in Equation (7.62) to Equation (7.22). For discussion of the results, it is customary to non-dimensionalise $S^\infty(a_0)$ with K^∞. The resulting non-dimensional dynamic-stiffness coefficient, a complex variable, is decomposed into the real part $k(a_0)$, a spring coefficient, and the imaginary part $a_0 c(a_0)$, with the damping coefficient $c(a_0)$:

$$S^\infty(a_0) = K^\infty[k(a_0) + i a_0 c(a_0)] \tag{7.68}$$

In the case of hysteretic material damping, as already mentioned, Equation (7.67) yields $S^\infty(a_0^*)$, which can also be represented as a function of a_0, $S^\infty(a_0)$. The latter can then be decomposed as specified in Equation (7.68). This permits $k(a_0)$ and $c(a_0)$ to be examined as a function of the corresponding real dimensionless frequency a_0 also in the case of material damping.

The asymptotic expansion for high (dimensionless) frequency of the dynamic-stiffness coefficient for the unbounded case is addressed. $S^\infty(\omega, \xi)$ in Equation (7.61) is a function of a or also of $\bar{\xi}$ (Equation (7.10)). The asymptotic expansion for $\bar{\xi} \to \infty$ yields

$$S^\infty(\omega, \xi) = S^\infty(\bar{\xi}) \approx i\bar{\xi} C_\infty + K_\infty + \frac{A_1}{i\bar{\xi}} \tag{7.69}$$

with the dashpot coefficient (subscript ∞ for $\bar{\xi} \to \infty$)

$$C_\infty = \sqrt{E^0 M^0} = \frac{1}{3}\sqrt{\Delta_x^2 + \Delta_y^2}\,\rho c_s \tag{7.70}$$

the spring coefficient

$$K_\infty = \frac{E^0}{2} - E^1 = G\frac{x_2\Delta_x + y_2\Delta_y}{2(x_1 y_2 - x_2 y_1)} \tag{7.71}$$

and

$$A_1 = \frac{1}{2}\sqrt{\frac{E^0}{M^0}}\left(E^2 - \frac{E^0}{4}\right) = \frac{1}{2}Gc_s\frac{\sqrt{\Delta_x^2 + \Delta_y^2(\bar{x}^2 + \bar{y}^2)}}{(x_1y_2 - x_2y_1)^2} \tag{7.72}$$

Note that the asymptotic value of the spring coefficient K_∞ (subscript ∞) in Equation (7.71) is not equal to the static-stiffness coefficient K^∞ (superscript ∞) in Equation (7.22). The asymptotic expansion of Equation (7.69) also applies for the side $1-2$ with $\xi = 1$, resulting with $\bar{\xi} = \omega$ in

$$S^\infty(\omega) \approx \mathrm{i}\omega C_\infty + K_\infty + \frac{1}{\mathrm{i}\omega}A_1 \tag{7.73}$$

The coefficients C_∞, K_∞, A_1 are not modified (Equations (7.70), (7.71), (7.72)). For the case of hysteretic material damping, Equation (7.73) still applies, with the coefficients on the right-hand side equal to

$$C_\infty^* = C_\infty\sqrt{1 + 2\mathrm{i}\zeta} \tag{7.74a}$$

$$K_\infty^* = K_\infty(1 + 2\mathrm{i}\zeta) \tag{7.74b}$$

$$A_1^* = A_1(1 + 2\mathrm{i}\zeta)^{\frac{3}{2}} \tag{7.74c}$$

Based on the asymptotic high-frequency expansion of the dynamic stiffness, it is possible to calculate the corresponding displacement. Substituting Equation (7.69) into Equation (7.11) yields

$$E^0\bar{\xi}u_2(\bar{\xi})_{,\bar{\xi}} + \mathrm{i}\bar{\xi}C_\infty u_2(\bar{\xi}) + (E^1 + K_\infty)u_2(\bar{\xi}) + \frac{1}{\mathrm{i}\bar{\xi}}A_1u_2(\bar{\xi}) = 0 \tag{7.75}$$

The solution of this linear first-order ordinary differential equation equals

$$u_2(\bar{\xi}) = c\bar{\xi}^{-\frac{E^1+K_\infty}{E^0}}\mathrm{e}^{-\mathrm{i}\frac{C_\infty}{E^0}\bar{\xi}}\mathrm{e}^{-\mathrm{i}\frac{A_1}{E^0\bar{\xi}}} \tag{7.76}$$

with the power of $\bar{\xi}$, $-(E^1 + K_\infty)/E^0 = -0.5$, being negative, leading to a decrease of the amplitude towards infinity.

Setting $A_1 = 0$ in Equation (7.75), i.e. using only the first two terms in the high-frequency asymptotic expansion, yields for the displacement amplitude in the far field

$$E^0\bar{\xi}u_2(\bar{\xi})_{,\bar{\xi}} + \mathrm{i}\bar{\xi}C_\infty u_2(\bar{\xi}) + (E^1 + K_\infty)u_2(\bar{\xi}) = 0 \tag{7.77}$$

8 Statics

In this and the following chapters, the Model Problem with one degree of freedom and thus governed by scalar equations, is solved by applying the general procedures of the matrix equations, which are present for several degrees of freedom.

8.1 VANISHING BODY LOAD

For vanishing body load the scaled boundary finite element equation in displacement for statics (Equation (7.12)) is a homogeneous second-order differential equation (Euler–Cauchy)

$$E^0 \xi^2 u_2(\xi)_{,\xi\xi} + E^0 \xi u_2(\xi)_{,\xi} - E^2 u_2(\xi) = 0 \tag{8.1}$$

It is transformed into two first-order differential equations with two unknowns. This is consistent with the solution procedure in dynamics for matrix equations (section 12.1). The dependent variables are defined as

$$\{X(\xi)\} = \begin{Bmatrix} u_2(\xi) \\ Q_2(\xi) \end{Bmatrix} \tag{8.2}$$

with the internal nodal force $Q_2(\xi)$ (Equation (5.51)). Equation (8.1) is formulated as

$$\xi\{X(\xi)\}_{,\xi} = -[Z]\{X(\xi)\} \tag{8.3}$$

with the coefficient matrix

$$[Z] = \begin{bmatrix} \dfrac{E^1}{E^0} & -\dfrac{1}{E^0} \\ -E^2 + \dfrac{(E^1)^2}{E^0} & -\dfrac{E^1}{E^0} \end{bmatrix} \tag{8.4}$$

To transform the coefficient matrix $[Z]$, the eigenvalue problem is formulated

$$[Z][\Phi] = -[\Phi]\lceil\Lambda\rfloor \tag{8.5}$$

where the eigenvalues are equal to

$$\lceil \Lambda \rfloor = \begin{bmatrix} -\lambda & \\ & \lambda \end{bmatrix} \tag{8.6}$$

with

$$\lambda = -\sqrt{\frac{E^2}{E^0}} = -\frac{1}{2}\sqrt{1 + \frac{12(\bar{x}^2 + \bar{y}^2)}{\Delta_x^2 + \Delta_y^2}} \tag{8.7}$$

and the eigenvectors are

$$[\Phi] = \begin{bmatrix} \Phi_{11} & \Phi_{12} \\ \Phi_{21} & \Phi_{22} \end{bmatrix} = \begin{bmatrix} 1 & 1 \\ E^1 + \sqrt{E^0 E^2} & E^1 - \sqrt{E^0 E^2} \end{bmatrix} \tag{8.8}$$

Substituting Equations (8.7) and (8.8), Equation (8.5) is indeed satisfied. Introducing the variable $\{W(\xi)\}$ defined as

$$\{X(\xi)\} = [\Phi]\{W(\xi)\} \tag{8.9}$$

substituting into Equation (8.3), using Equation (8.4) and premultiplying by $[\Phi]^{-1}$ yield

$$\xi\{W(\xi)\}_{,\xi} = \lceil \Lambda \rfloor \{W(\xi)\} \tag{8.10}$$

Its solution equals

$$\{W(\xi)\} = \begin{bmatrix} \xi^{-\lambda} & \\ & \xi^{\lambda} \end{bmatrix} \begin{Bmatrix} c_1 \\ c_2 \end{Bmatrix} \tag{8.11}$$

with the integration constants c_1, c_2. Substituting Equation (8.11) into Equation (8.9) with Equation (8.2) leads to

$$u_2(\xi) = c_1\Phi_{11}\xi^{-\lambda} + c_2\Phi_{12}\xi^{\lambda} = c_1\xi^{-\lambda} + c_2\xi^{\lambda} \tag{8.12a}$$

$$Q_2(\xi) = c_1\Phi_{21}\xi^{-\lambda} + c_2\Phi_{22}\xi^{\lambda} = c_1(E^1 + \sqrt{E^0 E^2})\xi^{-\lambda} + c_2(E^1 - \sqrt{E^0 E^2})\xi^{\lambda} \tag{8.12b}$$

For the wedge, $u_2(\xi)$ must remain finite at $\xi = 0$. As λ is negative (Equation (8.7)), this leads to $c_2 = 0$ and

$$u_2(\xi) = c\Phi_{11}\xi^{-\lambda} = c\xi^{-\lambda} \tag{8.13a}$$

$$Q_2(\xi) = c\Phi_{21}\xi^{-\lambda} = c(E^1 + \sqrt{E^0 E^2})\xi^{-\lambda} \tag{8.13b}$$

Based on Equation (5.52), the nodal force $R_2(\xi) = Q_2(\xi)$. The static-stiffness coefficient follows as (Equation (5.53) formulated for $\omega = 0$)

$$K^{\mathrm{b}} = \frac{R_2(\xi)}{u_2(\xi)} = \frac{\Phi_{21}}{\Phi_{11}} = E^1 + \sqrt{E^0 E^2} \tag{8.14}$$

For the truncated semi-infinite wedge, $u_2 \rightarrow 0$ for $\xi \rightarrow \infty$, yielding $c_1 = 0$. Equation (5.52) leads to $R_2(\xi) = -Q_2(\xi)$. The static-stiffness coefficient equals (Equation (5.53))

$$K^{\infty} = \frac{R_2(\xi)}{u_2(\xi)} = -\frac{\Phi_{22}}{\Phi_{12}} = -E^1 + \sqrt{E^0 E^2} \tag{8.15}$$

Alternatively, the static-stiffness coefficients can also be determined from Equations (7.13) and (7.14) using the solution procedure of the Riccati equation. The eigenvalue problem (Equation (8.5)) is solved. K^{b} and K^{∞} then follow from Equations (8.14) and (8.15) using the eigenvectors.

8.2 BODY LOAD

For a body load, the scaled boundary finite element equation (Equation (7.23)) is a non-homogeneous differential equation

$$E^0 \xi^2 u_2(\xi),_{\xi\xi} + E^0 \xi u_2(\xi),_{\xi} - E^2 u_2(\xi) + \xi^2 F_2^{\mathrm{b}}(\xi) = 0 \tag{8.16}$$

It is processed analogously, applying the method of variation of parameters. The corresponding first-order differential equations equal

$$\xi\{X(\xi)\},_{\xi} = -[Z]\{X(\xi)\} - \begin{Bmatrix} 0 \\ \xi^2 F_2^{\mathrm{b}}(\xi) \end{Bmatrix} \tag{8.17}$$

with $\{X(\xi)\}$ in Equation (8.2) and $[Z]$ in Equation (8.4). The transformed variables $\{W(\xi)\}$ are again introduced (Equation (8.9)). Substituting into Equation (8.17) yields

$$\xi\{W(\xi)\},_{\xi} = \lceil\Lambda\rfloor\{W(\xi)\} - [A]\begin{Bmatrix} 0 \\ \xi^2 F_2^{\mathrm{b}}(\xi) \end{Bmatrix} \tag{8.18}$$

where

$$[A] = \begin{bmatrix} A_{11} & A_{12} \\ A_{21} & A_{22} \end{bmatrix} = [\Phi]^{-1} = \frac{1}{2\sqrt{E^0 E^2}} \begin{bmatrix} -E^1 + \sqrt{E^0 E^2} & 1 \\ E^1 + \sqrt{E^0 E^2} & -1 \end{bmatrix} \tag{8.19}$$

The solution of Equation (8.18) is postulated as (Equation (8.11))

$$\{W(\xi)\} = \begin{bmatrix} \xi^{-\lambda} & \\ & \xi^{\lambda} \end{bmatrix} \begin{Bmatrix} c_1(\xi) \\ c_2(\xi) \end{Bmatrix} \tag{8.20}$$

Note that the integration constants are replaced by $c_1(\xi)$, $c_2(\xi)$, which are functions of ξ. Substituting Equation (8.20) into Equation (8.18) results in

$$c_1(\xi),_{\xi} = -\xi^{\lambda+1} A_{12} F_2^{\mathrm{b}}(\xi) \tag{8.21a}$$

$$c_2(\xi)_{,\xi} = -\xi^{-\lambda+1} A_{22} F_2^{\mathrm{b}}(\xi) \tag{8.21b}$$

Their solutions equal

$$c_1(\xi) = -\int \xi^{\lambda+1} A_{12} F_2^{\mathrm{b}}(\xi) \mathrm{d}\xi \tag{8.22a}$$

$$c_2(\xi) = -\int \xi^{-\lambda+1} A_{22} F_2^{\mathrm{b}}(\xi) \mathrm{d}\xi \tag{8.22b}$$

Substituting $F_2^{\mathrm{b}}(\xi)$ (Equation (7.5)) and A_{12}, A_{22} (Equation (8.19)) yields

$$c_1(\xi) = c_1 - \frac{x_1 y_2 - x_2 y_1}{4\sqrt{E^0 E^2}\,(\lambda + d + 2)} p_0 \xi^{\lambda+d+2} \tag{8.23a}$$

$$c_2(\xi) = c_2 + \frac{x_1 y_2 - x_2 y_1}{4\sqrt{E^0 E^2}\,(-\lambda + d + 2)} p_0 \xi^{-\lambda+d+2} \tag{8.23b}$$

From Equations (8.2), (8.9), (8.20) and (8.23),

$$u_2(\xi) = c_1 \Phi_{11} \xi^{-\lambda} + c_2 \Phi_{12} \xi^{\lambda} - \frac{x_1 y_2 - x_2 y_1}{2\left((d+2)^2 E^0 - E^2\right)} p_0 \xi^{d+2} \tag{8.24a}$$

$$Q_2(\xi) = c_1 \Phi_{21} \xi^{-\lambda} + c_2 \Phi_{22} \xi^{\lambda} - \frac{(x_1 y_2 - x_2 y_1)((d+2)E^0 + E^1)}{2((d+2)^2 E^0 - E^2)} p_0 \xi^{d+2} \tag{8.24b}$$

apply.

For the wedge, $u_2(\xi)$ must remain finite at $\xi = 0$ which leads as $\lambda < 0$ to $c_2 = 0$. $d > -2$ must be satisfied (Equation (8.24))

$$u_2(\xi) = c\xi^{-\lambda} - \frac{x_1 y_2 - x_2 y_1}{2((d+2)^2 E^0 - E^2)} p_0 \xi^{d+2} \tag{8.25a}$$

$$Q_2(\xi) = c(E^1 + \sqrt{E^0 E^2})\xi^{-\lambda} - \frac{(x_1 y_2 - x_2 y_1)((d+2)E^0 + E^1)}{2((d+2)^2 E^0 - E^2)} p_0 \xi^{d+2} \tag{8.25b}$$

Based on Equation (5.52), the nodal force $R_2(\xi) = Q_2(\xi)$. Eliminating c from Equation (8.25) yields

$$R_2(\xi) = (E^1 + \sqrt{E^0 E^2}) u_2(\xi) - \frac{(x_1 y_2 - x_2 y_1)\sqrt{E^0}}{2\left((d+2)\sqrt{E^0} + \sqrt{E^2}\right)} p_0 \xi^{d+2} \tag{8.26}$$

The coefficient of $u_2(\xi)$ is equal to the static-stiffness coefficient K^{b} (Equation (8.14)). Comparing Equations (8.26) and (5.53) results in the nodal load due to the body load:

$$R_2^{\mathrm{F}}(\xi) = \frac{(x_1 y_2 - x_2 y_1)\sqrt{E^0}}{2\left((d+2)\sqrt{E^0} + \sqrt{E^2}\right)} p_0 \xi^{d+2} \tag{8.27}$$

For the truncated semi-infinite wedge, $u_2 \to 0$ for $\xi \to \infty$ in Equation (8.24a) leads to $c_1 = 0$. $d < -2$ must be satisfied (Equation (8.24))

$$u_2(\xi) = c\xi^{\lambda} - \frac{x_1 y_2 - x_2 y_1}{2((d+2)^2 E^0 - E^2)} p_0 \xi^{d+2} \tag{8.28a}$$

$$Q_2(\xi) = c(E^1 - \sqrt{E^0 E^2})\xi^{\lambda} - \frac{(x_1 y_2 - x_2 y_1)((d+2)E^0 + E^1)}{2((d+2)^2 E^0 - E^2)} p_0 \xi^{d+2} \tag{8.28b}$$

apply. Equation (5.52) leads to $R_2(\xi) = -Q_2(\xi)$. Eliminating c from Equation (8.28) yields

$$R_2(\xi) = (-E^1 + \sqrt{E^0 E^2}) u_2(\xi) - \frac{(x_1 y_2 - x_2 y_1)\sqrt{E^0}}{2(-(d+2)\sqrt{E^0} + \sqrt{E^2})} p_0 \xi^{d+2} \tag{8.29}$$

The coefficient of $u_2(\xi)$ is equal to K^{∞} (Equation (8.15)). Comparing Equations (8.29) and (5.53) results in the nodal load due to the body load:

$$R_2^{\mathrm{F}}(\xi) = \frac{(x_1 y_2 - x_2 y_1)\sqrt{E^0}}{2\left(-(d+2)\sqrt{E^0} + \sqrt{E^2}\right)} p_0 \xi^{d+2} \tag{8.30}$$

This analytical procedure is equivalent to that in Chapter 7.

9 Mass of Wedge

For the dynamic analysis of a bounded medium, the mass matrix is required in addition to the static-stiffness matrix.

The mass coefficient M^b of the wedge with respect to node 2 on the boundary (Figure 7.1a) is defined using the low-frequency expansion of the dynamic-stiffness coefficient as (Equation (5.65))

$$S^b(\omega) = K^b - \omega^2 M^b + O(\omega^4) \tag{9.1}$$

Substituting Equation (9.1) in Equation (7.6) leads to

$$\frac{1}{E^0}\left(K^b\right)^2 - 2\frac{E^1}{E^0}K^b + \frac{(E^1)^2}{E^0} - E^2 + \omega^2\left(2\left(\frac{-K^b + E^1}{E^0} - 1\right)M^b + M^0\right) + O(\omega^4) = 0 \tag{9.2}$$

The constant term is used to determine the static-stiffness coefficient K^b, and thus vanishes (Equation (7.13)). Setting the coefficient of ω^2 equal to zero yields

$$2\left(\frac{-K^b + E^1}{E^0} - 1\right)M^b + M^0 = 0 \tag{9.3}$$

The mass coefficient M^b thus follows from a *linear* equation. Substituting K^b (Equation (8.14)) leads to

$$M^b = \frac{\sqrt{E^0}M^0}{2\left(\sqrt{E^2} + \sqrt{E^0}\right)} \tag{9.4}$$

which is the same result as in Equation (7.56).

Alternatively, the mass coefficient M^b can be determined using as the shape function the analytical expression for the static displacement, as in standard finite elements. The static displacement $u_2(\xi)$ specified in Equation (8.13a) is written as

$$u_2(\xi) = \bar{u}\xi^{-\lambda} \tag{9.5}$$

where the integration constant is equal to the displacement on the boundary $u_2(\xi = 1) = \bar{u}$. The static displacement over the element equals

$$u(\xi,\eta) = N_V(\xi,\eta)\bar{u} \tag{9.6}$$

with the shape function $N_V(\xi,\eta)$ (subscript V for volume)

$$N_V(\xi,\eta) = \tfrac{1}{2}\xi^{-\lambda}(1+\eta) \tag{9.7}$$

as specified in Equation (7.37). Based on the term in the weighted-residual statement in dynamics for finite elements (Equation (1.86)), the mass coefficient equals (Equations (1.87) and (1.88))

$$M^b = \rho \int N_V(\xi,\eta)^T N_V(\xi,\eta) \mathrm{d}V \tag{9.8}$$

with $\mathrm{d}V$ as specified in Equation (5.18) as

$$\mathrm{d}V = \xi|J|\mathrm{d}\xi \mathrm{d}\eta \tag{9.9}$$

Based on Equations (5.16) and (7.4), Equation (9.9) is rewritten as

$$\mathrm{d}V = \frac{3M^0}{2\rho}\xi \mathrm{d}\xi \mathrm{d}\eta \tag{9.10}$$

Substituting Equations (9.10) and (9.7) into Equation (9.8) yields

$$M^b = \frac{3M^0}{2}\int_0^1 \xi^{-2\lambda}\xi\left(\int_{-1}^{+1}\left(\frac{1+\eta}{2}\right)^2 \mathrm{d}\eta\right)\mathrm{d}\xi \tag{9.11}$$

Performing the integrations results in

$$M^b = \frac{M^0}{2}\frac{1}{-\lambda+1} \tag{9.12}$$

Substituting λ from Equation (8.7) leads to

$$M^b = \frac{\sqrt{E^0}M^0}{2\left(\sqrt{E^2}+\sqrt{E^0}\right)} \tag{9.13}$$

10 High-frequency Asymptotic Expansion for Dynamic Stiffness of Truncated Semi-infinite Wedge

For an unbounded medium a boundary condition at infinity must be formulated to achieve a *unique solution*. This so-called *radiation condition* states that no energy is radiated from infinity towards the medium [S2]. The unbounded medium thus acts as an *energy sink*, dissipating energy, and never as an energy source. As discussed in connection with Equation (7.57), this corresponds to waves propagating towards infinity. It is not sufficient to state that the waves decay to zero towards infinity.

Infinity corresponds to the (dimensionless) radial coordinate $\xi \to \infty$. The governing independent variable is, however, not ξ, but the dimensionless frequency which is proportional to $\bar{\xi} = \omega\xi$ (Equation (7.10)), as discussed in section 5.4. The limit $\xi \to \infty$ can also be achieved with $\omega \to \infty$, both leading to $\bar{\xi} \to \infty$. Thus, the high-frequency limit of the dynamic-stiffness coefficient is addressed to formulate the radiation condition. An asymptotic expansion is formulated. In addition, a starting value for the dynamic-stiffness coefficient at a high but finite frequency can be determined, which permits a numerical integration of the scaled boundary finite element equation in dynamic stiffness (Equation (7.7)) for decreasing frequency to be performed (see section 11.1).

The dynamic-stiffness coefficient $S^{\infty}(\omega)$ at high frequency is expanded in a power series of $i\omega$ in descending order starting at one with, for example, three terms

$$S^{\infty}(\omega) \approx i\omega C_{\infty} + K_{\infty} + \frac{1}{i\omega} A_1 \tag{10.1}$$

with the constant dashpot coefficient C_{∞}, the constant spring coefficient K_{∞} and the constant A_1. Substituting Equation (10.1) into Equation (7.7) and setting the coefficients of the terms in descending order of the power of $i\omega$ equal to zero determines analytically the unknown coefficients in Equation (10.1) sequentially:

$$\begin{aligned}(i\omega)^2\left(\frac{1}{E^0}C_{\infty}^2 - M^0\right) + i\omega\frac{1}{E^0}C_{\infty}(2K_{\infty} + 2E^1 - E^0)\\ + \frac{1}{E^0}\left(2C_{\infty}A_1 + \left(K_{\infty} + E^1\right)^2 - E^0E^2\right) + O\left(\frac{1}{i\omega}\right) = 0\end{aligned} \tag{10.2}$$

C_∞ follows from the coefficient of $(\mathrm{i}\omega)^2$ solving a *quadratic* equation:

$$C_\infty = \sqrt{E^0 M^0} = \frac{1}{3}\sqrt{\Delta_x^2 + \Delta_y^2}\rho c_s \tag{10.3}$$

The *positive* square root is chosen leading to a positive dashpot coefficient, and thus to dissipation of energy in the dashpot. K_∞ is determined from the coefficient of $\mathrm{i}\omega$ in Equation (10.2), solving a *linear* equation

$$K_\infty = \frac{E^0}{2} - E^1 = G\frac{x_2\Delta_x + y_2\Delta_y}{2(x_1y_2 - x_2y_1)} \tag{10.4}$$

Analogously, from a *linear* equation

$$A_1 = \frac{1}{2}\sqrt{\frac{E^0}{M^0}}\left(E^2 - \frac{E^0}{4}\right) = \frac{1}{2}Gc_s\frac{\sqrt{\Delta_x^2 + \Delta_y^2}(\bar{x}^2 + \bar{y}^2)}{(x_1y_2 - x_2y_1)^2} \tag{10.5}$$

results. Again, C_∞, K_∞ and A_1 are the same as in Equations (7.70), (7.71) and (7.72).

For hysteretic material damping, Equation (10.1) still applies, but with the coefficients on the right-hand side C_∞, K_∞ and A_1 multiplied by $\sqrt{1 + 2\mathrm{i}\zeta}, 1 + 2\mathrm{i}\zeta$ and $(1 + 2\mathrm{i}\zeta)^{\frac{3}{2}}$, respectively (Equation (7.74)).

It is important to note that the constant dashpot coefficient C_∞ can be constructed from the finite element discretisation of the boundary $1 - 2$ without solving a quadratic equation. The response of the truncated semi-infinite wedge at the initial time is calculated directly. An infinitesimal area $\mathrm{d}S$ of the side $1 - 2$, which is a length in this two-dimensional case, is addressed. The truncated semi-infinite wedge is initially at rest. After applying the load r per unit length (and unit depth $\mathrm{d}z$) acting perpendicular to the plane during the first infinitesimal time $\mathrm{d}t$, the wave propagates perpendicularly to the side $1 - 2$ in the plane, the wave front is at a distance $c_s\mathrm{d}t$ (shear-wave velocity c_s), and the domain of influence of the excitation equals $c_s\mathrm{d}t\mathrm{d}S$. The law of conservation of momentum is applied for the first infinitesimal time $\mathrm{d}t$. The initial momentum vanishes. The momentum (mass times velocity) at $\mathrm{d}t$ equals $\rho c_s\mathrm{d}t\mathrm{d}S\dot{u}$ (mass density ρ, velocity $\dot{u}$ at $\mathrm{d}t$). The law stating that the load integrated over time (load times $\mathrm{d}t$) is equal to the difference of the momentum at time $\mathrm{d}t$ and at zero is formulated as

$$r\mathrm{d}S\mathrm{d}t = \rho c_s\mathrm{d}t\mathrm{d}S\dot{u} \tag{10.6}$$

which yields

$$r = \rho c_s\dot{u} \tag{10.7}$$

The initial response of the side $1 - 2$ can thus be modelled by a dashpot with the coefficient per unit length ρc_s which is called the *impedance*. These dashpots which act perpendicular to the plane are shown in Figure 10.1.

Introducing a Dirac delta function $\delta(t)$, as described in Appendix B, Equation (10.7) can be written as

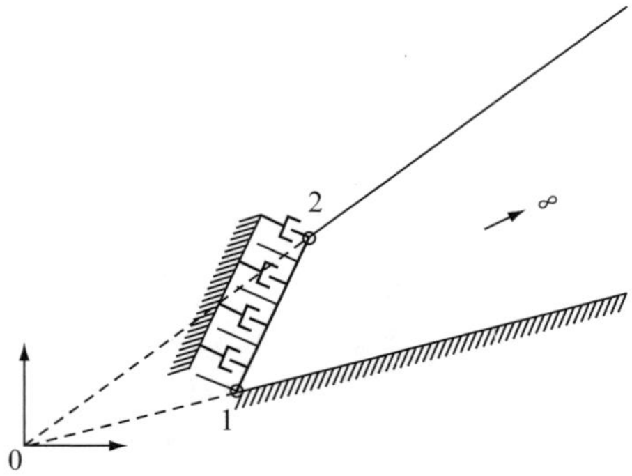

Figure 10.1 Initial response modelled by distributed dashpots

$$r(t) = \rho c_s \int_0^t \dot{\delta}(t-\tau)u(\tau)\mathrm{d}\tau \tag{10.8}$$

Comparing with Equation (B.47) yields the corresponding local response function:

$$s^{\infty}(t) = \rho c_s \dot{\delta}(t) \tag{10.9}$$

The Fourier transformation of Equation (10.9) yields, with the Fourier transform of $\dot{\delta}(t)$ equal to iω (Equations (B.38) and (B.39))

$$s^{\infty}(\omega) = \mathrm{i}\omega\rho c_s \tag{10.10}$$

This result can also be deduced from Equation (10.7) substituting $\dot{u}(\omega) = \mathrm{i}\omega u(\omega)$.

To determine the dynamic-stiffness coefficient corresponding to node 2, virtual work concepts are applied. Integrating on the side 1 − 2 with a linear shape function and with the length of the line 1 − 2 following from Equation (5.19) (with $\xi = 1$)

$$S^{\infty}(\omega) = s^{\infty}(\omega)\frac{1}{2}\sqrt{\Delta_x^2 + \Delta_y^2}\int_{-1}^{1} \tfrac{1}{4}(1+\eta)^2 \mathrm{d}\eta \tag{10.11}$$

Substituting Equation (10.10) into Equation (10.11) leads to

$$S^{\infty}(\omega) = \mathrm{i}\omega\frac{1}{3}\sqrt{\Delta_x^2 + \Delta_y^2}\rho c_s \tag{10.12}$$

where the coefficient of iω is equal to C_{∞} (Equation (10.3)).

The transformation from the time to the frequency domain in the second-last paragraph is better understood after studying section 11.2.

11 Numerical Solution of Dynamic Stiffness, Unit-impulse Response and Displacement of Truncated Semi-infinite Wedge

The governing ordinary differential equations for the semi-infinite truncated wedge can also be solved numerically in the ξ-direction. Thus, a numerical procedure is applied in both directions. This is meaningful, as the transformation to the scaled boundary coordinates permits the radiation condition for $\xi \to \infty$ to be introduced rigorously.

To be able to enforce the boundary conditions at infinity and on side $1-2$, the scaled boundary finite element equation in displacement (Equation (7.2)) is not solved directly. It is replaced by two first-order differential equations, the scaled boundary finite element equation in dynamic stiffness (Equation (7.7)), and the nodal force-displacement relationship (Equation (7.11)). When solving the scaled boundary finite element equation in dynamic stiffness, the radiation condition is incorporated. This first step, discussed in section 11.1, yields the dynamic-stiffness coefficient. In the second step, addressed in section 11.3, the displacement follows from the nodal force-displacement relationship with the known dynamic-stiffness coefficient. The scaled boundary finite element equation in dynamic stiffness can also be formulated and solved numerically in the time domain, resulting in the unit-impulse response function. This is developed in section 11.2.

As an example, the truncated semi-infinite wedge with vanishing body load shown in Figure 11.1 is addressed in this chapter. The side 2-∞ represents the free surface and 1-∞ the fixed boundary, both extending to infinity. The side $1-2$ representing the boundary can be interpreted as the structure-medium interface. The following parameters are selected: $x_1 = r_0,\ y_1 = -r_0/\sqrt{5},\ x_2 = r_0,\ y_2 = 0$, where r_0 denotes the characteristic length of the boundary $1-2$ (Equations (7.48) and (7.49)).

11.1 DYNAMIC STIFFNESS

For a discussion on the boundary condition at infinity, the radiation condition, the first two paragraphs of Chapter 10 should be consulted.

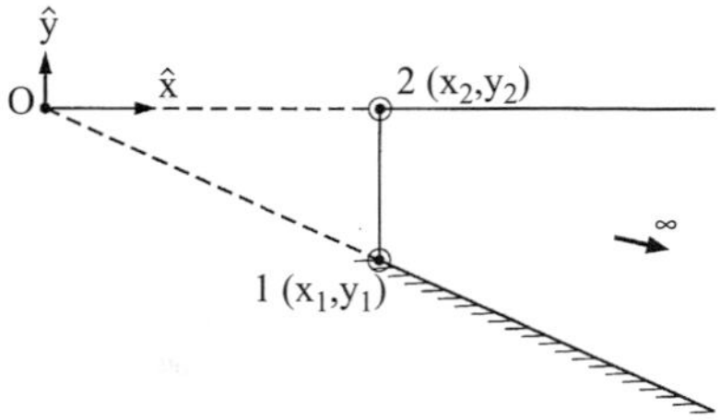

Figure 11.1 Truncated semi-infinite wedge used for numerical example (Reprinted from *Computers and Structures*, Vol. 78, Song and Wolf, 'The scaled boundary finite element method – a primer: solution procedures', pp. 211–225, © (2000), with permission from Elsevier Science)

The scaled boundary finite element equation in dynamic stiffness is a nonlinear first-order ordinary differential equation (Equation (7.7)) with $S^{\infty}(\omega)$ as the dependent variable and ω as the independent variable. The boundary condition required to be able to start the numerical algorithm is calculated from the high-frequency asymptotic expansion (Equation (10.1)) at a high but finite ω_h

$$S^{\infty}(\omega_h) = \mathrm{i}\omega_h C_{\infty} + K_{\infty} + \frac{1}{\mathrm{i}\omega_h} A_1 \tag{11.1}$$

with the constants C_{∞}, K_{∞} and A_1, specified in Equations (10.3), (10.4), (10.5). The radiation condition is satisfied. Equation (7.7) is then solved using a standard numerical integration procedure for decreasing ω. A fourth-order Runge–Kutta scheme [P3] is applied. At the beginning, for large ω where the variables vary smoothly, an adaptive integration step size is determined. Later on, for smaller ω, a fixed integration step size is selected.

The dynamic-stiffness coefficient $S^{\infty}(a_0)$ is calculated starting at $\omega_h = 8c_s/r_0$. It is then non-dimensionalised with $K^{\infty} = 7/(6\sqrt{5})G$ (Equation (7.22)), and finally, decomposed as indicated in Equation (7.68) in the dimensionless spring coefficient $k(a_0)$ and the dimensionless damping coefficient $c(a_0)$, with a_0 defined as in Equation (7.63). The numerical solution $S^{\infty}(a_0)$ agrees well with the analytical solution (Equation (7.62)) (Figure 11.2). The error introduced at ω_h through the high-frequency asymptotic expansion diminishes for decreasing ω.

The analysis is repeated introducing constant hysteretic material damping $\zeta = 0.05$ [W5]. Equation (7.7) with E^{0*}, E^{1*}, E^{2*} replacing E^0, E^1, E^2, respectively, is solved numerically. As $S^{\infty}(\omega)$ is also complex for the case of no material damping, the same algorithm is used. The same starting value ω_h is used. The decomposition of Equation (7.68) with a real a_0 (Equation (7.63)) applies. Again, good agreement with the analytical solution of Equation (7.67) occurs (Figure 11.3).

11.2 UNIT-IMPULSE RESPONSE

To be able to determine numerically the unit-impulse response function in the time domain, the scaled boundary finite element equation in dynamic stiffness is first

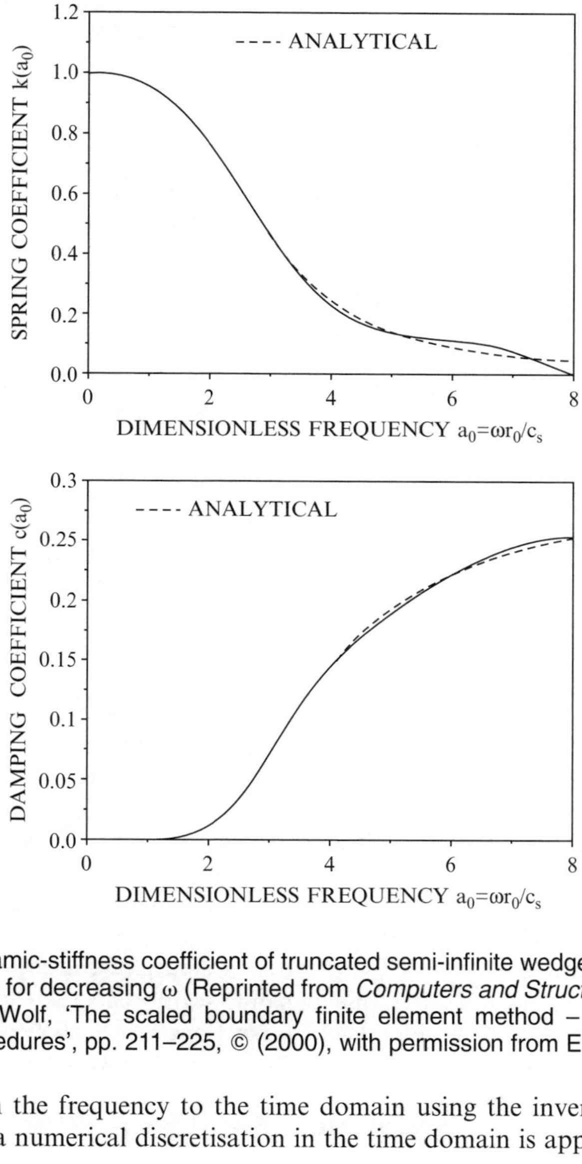

Figure 11.2 Dynamic-stiffness coefficient of truncated semi-infinite wedge calculated numerically for decreasing ω (Reprinted from *Computers and Structures*, Vol. 78, Song and Wolf, 'The scaled boundary finite element method – a primer: solution procedures', pp. 211–225, © (2000), with permission from Elsevier Science)

transformed from the frequency to the time domain using the inverse Fourier transformation. Then a numerical discretisation in the time domain is applied.

11.2.1 Inverse Fourier Transformation

Appendix B discusses the inverse Fourier transformation of the dynamic-stiffness coefficient $S^{\infty}(\omega)$ to the unit-impulse response coefficient $S^{\infty}(t)$.

The nodal force-displacement relationship for the truncated semi-infinite wedge (Figure 7.1b) on the side $1-2(\xi = 1)$ with $u_0 = u_2(\xi = 1)$ is written in the frequency domain as (Equation (5.53))

$$R_0(\omega) = S^{\infty}(\omega)u_0(\omega) \tag{11.2}$$

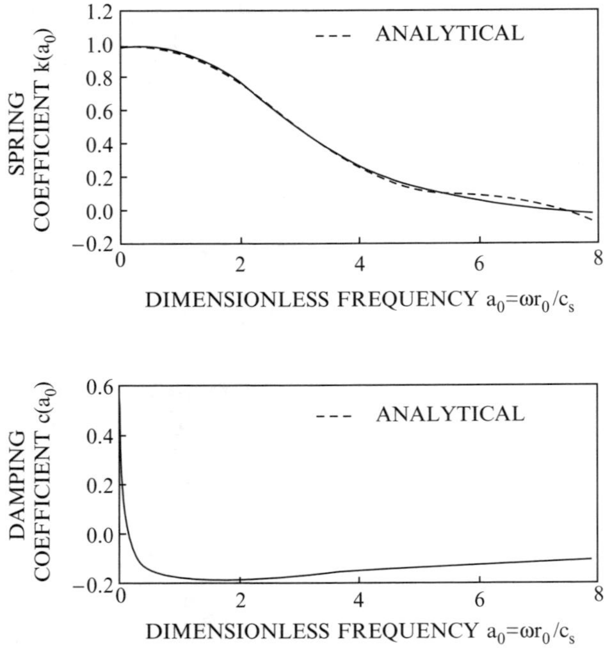

Figure 11.3 Dynamic-stiffness coefficient of truncated semi-infinite wedge with hysteretic damping calculated numerically for decreasing ω (Reproduced from *Proceedinds Computer Methods and Advances in Geomechanics*, © (2001), with permission from A. A. Balkema)

$S^{\infty}(\omega)$ is decomposed as (Equation (B.38))

$$S^{\infty}(\omega) = \mathrm{i}\omega C_{\infty} + K_{\infty} + S_r^{\infty}(\omega) \tag{11.3}$$

with the same dashpot coefficient C_{∞} and the same spring coefficient K_{∞} as in Equation (10.1). The remaining regular part of the dynamic-stiffness coefficient $S_r^{\infty}(\omega)$ (subscript r for regular) is square integrable. The inverse Fourier transformation of Equation (11.2) leads to the unit-impulse response coefficient in time (Equation (B. 39))

$$S^{\infty}(t) = C_{\infty}\dot{\delta}(t) + K_{\infty}\delta(t) + S_r^{\infty}(t) \tag{11.4}$$

with the Dirac delta function $\delta(t)$. $S_r^{\infty}(t)$ and $S_r^{\infty}(\omega)$ form a Fourier transformation pair (Equation (B.40)). Substituting Equation (11.3) into Equation (11.1) and performing the inverse Fourier transformation yields (Equation (B.49))

$$R(t) = C_{\infty}\dot{u}(t) + K_{\infty}u(t) + \int_0^t S_r^{\infty}(t-\tau)u(\tau)\mathrm{d}\tau \tag{11.5}$$

The first two terms on the right-hand side represent the instantaneous response and the convolution integral with the regular part $S_r^{\infty}(t)$ the lingering part.

Starting from the scaled boundary finite element equation in dynamic stiffness (Equation (7.7)), the corresponding scaled boundary finite element equation in the regular part of the dynamic stiffness can be formulated. Substituting Equation (11.3) into Equation (7.7) and using Equations (10.3) and (10.4) yields

$$\frac{1}{E^0}(S_r^{\infty}(\omega))^2 + \mathrm{i}\omega\frac{2}{E^0}C_{\infty}S_r^{\infty}(\omega) + \frac{2}{E^0}(K_{\infty} + E^1)S_r^{\infty}(\omega) \\ - \omega S_r^{\infty}(\omega)_{,\omega} + \frac{1}{E^0}(K_{\infty} + E^1)^2 - E^2 = 0 \tag{11.6}$$

To avoid the $\delta(t)$ function, and thus to simplify the numerical discretisation, the abbreviation

$$V_r^{\infty}(\omega) = \frac{1}{\mathrm{i}\omega}S_r^{\infty}(\omega) \tag{11.7}$$

is introduced. In the time domain, Equation (11.7) corresponds to (Equation (B.42))

$$V_r^{\infty}(t) = \int_0^t S_r^{\infty}(\tau)\mathrm{d}\tau \tag{11.8}$$

or

$$S_r^{\infty}(t) = \dot{V}_r^{\infty}(t) \tag{11.9}$$

The scaled boundary finite element equation in the time domain is developed for $V_r^{\infty}(t)$. The regular part of the unit-impulse response coefficient $S_r^{\infty}(t)$ is then calculated numerically as the time derivative of $V_r^{\infty}(t)$ (Equation (11.9))

Dividing Equation (11.6) by $(\mathrm{i}\omega)^2$ and substituting Equation (11.7) results in

$$\frac{1}{E^0}(V_r^{\infty}(\omega))^2 + \frac{2}{E^0}C_{\infty}V_r^{\infty}(\omega) + \frac{2}{E^0}(K_{\infty} + E^1)\frac{1}{\mathrm{i}\omega}V_r^{\infty}(\omega) \\ - \frac{1}{\mathrm{i}\omega}V_r^{\infty}(\omega) - V_r^{\infty}(\omega)_{,\,\mathrm{i}\omega} + \frac{1}{(\mathrm{i}\omega)^2}\left(\frac{1}{E^0}(K_{\infty} + E^1)^2 - E^2\right) = 0 \tag{11.10}$$

Applying the inverse Fourier transformation to Equation (11.10) yields (Equations (B.41) to (B.45))

$$\frac{1}{E^0}\int_0^t V_r^{\infty}(t-\tau)V_r^{\infty}(\tau)\mathrm{d}\tau + \frac{2}{E^0}C_{\infty}V_r^{\infty}(t) + \frac{2}{E^0}\left(K_{\infty} + E^1\right)\int_0^t V_r^{\infty}(\tau)\mathrm{d}\tau \\ - \int_0^t V_r^{\infty}(\tau)\mathrm{d}\tau + tV_r^{\infty}(t) + t(\frac{1}{E^0}(K_{\infty} + E^1)^2 - E^2)\mathrm{H}(t) = 0 \tag{11.11}$$

with the Heaviside step function $\mathrm{H}(t)$. This represents the *scaled boundary finite element equation in the time domain for* $V_r^{\infty}(t)$.

The high-frequency asymptotic expansion of the dynamic-stiffness coefficient (Equation (10.1)) corresponds to the *early-time asymptotic expansion of the unit-impulse response coefficient*. Performing the inverse Fourier transformation of Equation (10.1) yields, for this expansion (Equation (B.44)),

$$S^{\infty}(t) \approx C_{\infty}\dot{\delta}(t) + K_{\infty}\delta(t) + A_1 \mathrm{H}(t) \tag{11.12}$$

11.2.2 Time Discretisation

The regular part of the unit-impulse response coefficient at $t = 0$ $S_r^{\infty}(0)$ is obtained by comparing Equations (11.4) and (11.12) formulated at $t = 0$ as

$$S_r^{\infty}(0) = A_1 \tag{11.13}$$

As $S_r^{\infty}(0)$ is finite, formulating Equation (11.9) for $t = 0$ yields

$$V_r^{\infty}(0) = 0 \tag{11.14}$$

After discretisation with respect to time, Equation (11.11) yields an equation for $V_r^{\infty}{}_n$ at each time station n ($n \geq 1$). It is assumed that $V_r^{\infty}{}_n$ is piece-wise constant over the time $(n - 0.5)\Delta t \leq t < (n + 0.5)\Delta t$. At time station $n = 0$, $V_r^{\infty}{}_0 = 0$ (Equation (11.14)), which is assumed to apply for $0 \leq t < 0.5\Delta t$. The integral terms in Equation (11.11) are discretised as

$$\int_0^{n\Delta t} V_r^{\infty}(\tau)\mathrm{d}\tau = \Delta t \sum_{j=1}^{n-1} V_r{}^{\infty}{}_j + 0.5\Delta t V_r{}^{\infty}{}_n \tag{11.15}$$

$$\int_0^{n\Delta t} V_r^{\infty}(n\Delta t - \tau)V_r^{\infty}(\tau)\mathrm{d}\tau = \Delta t \sum_{j=1}^{n-1} V_r{}^{\infty}{}_j V_r{}^{\infty}{}_{n-j} \tag{11.16}$$

Substituting Equations (11.15) and (11.16) into Equation (11.11) formulated for $t = n\Delta t$ results in a *linear* equation for $V_r^{\infty}{}_n$

$$\begin{aligned}\left(\frac{1}{E^0}\frac{2}{\Delta t}C_{\infty} + \frac{K_{\infty} + E^1}{E^0} + n - 1\right)V_r{}^{\infty}{}_n = &-\frac{1}{E^0}\sum_{j=1}^{n-1} V_r^{\infty}{}_j V_r^{\infty}{}_{n-j} \\ &- \left(\frac{2(K_{\infty} + E^1)}{E^0} - 1\right)\sum_{j=1}^{n-1} V_r{}^{\infty}{}_j + n\left(E^2 - \frac{(K_{\infty} + E^1)^2}{E^0}\right)\end{aligned} \tag{11.17}$$

The regular part of the unit-impulse response coefficient $S_r^{\infty}{}_n$ then follows as (Equation (11.9))

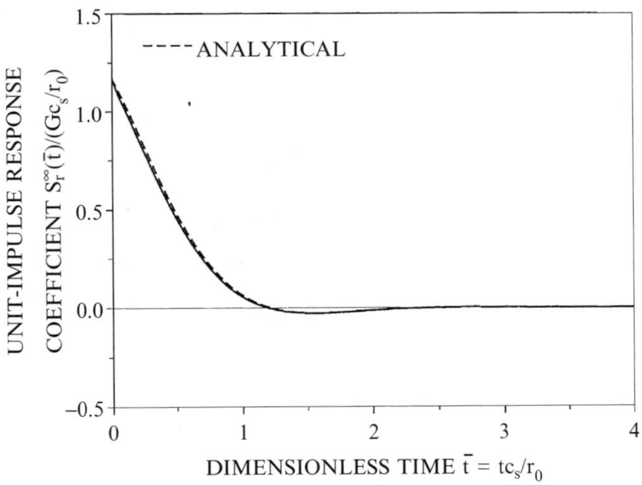

Figure 11.4 Regular part of unit-impulse response coefficient of truncated semi-infinite wedge (Reprinted from *Computers and Structures*, Vol. 78, Song and Wolf, 'The scaled boundary finite element method – a primer: solution procedures', pp. 211–225. © (2000), with permission from Elsevier Science)

$$S^{\infty}_{r\ n} = \frac{1}{\Delta t}(V^{\infty}_{r\ n} - V^{\infty}_{r\ n-1}) \tag{11.18}$$

For $\Delta t = 0.05 r_0/c_s$, $S_r^{\infty}(t)$, calculated step by step numerically using Equations (11.17) and (11.18), is plotted as a function of the dimensionless time $\bar{t} = tc_s/r_0$ in Figure 11.4. Excellent agreement results with the regular part of the analytical solution in Equation (7.62) transformed into the time domain.

11.3 DISPLACEMENT

To determine the displacements in the interior of the truncated semi-infinite wedge, the dynamic-stiffness coefficient is calculated first. In section 11.1 $S^{\infty}(\omega)$ follows from solving Equation (7.7) numerically. $S^{\infty}(a_0)$ is plotted in Figure 11.2 as a function of $a_0 = \omega r_0/c_s$, i.e. on side $1 - 2$ with $\xi = 1$. As discussed in connection with Equation (5.60), the dynamic-stiffness coefficient is a function of the product $\bar{\xi} = \omega\xi$ (or of $a = \omega\xi r_0/c_s$). This means that $S^{\infty}(a_0)$ determined from Equation (7.7) with $\xi = 1$ applies for any ξ $(1 \leq \xi < \infty)$ selecting the appropriate ξ (or a) and is denoted as $S^{\infty}(\bar{\xi})$. The nodal force–displacement relationship (Equation (7.11)), a linear first-order ordinary differential equation in $u_2(\bar{\xi})$ with $\bar{\xi}$ as the independent variable, is then solved numerically with the known $S^{\infty}(\bar{\xi})$. The boundary condition on line $1 - 2(\xi = 1)$ is known, which can be expressed in the displacement amplitude. Either the displacement amplitude $u_2(\bar{\xi} = \omega)$ is known, or if the nodal force amplitude $R_2(\omega)$ is specified,

$$u_2(\bar{\xi} = \omega) = \frac{1}{S^{\infty}(\omega)} R_2(\omega) \tag{11.19}$$

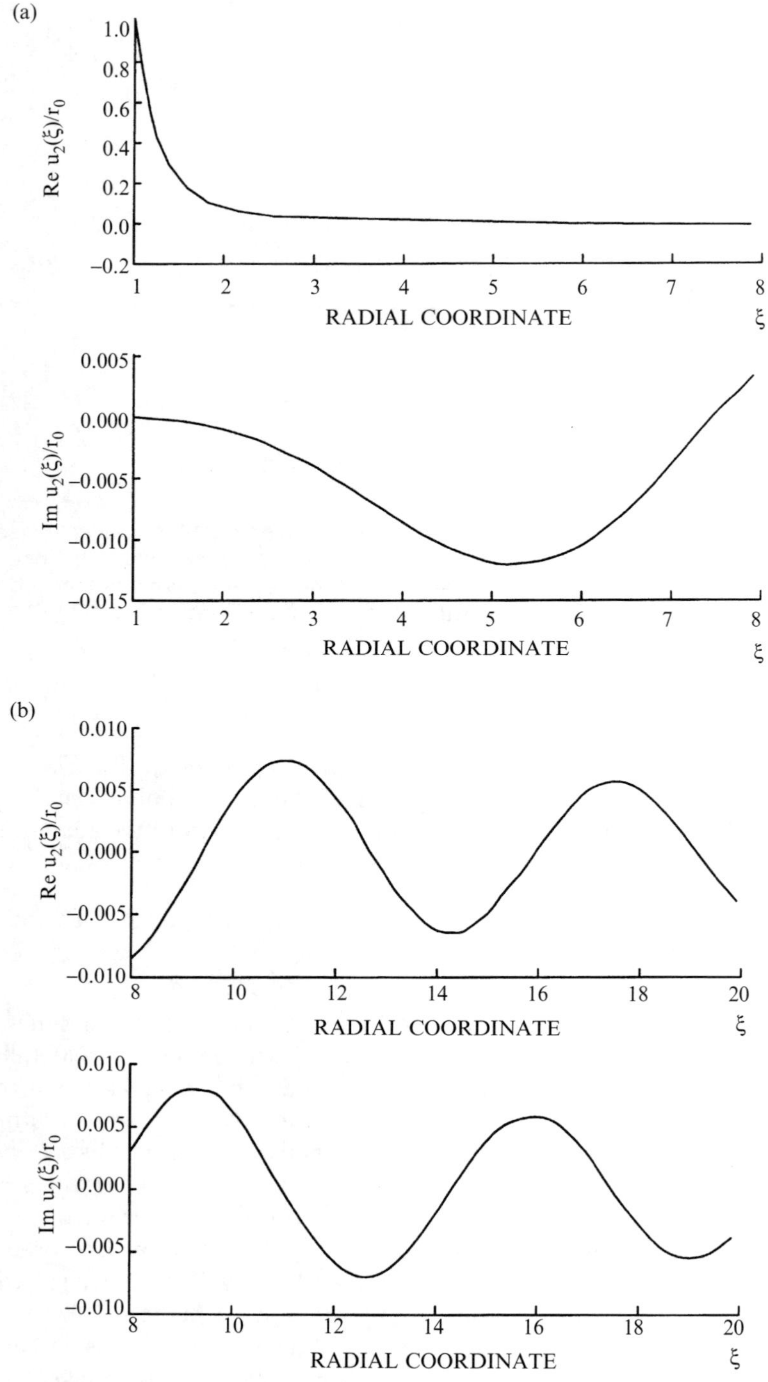

Figure 11.5 Displacement amplitude in radial direction of truncated semi-infinite wedge calculated numerically (a) $1 \leq \xi \leq 8$, (b) far field $8 \leq \xi \leq 20$ (Reproduced from *Proceedings Computer Methods and Advances in Geomechanics*, © (2001), with permission from A.A. Balkema)

applies. Equation (7.11) is solved from $\bar{\xi} = \omega$ for increasing $\bar{\xi}$. A forward Euler scheme can be applied. The same integration step size as when calculating $S^\infty(\bar{\xi})$ from Equation (7.7) is used, as $S^\infty(\bar{\xi})$ must be known. This yields the dependent variable $u_2(\bar{\xi})$.

In the high-frequency range, actually for large $\bar{\xi}$, the asymptotic expansion of the dynamic-stiffness coefficient $S^\infty(\bar{\xi})$ (Equation (7.69)) can be substituted, leading to Equation (7.75). The boundary condition $u_2(\bar{\xi}_e)$ at the exterior coordinate $\bar{\xi}_e$ is determined from integrating numerically Equation (7.11), as discussed above. A numerical scheme can also be used to solve Equation (7.75) for increasing $\bar{\xi}$, yielding the far-field displacement amplitudes.

As an example [W5], the displacement amplitude $u_2(\bar{\xi})$ is calculated with the boundary condition at node 2 with $\xi = 1$ $u_2(\bar{\xi} = \omega)/r_0 = 1$. The frequency ω equals c_s/r_0, i.e. $a_0 = 1$ is selected. As discussed in section 11.1, the high-frequency expansion of the dynamic-stiffness coefficient is applied at $\omega_h = 8c_s/r_0$ and $S^\infty(\omega)$ $(= S^\infty(\bar{\xi}))$ is calculated numerically from Equation (7.7) for decreasing $\bar{\xi}$ down to $\bar{\xi} = \omega$ in the first step. Equation (7.11) is then integrated numerically for increasing $\bar{\xi}$ up to $\bar{\xi} = 8c_s/r_0$ $(\xi = 8)$ with a step size $\Delta\bar{\xi} = 0.2$ in the second step enforcing the specified boundary condition $u_2(\bar{\xi} = \omega)$ at $\xi = 1$. The real and imaginary parts of $u_2(\xi)$ are plotted for $1 \leq \xi \leq 8$ in Figure 11.5a. The results coincide with the analytical solution (Equation (7.59)) from a practical point of view. The result at $\bar{\xi}_e = 8c_s/r_0$ $(\xi_e = 8)$ represents the boundary condition for the third step where Equation (7.75) is integrated numerically for increasing $\bar{\xi}$. Again, agreement with the analytical solution for $8 \leq \xi \leq 20$ shown in Figure 11.5b exists. Instead of solving Equation (7.75) numerically, an analytical solution (Equation (7.76)) is available for the scalar case.

12 Analytical Solution in Frequency Domain

A key concept and significant advantage of the scaled boundary finite element method consists of the transformation from partial differential equations to ordinary differential equations, which can be solved to a large extent analytically. After determining the integration constants in the general solution by enforcing boundary conditions which are satisfied exactly, this analytical procedure permits, in general, a closed-form expression of the results to be formulated, yielding an in-depth understanding of the physical behaviour. In addition, results can be calculated selectively, e.g. the dynamic-stiffness coefficient for a specific frequency or the displacement in a specific point. This is in contrast to a numerical procedure where, for example, the dynamic-stiffness coefficient for all frequencies from a very large value down to the specific frequency (section 11.1), or the displacements in all points located from the boundary to the specific point (section 11.3) must be determined.

In statics (Chapter 8), analytical solutions are readily available. This includes body loads. This chapter extends the analytical solution to dynamics in the frequency domain for vanishing body loads. The formulation using series becomes significantly more complicated. Both the wedge and the truncated semi-infinite wedge are addressed.

12.1 DISPLACEMENT

For vanishing body load the scaled boundary finite element equation in displacement amplitude $u_2(\xi)$ equals (Equation (7.2))

$$E^0\xi^2u_2(\xi),_{\xi\xi} + E^0\xi u_2(\xi),_{\xi} - E^2u_2(\xi) + \omega^2\xi^2M^0u_2(\xi) = 0 \tag{12.1}$$

Introducing the independent variable

$$\bar{\xi} = \omega\xi \tag{12.2}$$

which is proportional to the dimensionless frequency (Equation (7.48)) appearing in the argument of the analytical solution of the displacement amplitude (Equation (7.51)) and the dependent variable

$$\{X(\bar{\xi})\} = \begin{Bmatrix} u_2(\bar{\xi}) \\ Q_2(\bar{\xi}) \end{Bmatrix} \tag{12.3}$$

with the internal nodal force $Q_2(\bar{\xi})$ (Equation (5.51)). Equation (12.1) is transformed into

$$\bar{\xi}\{X(\bar{\xi})\},_{\bar{\xi}} = -[Z]\{X(\bar{\xi})\} - \bar{\xi}^2 \begin{bmatrix} 0 & 0 \\ M^0 & 0 \end{bmatrix} \{X(\bar{\xi})\} \tag{12.4}$$

The coefficient matrix $[Z]$ is specified in Equation (8.4).

An independent set of solutions grouped in the matrix $[X(\bar{\xi})]$ is determined. The general solution $\{X(\bar{\xi})\}$ can be expressed as

$$\{X(\bar{\xi})\} = [X(\bar{\xi})]\{c\} \tag{12.5}$$

with the integration constants $\{c\}$. $[X(\bar{\xi})]$ satisfies Equation (12.4)

$$\bar{\xi}[X(\bar{\xi})],_{\bar{\xi}} = -[Z][X(\bar{\xi})] - \bar{\xi}^2 \begin{bmatrix} 0 & 0 \\ M^0 & 0 \end{bmatrix} [X(\bar{\xi})] \tag{12.6}$$

Introducing the variable $[W(\bar{\xi})]$ defined as

$$[X(\bar{\xi})] = [\Phi][W(\bar{\xi})] \tag{12.7}$$

with the eigenvectors $[\Phi]$ of $[Z]$ defined in Equation (8.8) yields

$$\bar{\xi}[W(\bar{\xi})],_{\bar{\xi}} = \lceil \Lambda \rfloor [W(\bar{\xi})] - \bar{\xi}^2[M][W(\bar{\xi})] \tag{12.8}$$

with the eigenvalues $\lceil \Lambda \rfloor$ defined in Equation (8.6). $[M]$ equals

$$[M] = [\Phi]^{-1} \begin{bmatrix} 0 & 0 \\ M^0 & 0 \end{bmatrix} [\Phi] = \frac{M^0}{2\sqrt{E^0 E^2}} \begin{bmatrix} 1 & 1 \\ -1 & -1 \end{bmatrix} \tag{12.9}$$

Proceeding as in section 10, Chapter 14 of Ref [G1], the solution of Equation (12.8) is written as the product of two square matrix functions in $\bar{\xi}$

$$[W(\bar{\xi})] = [A(\bar{\xi})][Y(\bar{\xi})] \tag{12.10}$$

The first matrix $[A(\bar{\xi})]$ is formulated as a power series in $\bar{\xi}$

$$[A(\bar{\xi})] = [A_0] + \bar{\xi}^2[A_1] + \bar{\xi}^4[A_2] + \ldots + \bar{\xi}^{2k}[A_k] + \ldots \tag{12.11}$$

with

$$[A_0] = [I] \tag{12.12}$$

and the other coefficient matrices $[A_k]\, k = 1, 2, \ldots$ to be determined. The second matrix $[Y(\bar{\xi})]$ in Equation (12.10) satisfies the ordinary differential equations

$$\bar{\xi}[Y(\bar{\xi})]_{,\bar{\xi}} = [P^*(\bar{\xi})][Y(\bar{\xi})] \tag{12.13}$$

where $[P^*(\bar{\xi})]$ is again a power series in $\bar{\xi}$

$$[P^*(\bar{\xi})] = [P_0^*] + \bar{\xi}^2[P_1^*] + \bar{\xi}^4[P_2^*] + \ldots + \bar{\xi}^{2k}[P_k^*] + \ldots \tag{12.14}$$

The coefficient matrices $[P_k^*]$ $k = 0, 1, \ldots$ will be chosen in the simplest form possible.

The derivative $[W(\bar{\xi})]_{,\bar{\xi}}$ equals (Equation (12.10))

$$\begin{aligned}[W(\bar{\xi})]_{,\bar{\xi}} &= [A(\bar{\xi})]_{,\bar{\xi}}[Y(\bar{\xi})] + [A(\bar{\xi})][Y(\bar{\xi})]_{,\bar{\xi}} \\ &= \left([A(\bar{\xi})]_{,\bar{\xi}} + \frac{1}{\bar{\xi}}[A(\bar{\xi})][P^*(\bar{\xi})]\right)[Y(\bar{\xi})]\end{aligned} \tag{12.15}$$

where Equation (12.13) is substituted. Substituting Equation (12.10) and (12.15) into Equation (12.8), and post-multiplying by $[Y(\bar{\xi})]^{-1}$ results in

$$\bar{\xi}[A(\bar{\xi})]_{,\bar{\xi}} - \lceil\Lambda\rfloor[A(\bar{\xi})] + [A(\bar{\xi})][P^*(\bar{\xi})] + \bar{\xi}^2[M][A(\bar{\xi})] = 0 \tag{12.16}$$

The unknown coefficient matrices $[A_k]$, $[P_k^*]$ will be determined from this equation.

Substituting Equations (12.11) and (12.14) into Equation (12.16) leads to a power series in $\bar{\xi}$

$$\begin{aligned}&[P_0^*] - \lceil\Lambda\rfloor + \bar{\xi}^2(-\lceil\Lambda\rfloor[A_1] + [A_1]\lceil\Lambda\rfloor + 2[A_1] + [P_1^*] + [M]) \\ &\quad + \ldots + \bar{\xi}^{2k}(-\lceil\Lambda\rfloor[A_k] + [A_k]\lceil\Lambda\rfloor + 2k[A_k] + [P_k^*] + [C_k]) + \ldots = 0\end{aligned} \tag{12.17}$$

where

$$[C_k] = [M][A_{k-1}] + \sum_{i=1}^{k-1}[A_i][P_{k-i}^*] \tag{12.18}$$

Note that $[C_k]$ is determined from $[A_i]$ and $[P_i^*]$ $(i = 0, 1, \ldots k-1)$.

To satisfy Equation (12.17), all the coefficients of the power series in $\bar{\xi}$ must vanish. This yields for the constant term (Equation (8.7))

$$[P_0^*] = \lceil\Lambda\rfloor = \begin{bmatrix} -\lambda & \\ & \lambda \end{bmatrix} = \sqrt{\frac{E^2}{E^0}}\begin{bmatrix} 1 & \\ & -1 \end{bmatrix} \tag{12.19}$$

and for the quadratic term $(k = 1)$

$$\lceil\Lambda\rfloor[A_1] - [A_1]\lceil\Lambda\rfloor - 2[A_1] = [P_1^*] + [M] \tag{12.20}$$

which is written as four scalar equations

$$-2[A_1]_{11} = [P_1^*]_{11} + \frac{M^0}{2\sqrt{E^0 E^2}} \tag{12.21a}$$

$$-2(\lambda + 1)[A_1]_{12} = [P_1^*]_{12} + \frac{M^0}{2\sqrt{E^0 E^2}} \tag{12.21b}$$

$$2(\lambda - 1)[A_1]_{21} = [P_1^*]_{21} - \frac{M^0}{2\sqrt{E^0 E^2}} \tag{12.21c}$$

$$-2[A_1]_{22} = [P_1^*]_{22} - \frac{M^0}{2\sqrt{E^0 E^2}} \tag{12.21d}$$

To demonstrate the general procedure, $\lambda = -1$ is chosen. This is satisfied (Equation (8.7)) for the wedge and truncated semi-infinite wedge shown in Figure 12.1 with the coordinates $x_1 = 0$, $x_2 = -y_1 = \sqrt{2}r_0$, $y_2 = 0$. Equation (12.21) permits the four coefficients of $[A_1]$ to be determined for the simplest possible choice of the coefficients of $[P_1^*]$. For $[P_1^*]_{11} = [P_1^*]_{21} = [P_1^*]_{22} = 0$ the coefficients $[A_1]_{11}$, $[A_1]_{21}$, $[A_1]_{22}$ follow from Equations (12.21a), (12.21c) and (12.21d). As the coefficient $-2(\lambda + 1)$ of $[A_1]_{12}$ in Equation (12.21b) vanishes, the right-hand side must also be zero for a solution which determines $[P_1^*]_{12}$. $[A_1]_{12}$ can be selected arbitrarily, for instance zero. This yields

$$[A_1] = \frac{M^0}{4\sqrt{E^0 E^2}} \begin{bmatrix} -1 & 0 \\ 0.5 & 1 \end{bmatrix} \tag{12.22}$$

$$[P_1^*] = -\frac{M^0}{2\sqrt{E^0 E^2}} \begin{bmatrix} 0 & 1 \\ 0 & 0 \end{bmatrix} \tag{12.23}$$

For subsequent terms, Equation (12.17) yields

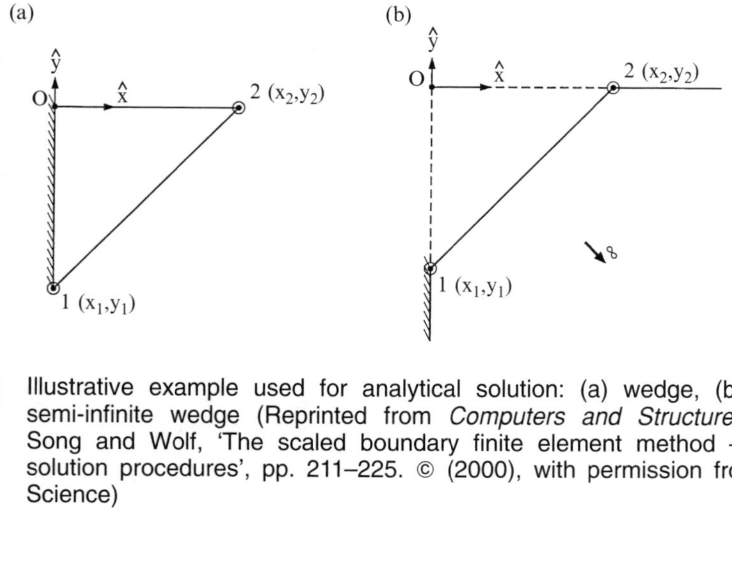

Figure 12.1 Illustrative example used for analytical solution: (a) wedge, (b) truncated semi-infinite wedge (Reprinted from *Computers and Structures*, Vol. 78, Song and Wolf, 'The scaled boundary finite element method – a primer: solution procedures', pp. 211–225. © (2000), with permission from Elsevier Science)

$$\lceil\Lambda\rfloor[A_k] - [A_k]\lceil\Lambda\rfloor - 2k[A_k] = [P_k^*] + [C_k] \quad k = 2, 3, \ldots \tag{12.24}$$

The scalar equations corresponding to Equation (12.24) with $\lambda = -1$ equal

$$-2k[A_k]_{11} = [P_k^*]_{11} + [C_k]_{11} \tag{12.25a}$$

$$2(1-k)[A_k]_{12} = [P_k^*]_{12} + [C_k]_{12} \tag{12.25b}$$

$$-2(1+k)[A_k]_{21} = [P_k^*]_{21} + [C_k]_{21} \tag{12.25c}$$

$$-2k[A_k]_{22} = [P_k^*]_{22} + [C_k]_{22} \tag{12.25d}$$

For $k \geq 2, [P_k^*] = 0$ can be selected, and $[A_k]$ follows from Equation (12.25) with $[C_k]$ defined in Equation (12.18). $[A(\bar{\xi})]$ in Equation (12.11) up to $k = 1$ equals (Equations (12.12) and (12.22))

$$[A(\bar{\xi})] = \begin{bmatrix} 1 & 0 \\ 0 & 1 \end{bmatrix} + \bar{\xi}^2 \frac{M^0}{4\sqrt{E^0 E^2}} \begin{bmatrix} -1 & 0 \\ 0.5 & 1 \end{bmatrix} + \ldots \tag{12.26}$$

$[P^*(\bar{\xi})]$ in Equation (12.14) consisting of only the first two terms is formulated as (Equations (12.19) and (12.23))

$$[P^*(\bar{\xi})] = \begin{bmatrix} 1 & 0 \\ 0 & -1 \end{bmatrix} - \bar{\xi}^2 \frac{M^0}{2\sqrt{E^0 E^2}} \begin{bmatrix} 0 & 1 \\ 0 & 0 \end{bmatrix} \tag{12.27}$$

$[Y(\bar{\xi})]$ follows from solving the differential equation (Equation (12.13)), which is written in vector form of $\{Y(\bar{\xi})\}$ as

$$\bar{\xi}\{Y(\bar{\xi})\}_{,\bar{\xi}} = [P^*(\bar{\xi})]\{Y(\bar{\xi})\} \tag{12.28}$$

The two scalar equations are

$$\bar{\xi} Y_1(\bar{\xi})_{,\bar{\xi}} = Y_1(\bar{\xi}) - \frac{M^0}{2\sqrt{E^0 E^2}} \bar{\xi}^2 Y_2(\bar{\xi}) \tag{12.29a}$$

$$\bar{\xi} Y_2(\bar{\xi})_{,\bar{\xi}} = -Y_2(\bar{\xi}) \tag{12.29b}$$

The solution of Equation (12.29b) equals

$$Y_2(\bar{\xi}) = c_2 \frac{1}{\bar{\xi}} \tag{12.30}$$

with the integration constant c_2. Substituting Equation (12.30) into Equation (12.29a) yields the non-homogeneous differential equation

$$\bar{\xi} Y_1(\bar{\xi})_{,\bar{\xi}} = Y_1(\bar{\xi}) - c_2 \frac{M^0}{2\sqrt{E^0 E^2}} \bar{\xi} \tag{12.31}$$

Applying the method of variation of parameters, the solution is postulated using the homogenous solution as

$$Y_1(\bar{\xi}) = \bar{\xi} c_1(\bar{\xi}) \tag{12.32}$$

Note that $c_1(\bar{\xi})$ is a function of $\bar{\xi}$. Substituting Equation (12.32) into Equation (12.31) results in

$$c_1(\bar{\xi})_{,\bar{\xi}} = -c_2 \frac{M^0}{2\sqrt{E^0 E^2}} \frac{1}{\bar{\xi}} \tag{12.33}$$

Its solution equals

$$c_1(\bar{\xi}) = c_1 - c_2 \frac{M^0}{2\sqrt{E^0 E^2}} \ln\bar{\xi} \tag{12.34}$$

Using Equations (12.34), (12.32) and (12.30) leads to the solution of Equation (12.28)

$$\begin{Bmatrix} Y_1(\bar{\xi}) \\ Y_2(\bar{\xi}) \end{Bmatrix} = [Y(\bar{\xi})] \begin{Bmatrix} c_1 \\ c_2 \end{Bmatrix} \tag{12.35}$$

with

$$[Y(\bar{\xi})] = \begin{bmatrix} \bar{\xi} & -\dfrac{M^0}{2\sqrt{E^0 E^2}} \bar{\xi} \ln\bar{\xi} \\ 0 & \dfrac{1}{\bar{\xi}} \end{bmatrix} \tag{12.36}$$

Substituting Equations (12.36), (12.26) and (12.10) into Equation (12.7) yields

$$[X(\bar{\xi})] = \begin{bmatrix} X_{11}(\bar{\xi}) & X_{12}(\bar{\xi}) \\ X_{21}(\bar{\xi}) & X_{22}(\bar{\xi}) \end{bmatrix} = \begin{bmatrix} \Phi_{11} & \Phi_{12} \\ \Phi_{21} & \Phi_{22} \end{bmatrix} \begin{bmatrix} A_{11}(\bar{\xi}) & A_{12}(\bar{\xi}) \\ A_{21}(\bar{\xi}) & A_{22}(\bar{\xi}) \end{bmatrix} \begin{bmatrix} Y_{11}(\bar{\xi}) & Y_{12}(\bar{\xi}) \\ Y_{21}(\bar{\xi}) & Y_{22}(\bar{\xi}) \end{bmatrix} \tag{12.37}$$

The amplitudes of the displacement and internal nodal force follow from Equations (12.3) and (12.5) as

$$\begin{Bmatrix} u_2(\bar{\xi}) \\ Q_2(\bar{\xi}) \end{Bmatrix} = \begin{bmatrix} X_{11}(\bar{\xi}) & X_{12}(\bar{\xi}) \\ X_{21}(\bar{\xi}) & X_{22}(\bar{\xi}) \end{bmatrix} \begin{Bmatrix} c_1 \\ c_2 \end{Bmatrix} \tag{12.38}$$

For the wedge, the boundary condition of a finite displacement at $\bar{\xi} = 0$ yields $c_2 = 0$ to eliminate the effect of $Y_{22}(\bar{\xi}) = 1/\bar{\xi}$ (Equations (12.36), (12.37) and (12.38))

$$\begin{Bmatrix} u_2(\bar{\xi}) \\ Q_2(\bar{\xi}) \end{Bmatrix} = c \begin{Bmatrix} \bar{\xi} - \dfrac{M^0}{8\sqrt{E^0 E^2}} \bar{\xi}^3 \\ E^1 \left(\bar{\xi} - \dfrac{M^0}{8\sqrt{E^0 E^2}} \bar{\xi}^3 \right) + \sqrt{E^0 E^2} \left(\bar{\xi} - \dfrac{3M^0}{8\sqrt{E^0 E^2}} \bar{\xi}^3 \right) \end{Bmatrix} \tag{12.39}$$

12.2 DYNAMIC STIFFNESS OF WEDGE

The dynamic-stiffness coefficient of the wedge is defined as the ratio of the nodal force amplitude on the boundary $\xi = 1$ $R_0 = Q_2(\xi = 1)$ to the displacement amplitude $u_0 = u_2(\xi = 1)$ (Equation (12.39))

$$S^{\mathrm{b}}(\omega) = \frac{R_0}{u_0} = E^1 + \sqrt{E^0 E^2}\,\frac{1 - \dfrac{3M^0}{8\sqrt{E^0 E^2}}\omega^2}{1 - \dfrac{M^0}{8\sqrt{E^0 E^2}}\omega^2} \tag{12.40}$$

Note that on the side $1-2$, $\bar{\xi} = \omega$ applies. The low-frequency expansion of Equation (12.40) yields Equation (9.1) for $\lambda = -1$ $(E^0 = E^2)$ with K^{b} and M^{b} specified in Equations (8.14) and (9.4).

12.3 DYNAMIC STIFFNESS OF TRUNCATED SEMI-INFINITE WEDGE

The integration constants c_1, c_2 in Equation (12.38) must be determined to satisfy the radiation condition at infinity of the truncated semi-infinite wedge. To be able to evaluate $[X(\xi \to \infty)]$, an infinite number of terms in $[A(\bar{\xi})]$ $(k \to \infty$ in Equation (12.11)) would be required, which is obviously not possible. To overcome this difficulty, an asymptotic expansion for $S^{\infty}(\omega, \xi)$ for large ξ satisfying the radiation condition is constructed similarly to that for high frequency. The scaled boundary finite element equation in dynamic stiffness for the truncated semi-infinite wedge specified in Equation (5.62) is formulated with $\bar{\xi} S^{\infty}(\bar{\xi})_{,\bar{\xi}} = \omega S^{\infty}(\omega)_{,\omega}$ as

$$\frac{1}{E^0}(S^{\infty}(\bar{\xi}))^2 + 2\frac{E^1}{E^0}S^{\infty}(\bar{\xi}) - \bar{\xi}S^{\infty}(\bar{\xi})_{,\bar{\xi}} + \frac{(E^1)^2}{E^0} - E^2 + \bar{\xi}^2 M^0 = 0 \tag{12.41}$$

The asymptotic expansion of $S^{\infty}(\bar{\xi})$ for large $\bar{\xi} = \omega\xi$ is defined as

$$S^{\infty}(\bar{\xi}) \approx \mathrm{i}\bar{\xi}C_{\infty} + K_{\infty} + \frac{1}{\mathrm{i}\bar{\xi}}A_1 \tag{12.42}$$

Comparing Equations (12.41), (12.42) formulated for $\xi = 1$ with Equations (7.7), (10.1), it follows that the coefficients $C_{\infty}, K_{\infty}, A_1$ in Equation (12.42) are identical to those in Equation (10.1) calculated in Equations (10.3), (10.4) and (10.5). Another similar boundary is introduced at ξ_e (Figure 12.2) to obtain a sufficiently large $\bar{\xi}_e = \omega\xi_e$, permitting the asymptotic expansion $S^{\infty}(\omega\xi_e) = S^{\infty}_e(\omega)$ (subscript e for exterior) to be used as the boundary condition (Equation (12.42))

$$S^{\infty}_e(\omega) = \mathrm{i}\omega\xi_e C_{\infty} + K_{\infty} + \frac{1}{\mathrm{i}\omega\xi_e}A_1 \tag{12.43}$$

The domain between the two similar boundaries at ξ_i and ξ_e, a trapezoid, is addressed. Formulating Equation (12.38) for the two boundaries yields

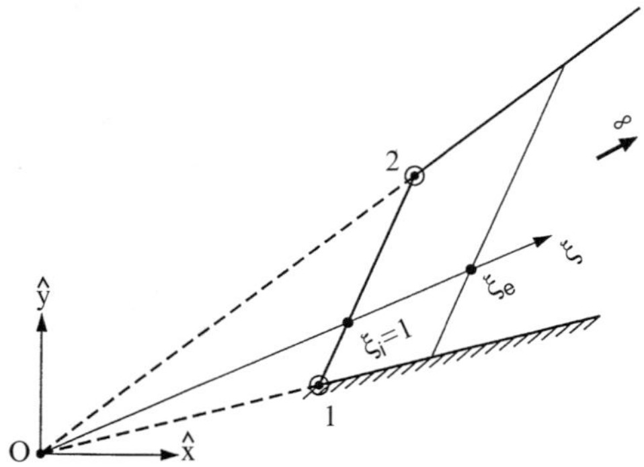

Figure 12.2 Truncated semi-infinite wedge on exterior of side 1 – 2 corresponding to $\xi_i \leq \xi < \infty$ (Reprinted from *Computers and Structures*, Vol. 78, Song and Wolf, 'The scaled boundary finite element method – a primer: solution procedures', pp. 211–225. © (2000), with permission from Elsevier Science)

$$\begin{Bmatrix} u_i \\ Q_i \end{Bmatrix} = \begin{Bmatrix} u_2(\bar{\xi}_i) \\ Q_2(\bar{\xi}_i) \end{Bmatrix} = \begin{bmatrix} X_{11}(\bar{\xi}_i) & X_{12}(\bar{\xi}_i) \\ X_{21}(\bar{\xi}_i) & X_{22}(\bar{\xi}_i) \end{bmatrix} \begin{Bmatrix} c_1 \\ c_2 \end{Bmatrix} \tag{12.44}$$

$$\begin{Bmatrix} u_e \\ Q_e \end{Bmatrix} = \begin{Bmatrix} u_2(\bar{\xi}_e) \\ Q_2(\bar{\xi}_e) \end{Bmatrix} = \begin{bmatrix} X_{11}(\bar{\xi}_e) & X_{12}(\bar{\xi}_e) \\ X_{21}(\bar{\xi}_e) & X_{22}(\bar{\xi}_e) \end{bmatrix} \begin{Bmatrix} c_1 \\ c_2 \end{Bmatrix} \tag{12.45}$$

with $\bar{\xi}_i = \omega\xi_i$ and $\bar{\xi}_e = \omega\bar{\xi}_e$. For the boundaries ξ_i and ξ_e which are negative and positive faces, respectively, Equation (5.52):

$$P_i = -Q_i \tag{12.46}$$

$$P_e = Q_e \tag{12.47}$$

applies with the nodal force amplitudes P_i, P_e of the trapezoid. Eliminating c_1, c_2 substituting Equations (12.46) and (12.47) and rearranging Equations (12.44) and (12.45) leads to

$$\begin{Bmatrix} P_i \\ P_e \end{Bmatrix} = \begin{bmatrix} S_{ii}(\omega) & S_{ie}(\omega) \\ S_{ei}(\omega) & S_{ee}(\omega) \end{bmatrix} \begin{Bmatrix} u_i \\ u_e \end{Bmatrix} \tag{12.48}$$

with the dynamic-stiffness matrix

$$\begin{bmatrix} S_{ii}(\omega) & S_{ie}(\omega) \\ S_{ei}(\omega) & S_{ee}(\omega) \end{bmatrix} = \begin{bmatrix} -X_{21}(\bar{\xi}_i) & -X_{22}(\bar{\xi}_i) \\ X_{21}(\bar{\xi}_e) & X_{22}(\bar{\xi}_e) \end{bmatrix} \begin{bmatrix} X_{11}(\bar{\xi}_i) & X_{12}(\bar{\xi}_i) \\ X_{11}(\bar{\xi}_e) & X_{12}(\bar{\xi}_e) \end{bmatrix}^{-1} \tag{12.49}$$

The trapezoid is a finite element cell with its dynamic-stiffness matrix calculated analytically (Equation (12.49). From the dynamic condensation equation derived in section 6.3 (Equation (6.60)) the dynamic-stiffness coefficient of the truncated semi-infinite wedge at $\xi_i = 1, S^{\infty}(\omega) = S_i^{\infty}(\omega)$, is related to that at ξ_e as

$$S^{\infty}(\omega) = S_{ii}(\omega) - S_{ie}(\omega)(S_e^{\infty}(\omega) + S_{ee}(\omega))^{-1}S_{ei}(\omega) \tag{12.50}$$

$S_e^{\infty}(\omega)$ is specified in Equation (12.43).

As already mentioned, the analytical procedure permits the calculation of the dynamic-stiffness coefficient directly for any specific ω, in contrast to the numerical procedure for decreasing ω in section 11.1. The truncated semi-infinite wedge shown in Figure 12.1b is examined. For the analysis of the trapezoid $k = 6$ in Equation (12.11) is selected. $\bar{\xi}_e = 3c_s/r_0$ is chosen to calculate the asymptotic expansion for $S_e^{\infty}(\omega)$ in Equation (12.43). The dynamic-stiffness coefficient $S^{\infty}(\omega)$ in Equation (12.50) is a function of ω. This analytical $S^{\infty}(\omega)$ using a series agrees well with the analytical solution using Hankel functions (Equation 7.62), denoted as analytical in Figure 12.3. $S^{\infty}(\omega)$ is decomposed as in Equation (7.68).

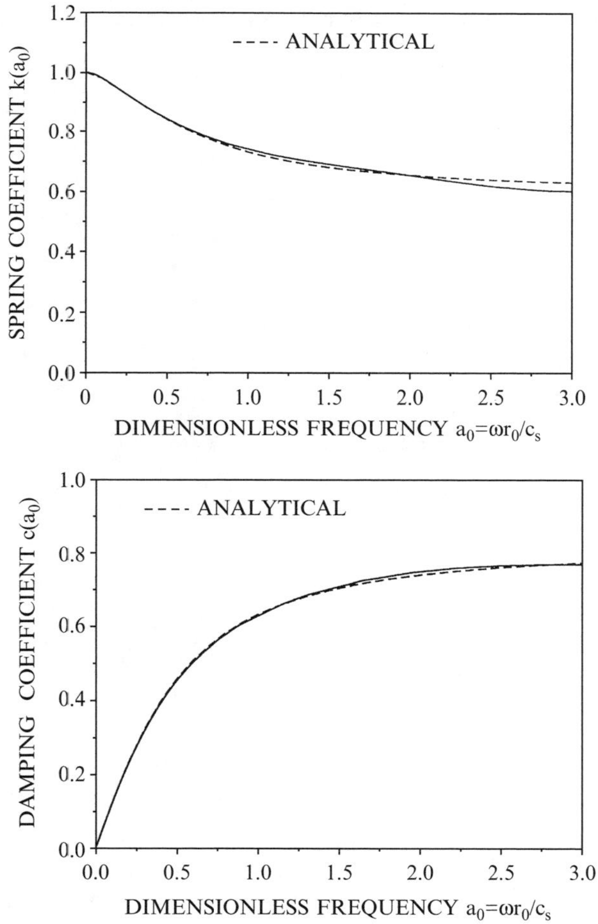

Figure 12.3 Dynamic-stiffness coefficient of truncated semi-infinite wedge calculated with series solution (Reprinted from *Computers and Structures*, Vol. 78 Song and Wolf, 'The scaled boundary finite element method – a primer: solution procedures', pp. 211–225. © (2000), with permission from Elsevier Science)

12.4 FAR-FIELD DISPLACEMENT

The displacement amplitudes in the far field, i.e. for $\bar{\xi} > \bar{\xi}_e$ where the high-frequency asymptotic expansion of the dynamic stiffness with the first two terms applies, can be determined. In this case, Equation (7.77) can be solved with a series expansion analogous to the procedure of section 12.1 used to address Equation (12.4). In this first-order differential equation, the boundary condition $u_2(\bar{\xi}_e)$ follows from the corresponding series expansion in section 12.1. (Of course, for this scalar case, an analytical solution, Equation (7.76) with $A_1 = 0$), can directly be formulated.)

Thus, it is possible to calculate analytically the response throughout the truncated semi-infinite wedge, avoiding numerical discretisation in the $\bar{\xi}$-direction. However, the formulation in the form of a series is, in the dynamic case, much more complicated than in the static case. In addition, as for the numerical solution, the high-frequency asymptotic expansion of the dynamic stiffness has to be introduced as an approximation.

13 Implementation

The Model Problem addressed in Part I is a simple example which is easy to follow and at the same time permits essentially all the features to be demonstrated. This applies to the two different derivations of the governing differential equations, and to their solution procedures. However, as only a single line finite element is selected in the modelisation, there is no assembly process of the coefficient matrices. In general, the side 1 – 2 of the wedge and the truncated semi-infinite wedge (Figure 4.1) will be subdivided into several line finite elements, introducing additional nodes with (displacement) degrees of freedom. This is shown for a more general case, for instance, in Figures 3.1 and 3.4. For each line finite element between two adjacent nodes, the coefficient matrices of order 2×2 $[E^0]$, $[E^1]$, $[E^2]$, $[M^0]$ and the nodal load vectors of order 2×1 $\{F\}$ follow as before using the corresponding nodal coordinates (Equations (5.42), (5.43), (5.45)). In the assemblage process, the coefficient matrices and nodal load vectors sharing the same degrees of freedom are added, which is analogous to the procedure in finite elements (section 1.4). The individual line finite elements are one-dimensional. The displacement boundary condition in node 1 is then enforced, leading to the final assembled coefficient matrices $[E^0]$, $[E^1]$, $[E^2]$, $[M^0]$ and nodal load vector $\{F\}$. As already mentioned, to simplify the nomenclature, the same symbols are used for the assembled quantities. When the side 1 – 2 is subdivided into m line finite elements, the order of the assembled coefficient matrices after enforcing the displacement boundary condition in node 1 will be $m \times m$. In the scaled boundary finite element equations in displacement (Equation (5.47)) and in dynamic stiffness (Equations (5.63), (5.64)), the unknowns $\{u(\xi)\}$ and $[S(\omega, \xi)]$ are of orders $m \times 1$ and $m \times m$, respectively.

Summarising, the implementation of the scaled boundary finite element method for the out-of-plane motion governed by the scalar wave equation in two dimensions with several degrees of freedom on the boundary (side 1 – 2) proceeds as follows. The general case of elastodynamics with the vector wave equation and diffusion addressed in Part II follows the same steps:

- If necessary, subdivide the total (bounded) domain into subdomains (super elements) with their own scaling centres, as illustrated in Figure 3.2. This can occur for two reasons (besides other aspects related to the behaviour of the scaling centre and on the adjacent side-faces). First, the boundary must be visible from its scaling centre. Secondly, in dynamics of a bounded domain, as the representation with the static-stiffness matrix $[K^b]$ and mass matrix $[M^b]$ is selected with degrees of freedom only on the boundary, a sufficient number of degrees of freedom must be present in the

Table 13.1 Addressed problems and results with chapter numbers (static stiffness of bounded medium $[K^{\mathrm{b}}]$, mass of bounded medium $[M^{\mathrm{b}}]$, static stiffness of unbounded medium $[K^{\infty}]$, internal displacement $\{u(\xi)\}$, nodal loads from body loads $\{R^{\mathrm{F}}\}$, dynamic stiffness of bounded medium $[S^{\mathrm{b}}(\omega)]$, dynamic stiffness of unbounded medium $[S^{\infty}(\omega)]$, unit-impulse response of unbounded medium $[S^{\infty}(t)]$, dashpot $[C_{\infty}]$, spring $[K_{\infty}]$, high frequency ω_h) (Reprinted from *Computers and Structures*, Vol. 78, Song and Wolf, 'The scaled boundary finite element method – a primer: solution procedures', pp. 211–225, © (2000), with permission from Elsevier Science)

	Bounded medium		Unbounded medium	
Statics	$[K^{\mathrm{b}}], \{u(\xi)\}, \{R^{\mathrm{F}}\}$ (Ch. 8)		$[K^{\infty}], \{u(\xi)\}, \{R^{\mathrm{F}}\}$ (Ch. 8)	
Frequency domain	$[M^{\mathrm{b}}]$ (Ch. 9)	$[S^{\mathrm{b}}(\omega)]$, $\{u(\xi)\}$ (Ch. 12.1, 12.2)	$[S^{\infty}(\omega_h)]$ (Ch. 10)	$[S^{\infty}(\omega)]$, $\{u(\xi)]$ (Ch. 11.1, 11.3, numerically) $[S^{\infty}(\omega)]$, $\{u(\xi)\}$ (Ch. 12.1, 12.3, 12.4)
Time domain			$[C_{\infty}]$, $[K_{\infty}]$ (Ch. 10)	$[S^{\infty}(t)]$ (Ch. 11.2, numerically)

interior of the domain to adequately represent the wave length (typically 4 – 10). This latter criterion is the same as must be enforced in the finite element method.

- The boundary is discretised spatially with surface finite elements, which can also be called scaled boundary elements. The same also applies to the interfaces between adjacent subdomains. This determines the degrees of freedom present in the model.
- For each surface finite element (for every subdomain) the coefficient matrices $[E^0]$, $[E^1]$, $[E^2]$, $[M^0]$ and nodal loads $\{F\}$ are calculated and assembled, enforcing the displacement boundary conditions.
- The scaled boundary finite element equation either in displacement or in dynamic stiffness can be formulated and solved, depending on the problem to be addressed. Table 13.1 identifies the results and the chapter numbers in Part I for static, frequency- and time-domain analyses of the wedge and truncated semi-infinite wedge, which also apply to general bounded and unbounded media. Note that the scaled boundary finite element equations are solved analytically, except for the procedures described in Chapter 11 for the dynamic analysis of an unbounded medium. The word analytical is understood in a broad sense where, for instance, the solution of an eigenvalue problem is included.

14 Conclusions

The reader should consult Chapters 2 and 3 again to be able to start developing the notions which will be addressed in detail for the general case of elastodynamics and diffusion in Part II. In the following, certain extensions to the Model Problem such as discussing the three-dimensional case with a curved boundary are mentioned.

14.1 DERIVATIONS

A novel boundary element method based on finite elements to analyse bounded and unbounded media is developed addressing the Model Problem (Chapter 4). Two derivations for this scaled boundary finite element method are presented.

In the first, the scaled boundary transformation-based derivation (Chapter 5), a scaling centre is selected. The distance from this origin to a point is represented by the radial coordinate ξ. Discretising the boundary with surface finite elements determines the other two coordinates η, ζ. The material properties can vary in the circumferential directions on the boundary. Applying the scaled boundary transformation to the geometry, the governing partial differential equations in the Cartesian coordinates are formulated in the local coordinates ξ, η, ζ. The boundary conditions are conveniently formulated in ξ, η, ζ. Using the weighted-residual technique of finite elements in the two circumferential directions η, ζ parallel to the boundary results in the scaled boundary finite element equation in displacement amplitudes, a system of linear second-order ordinary differential equations with the radial coordinate ξ as the independent variable (Equation (5.47)). The scaled boundary finite element equation can also be formulated in dynamic stiffness with the frequency ω as the independent variable (Equations (5.63, (5.64)). For loads, the scaled boundary finite element equation in displacement becomes non-homogeneous, and an additional term in form of nodal loads appears in the nodal force-displacement relationship.

In the second, the mechanically-based derivation (Chapter 6), a similarity centre which corresponds to the scaling centre is selected. A similar fictitious boundary is constructed by scaling the boundary. A finite-element cell is introduced between the two boundaries. Assembling the finite element cell and the medium yields a relationship linking the dynamic-stiffness matrices at the boundaries of the medium via the dynamic-stiffness matrix of the finite element cell. Another relationship of the dynamic-stiffness matrices at the boundaries follows from similarity. Performing the limit of the cell width towards zero analytically yields the scaled boundary finite element equation in dynamic stiffness (Equations (6.74), (6.75)), which can also be expressed in displacement (Equation (6.96)).

Only the boundary is discretised with curved surface finite elements. When modelling a semi-infinite medium, the free surface extending to infinity is not discretised. The same applies for straight crack faces in fracture mechanics.

14.2 SOLUTION PROCEDURES

The scaled boundary finite element equations can be solved analytically to a very large extent, making the novel computational procedure semi-analytical.

In statics (Chapter 8), a quadratic eigenvalue problem is solved. The analytical solution of the displacements consists of a finite number of power functions in the radial coordinate. By selecting the appropriate eigenvalues, it is guaranteed that the displacements remain finite within the domain for a bounded medium and towards infinity for an unbounded domain.

The mass matrix of a bounded medium (Chapter 9) is calculated analytically from the low-frequency expansion of the dynamic-stiffness matrix by solving linear equations.

The high-frequency asymptotic expansion for the dynamic-stiffness matrix of an unbounded medium (Chapter 10) satisfies the radiation condition at infinity exactly, which permits the dynamic-stiffness matrix to be calculated analytically at a high but finite frequency. This also leads to the unit-impulse response function at early times.

The analytical solution in the frequency domain of the scaled boundary finite element equation in displacement (Chapter 12) is based on a series expansion, and consists of two sets of functions with the corresponding integration constants. One set of integration constants is determined by the boundary condition (finite displacements) for vanishing radial coordinate for a bounded medium and by the boundary condition (radiation condition) for the radial coordinate approaching infinity for an unbounded medium. The other set follows from the condition to be satisfied at the boundary. In dynamics the analytical solution consists of an (infinite) series of the product of the radial coordinate and the frequency. The dynamic-stiffness matrix of an unbounded medium follows analytically, enforcing the high-frequency asymptotic expansion at a high but finite frequency.

As an alternative (Chapter 11), the linear second-order differential equation, the scaled boundary finite element equation in displacement, is not solved directly, but is replaced by two first-order differential equations, a nonlinear one, the scaled boundary finite element equation in dynamic stiffness, and a linear one, the nodal force-displacement relationship. A numerical procedure consisting of two steps is chosen. In the first step, the scaled boundary finite element equation in dynamic stiffness is solved for decreasing product of radial coordinate and frequency down to the boundary. The high-frequency expansion of the dynamic stiffness satisfying the radiation condition serves as the starting value (boundary condition). This boundary condition follows either from solving an eigenvalue problem, or directly from the radiation condition using impedances. This leads to the dynamic stiffness. In the second step, the nodal force-displacement relationship is integrated numerically for increasing product of radial coordinate and frequency enforcing the boundary condition at the boundary. This results in the displacements.

The unit-impulse response can also be calculated numerically (section 11.2). The scaled boundary finite element equation in unit-impulse response is derived performing the inverse Fourier transformation of the corresponding equation in dynamic stiffness and using the high-frequency asymptotic expansion. The unit-impulse response matrix is solved numerically for increasing time starting at time zero.

14.3 PROPERTIES AND ADVANTAGES

Most properties and advantages of the scaled boundary finite element method are described in Chapter 2. The key advantages of the finite element and boundary element methods are also listed.

The guts of the scaled boundary finite element method consists in transforming the *partial* differential equations to *ordinary* differential equations which can be solved *analytically*. In the circumferential directions (parallel to the boundary), where the behaviour is in general smooth, the weighted-residual approximation of finite elements applies, leading to convergence in the finite element sense. In the radial direction (either pointing from the boundary towards infinity for an unbounded medium or from the boundary towards the interior for a bounded medium), where either the radiation condition at infinity or stress singularities as in fracture mechanics are present, an analytical (exact) solution is obtained. As the side-faces are formed by radial coordinates, the corresponding boundary conditions are also enforced exactly. Effectively, the governing partial differential equations are weakened in the circumferential directions in a finite element manner, but remain strong in the radial direction. The novel procedure is thus a *semi-analytical fundamental solution-less boundary element method based on finite elements*. The best of both worlds is achieved: with respect to the numerical and analytical methods, and with respect to the finite element and boundary element methods.

The advantages of the scaled boundary finite element method are compared with those of the finite element and boundary element methods in Table 14.1. Most items are obvious after studying the Model Problem in the preceding chapters. The following comments are appropriate:

- As no fundamental solution is required, as an example of expanding the scope of application, anisotropic material is straightforwardly processed, as only the elasticity matrix $[D]$, for the out-of-plane motion the shear modules G is affected (coefficient matrices $[E^0], [E^1], [E^2]$, in Equation (5.42)). This is also valid for the finite element method.
- When a straight interface between two different materials passes through the scaling centre, no additional discretisation is required. It is sufficient to calculate the coefficient matrices $[E^0], [E^1], [E^2]$, (Equation (5.42)) and $[M^0]$ (Equation (5.43)) of the two scaled boundary finite elements on both sides of the interface with the corresponding material constants. Thus, the material properties, the elasticity matrix $[D(\eta, \zeta)]$ and the mass density $\rho(\eta, \zeta)$, can vary in the circumferential directions η, ζ on the boundary.
- The analytical solution in the radial direction permits a stress singularity to be calculated analytically based directly on its definition without prior knowledge of the nature of the singularity (Equation (7.44)).

Table 14.1 Advantages of scaled boundary finite element method compared with those of finite element and boundary element methods

	Finite element method	Boundary element method	Scaled boundary finite element method
Reduction of the spatial dimension by one as only the boundary is discretised with surface finite elements, reducing the data preparation and computational efforts		X	X
Analytical solution achieved inside domain			X
No fundamental solution required, expanding the scope of application and avoiding singular integrals	X		X
Radiation condition at infinity satisfied exactly when modelling unbounded (infinite or semi-infinite) media		X	X
No discretisation of free and fixed boundaries and interfaces between different materials			X
No approximation other than that of the surface finite elements on the boundary		X	X
Symmetric dynamic-stiffness and unit-impulse response matrices for unbounded media	X	(X)	X
Symmetric static-stiffness and mass matrices for bounded media (super element)	X	(X)	X
Body loads processed without additional domain discretisation and thus additional approximation	X		X
Straightforward calculation of stress concentrations and intensity factors based on their definition			X
No fictitious eigenfrequencies for unbounded media	X		X
Straightforward coupling by standard assemblage of structure discretised with finite elements with unbounded medium	X		X

- In the boundary element method, special procedures are necessary to be able to calculate the response of unbounded media at certain frequencies (which are equal to the natural frequencies of the corresponding bounded media). Such fictitious eigenfrequencies do not occur in the scaled boundary finite element method.
- In contrast to the boundary element method, the property matrices of an unbounded medium being symmetric can straightforwardly be coupled with those of the structure when formulating the equations to analyse unbounded medium-structure interaction based on the substructure method (Appendix C).

Appendix A – Solid Modelling

In a numerical analysis, the accurate and consistent representation of the geometry of the problem to be solved is essential. An exact description of a complex shape would require the specification of the location of an infinite number of points on the shape. To reduce the effort of definition to a reasonable amount, the coordinates in, for instance, a Cartesian system of a limited number of points on the shape called nodes are specified. The geometry of the elements between these nodes is approximated using interpolation functions defined in a local coordinate system. This concept of *solid modelling* is the same as that used to describe the variation of, for instance, displacements in finite elements or of displacements and surface tractions in boundary elements. In the local coordinate system, all operations such as differentions of the interpolation functions and integrations over the element must be systematic and easy to perform.

A thorough, easy-to-follow discussion can be found in Reference [B2].

The following discussion is restricted to plane two-dimensional elements with four nodes, which is sufficient to understand the derivation of the Model Problem in Part I. To simplify the nomenclature, the circumflex (hat)ˆdenoting the Cartesian coordinates is omitted.

An arbitrary quadrilateral element with the coordinates of the nodes 1–4 at the corners specified in the Cartesian coordinate system x, y with a bilinear interpolation is shown in the upper part of Figure A.1. The required calculations such as integrations to determine the property matrices would be awkward to perform in the x-y coordinate system. Thus, for all computations a *parent element* with a local intrinsic coordinate system ξ, η is introduced (lower part of Figure A.1). The square parent element is scaled such that $-1 \leq \xi \leq +1, \ -1 \leq \eta \leq +1$, leading to simple limits of integration. To be able to generate the quadrilateral element from the parent element, a transformation called map is introduced:

$$x = x(\xi, \eta)$$

$$y = y(\xi, \eta) \tag{A.1}$$

A point (ξ, η) in the parent element is mapped to a point (x, y) in the quadrilateral element with parameters ξ, η. For instance, the side 2–3 of the parent element corresponding to $\xi = 1$ and $-1 \leq \eta \leq +1$, $(1, \eta)$, results in $(x = x(1, \eta),\ y = y(1, \eta))$, describing a line with the parameter η in the x, y plane. The same parent element can be used to generate all elements of a mesh (as long as the latter are of the same type, i.e. quadrilateral).

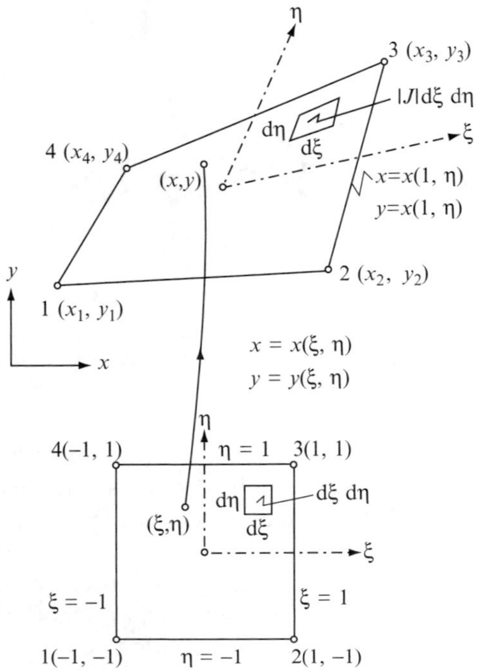

Figure A.1 Generation of quadrilateral finite element by mapping square parent element

To construct the map (Equation (A.1)), the coordinates of the nodes (x_i, y_i), $i = 1, \ldots, 4$ are interpolated in the local coordinates ξ, η. Defining

$$\{x\} = \begin{Bmatrix} x_1 \\ x_2 \\ x_3 \\ x_4 \end{Bmatrix} \quad \{y\} = \begin{Bmatrix} y_1 \\ y_2 \\ y_3 \\ y_4 \end{Bmatrix} \tag{A.2}$$

and with the *mapping functions* $[N(\xi, \eta)]$, as the interpolation functions are called,

$$[N(\xi, \eta)] = 1/4\,[(1-\xi)(1-\eta), \quad (1+\xi)(1-\eta), (1+\xi)(1+\eta), (1-\xi)(1+\eta)] \tag{A.3}$$

the mapping transformation equals

$$\begin{aligned} x(\xi, \eta) &= [N(\xi, \eta)]\{x\} \\ y(\xi, \eta) &= [N(\xi, \eta)]\{y\} \end{aligned} \tag{A.4}$$

$[N(\xi, \eta)]$ in Equation (A.3) describes bilinear functions in ξ, η, which subjected to the linear transformation in Equation (A.4) results in bilinear functions also in x, y.

A remark concerning nomenclature is appropriate. A function can be expressed in x, y or in ξ, η. For instance, when $u(x, y)$ is specified, substitution of $x = x(\xi, \eta), y = y\,(\xi,\eta)$ leads to the same physical quantity expressed in ξ, η, which is a different function.

Rigorous nomenclature would demand that another symbol be used, for instance $\tilde{u}(\xi, \eta)$. For simplicity, the same symbol is used, indicating the arguments in parenthesis, $u(\xi, \eta)$, when an ambiguity could arise. (The same concept is also used when a transformation from the time domain, $u(t)$, to the frequency domain, $u(\omega)$, is performed.)

To be able to derive equations for the property matrices of the arbitrary quadrilateral working with the parent element, the inverse transformation from (x,y) to (ξ, η) must be addressed. In particular, the derivatives with respect to the two coordinate systems have to be linked. Applying the chain rule to the function u, for example, yields

$$u,_{\xi} = u,_{x} x,_{\xi} + u,_{y} y,_{\xi} \tag{A.5}$$

and generalising leads to

$$\begin{Bmatrix} \dfrac{\partial}{\partial \xi} \\ \dfrac{\partial}{\partial \eta} \end{Bmatrix} = \begin{bmatrix} x,_{\xi} & y,_{\xi} \\ x,_{\eta} & y,_{\eta} \end{bmatrix} \begin{Bmatrix} \dfrac{\partial}{\partial x} \\ \dfrac{\partial}{\partial y} \end{Bmatrix} = [J(\xi, \eta)] \begin{Bmatrix} \dfrac{\partial}{\partial x} \\ \dfrac{\partial}{\partial y} \end{Bmatrix} \tag{A.6}$$

The so-called *Jacobian matrix* $[J(\xi, \eta)]$ follows straightforwardly from Equation (A.4). As derivatives with respect to x, y have to be expressed as a function of those with respect to ξ, η, Equation (A.6) is inverted. This results in

$$\begin{Bmatrix} \dfrac{\partial}{\partial x} \\ \dfrac{\partial}{\partial y} \end{Bmatrix} = \frac{1}{|J|} \begin{bmatrix} y,_{\eta} & -y,_{\xi} \\ -x,_{\eta} & x,_{\xi} \end{bmatrix} \begin{Bmatrix} \dfrac{\partial}{\partial \xi} \\ \dfrac{\partial}{\partial \eta} \end{Bmatrix} \tag{A.7}$$

with the determinant $|J| = |[J(\xi, \eta)]|$, called the *Jacobian*, equal to

$$|J| = x,_{\xi} y,_{\eta} - x,_{\eta} y,_{\xi} \tag{A.8}$$

For an acceptable mapping $|J| > 0$ must apply for all points. In this case, the inverse transformation $\xi = \xi(x, y), \eta = \eta(x, y)$ exists.

Integrations over the parent element require a relationship between the infinitesimal areas in the arbitrary quadrilateral element and the parent element (Figure A.1). At a point (ξ, η) of the parent element the infinitesimal area equals $d\xi d\eta$. The map of this infinitesimal area in the quadrilateral element at (ξ, η) with the side lengths $d\xi$ and $d\eta$ is shown in Figure A.2. The infinitesimal area is equal to the cross product of the two vectors along the local coordinate axes ξ, η with the components $[x,_{\xi} d\xi, y,_{\xi} d\xi]$ and $[x,_{\eta} d\eta, y,_{\eta} d\eta]$ yielding $(x,_{\xi} y,_{\eta} - x,_{\eta} y,_{\xi}) d\xi d\eta$. The expression in the parenthesis equals $|J|$ (Equation (A.8)). Thus, the infinitesimal area of the quadrilateral element formulated in the parent element at (ξ, η) in the $(\xi-, \eta-$ coordinate system equals $|J| d\xi d\eta$.

$$dx dy = |J| d\xi d\eta \tag{A.9}$$

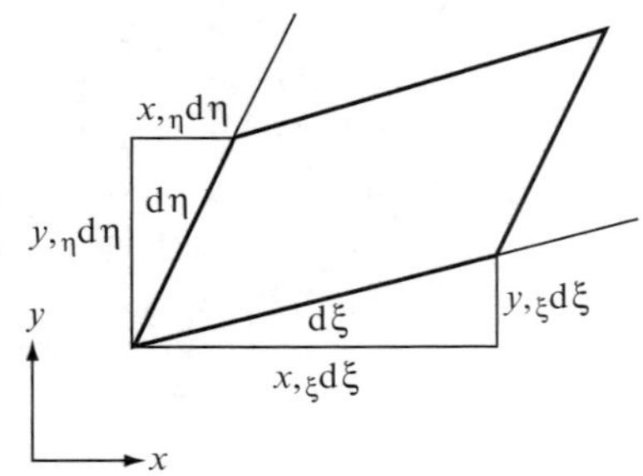

Figure A.2 Infinitesimal area of arbitrary quadrilateral

The integral I of a function $f(\xi, \eta)$ over the total quadrilateral element follows as

$$I = \int_{-1}^{1} \int_{-1}^{1} f(\xi, \eta) \, |J| \, d\xi d\eta \tag{A.10}$$

Summarising, the solid modelling outlined above achieves two goals. First, it permits an accurate definition of the geometry of a complex shape with a mesh by specifying the location of nodes and using interpolation functions, and secondly, it permits all pertinent computations to be performed in a parent element, which can then be mapped with a sequence of transformations to generate all elements of the mesh.

Appendix B – Harmonic Motion and Fourier Transformation

Several alternative ways exist to describe a harmonic motion of a specific frequency ω_j, which repeats itself infinitely many times. They are discussed in the following for a load P, but apply also for any function, for example a displacement u.

In the *first form*

$$P_j(t) = |P(\omega_j)| \cos[\omega_j t + \varphi(\omega_j)] \tag{B.1}$$

with $|P(\omega_j)|$ denoting the magnitude, a positive real value, and $\varphi(\omega_j)$, the phase angle covering the range from $-\pi$ to $+\pi$.

Expanding the cos-function leads to the *second form*

$$P_j(t) = A(\omega_j) \cos \omega_j t + B(\omega_j) \sin \omega_j t \tag{B.2}$$

with the positive or negative real values

$$\begin{aligned} A(\omega_j) &= |P(\omega_j)| \cos \varphi(\omega_j) \\ B(\omega_j) &= -|P(\omega_j)| \sin \varphi(\omega_j) \end{aligned} \tag{B.3}$$

As the *third* (intermediate) *form* a complex amplitude and an exponential function with an imaginary argument are introduced

$$P_j(t) = C(\omega_j)e^{i\omega_j t} + C^*(\omega_j)e^{-i\omega_j t} \tag{B.4}$$

with $C(\omega_j) = \mathrm{Re}C(\omega_j) + \mathrm{i}\,\mathrm{Im}C(\omega_j)$ being complex and an asterisk denoting the complex conjugate value $C^*(\omega_j) = \mathrm{Re}C(\omega_j) - \mathrm{i}\,\mathrm{Im}C(\omega_j)$. Substituting $e^{i\omega_j t} = \cos \omega_j t + i\sin\omega_j t$ results in

$$P_j(t) = 2\mathrm{Re}C(\omega_j) \cos \omega_j t - 2\mathrm{Im}C(\omega_j) \sin \omega_j t \tag{B.5}$$

which is in the same form as Equation (B.2) with

$$\begin{aligned} A(\omega_j) &= 2\mathrm{Re}C(\omega_j) \\ B(\omega_j) &= -2\mathrm{Im}C(\omega_j) \end{aligned} \tag{B.6}$$

Defining a complex amplitude $P(\omega_j) = \mathrm{Re}P(\omega_j) + \mathrm{iIm}P(\omega_j)$ the *fourth form* is constructed

$$P_j(t) = \mathrm{Re}[P(\omega_j)e^{i\omega_j t}] \tag{B.7}$$

Expanding leads to

$$P_j(t) = \mathrm{Re}P(\omega_j)\cos\omega_j t - \mathrm{Im}P(\omega_j)\sin\omega_j t \tag{B.8}$$

which is identical to Equation (B.2) with

$$\begin{aligned} A(\omega_j) &= \mathrm{Re}P(\omega_j) \\ B(\omega_j) &= -\mathrm{Im}P(\omega_j) \end{aligned} \tag{B.9}$$

The complex amplitude is thus given by

$$P(\omega_j) = |P(\omega_j)|\cos\varphi(\omega_j) + \mathrm{i}\,|P(\omega_j)|\sin\varphi(\omega_j) \tag{B.10}$$

It is important to note that the result is equal to the real part of the product of the complex amplitude $P(\omega_j)$ and $e^{i\omega_j t}$ and not just to the real part of $P(\omega_j)$ (Equation (B.7)). To simplify the nomenclature, $P_j(t)$ is set equal to the complex function $P(\omega_j)e^{i\omega_j t}$, and it is implicitly implied that at the end of the complex operation the real part only is regarded.

$$P_j(t) = P(\omega_j)e^{i\omega_j t} \tag{B.11}$$

This *complex form with a complex amplitude and an exponential function with an imaginary argument* provides a powerful and compact way of processing harmonic motion.

The constants of the four forms (first $|P(\omega_j)|$, $\varphi(\omega_j)$; second $A(\omega_j), B(\omega_j)$; third: $\mathrm{Re}C(\omega_j), \mathrm{Im}C(\omega_j)$; fourth: $\mathrm{Re}P(\omega_j), \mathrm{Im}P(\omega_j)$) can be calculated from one another using Equations (B.3), (B.6) and (B.9). In particular, the complex amplitude form follows from the magnitude-phase angle form using Equation (B.10), and going the other way

$$\begin{aligned} |P(\omega_j)| &= \sqrt{\mathrm{Re}P(\omega_j)^2 + \mathrm{Im}P(\omega_j)^2} \\ \varphi(\omega_j) &= \arctan\frac{\mathrm{Im}P(\omega_j)}{\mathrm{Re}P(\omega_j)} \end{aligned} \tag{B.12}$$

applies.

To demonstrate the advantages of working with the complex-amplitude form for harmonic motion, differentiation is first addressed. The derivative with respect to time of a function $u_j(t)$, which is regarded as a displacement in the following with the amplitude $u(\omega_j)$, is calculated as

$$u_j(t) = u(\omega_j)e^{i\omega_j t} \tag{B.13}$$

$$\dot{u}_j(t) = \mathrm{i}\omega u(\omega_j)\mathrm{e}^{\mathrm{i}\omega_j t} = \dot{u}(\omega_j)\mathrm{e}^{\mathrm{i}\omega_j t} \tag{B.14}$$

from which

$$\dot{u}(\omega_j) = \mathrm{i}\omega u(\omega_j) \tag{B.15}$$

follows. Differentiation thus amounts to multiplication of the amplitude by iω, which represents the amplitude of the velocity. Addressing the second derivative, the amplitude of the acceleration is equal to $-\omega^2$ times the amplitude of the displacement.

$$\ddot{u}(\omega_j) = -\omega^2 u(\omega_j) \tag{B.16}$$

Turning next to the operations for harmonic motion, the single-degree-of-freedom system with mass m, spring with coefficient k and dashpot with coefficient c is addressed. For a specified harmonic load $P_j(t)$ in the time domain corresponding to the frequency ω_j, the displacement $u_j(t)$ is to be calculated. The differential equation of motion in the time domain equals

$$m\ddot{u}_j(t) + c\dot{u}_j(t) + ku_j(t) = P_j(t) \tag{B.17}$$

Substituting Equations (B.11), (B.13), (B.14) and the corresponding expression for $\ddot{u}_j(t)$ with Equation (B.16) in Equation (B.17) and cancelling $\mathrm{e}^{\mathrm{i}\omega t}$ yields

$$(-\omega^2 m + \mathrm{i}\omega c + k)\, u(\omega_j) = P(\omega_j) \tag{B.18}$$

The coefficient on the left-hand side represents the complex dynamic-stiffness coefficient

$$S(\omega_j) = k - \omega^2 m + \mathrm{i}\omega c \tag{B.19}$$

After solving the algebraic equation for the unknown amplitude $u(\omega_j)$

$$u(\omega_j) = \frac{P(\omega_j)}{S(\omega_j)} \tag{B.20}$$

the displacement in the time domain equals

$$u_j(t) = u(\omega_j)\mathrm{e}^{\mathrm{i}\omega_j t} \tag{B.21}$$

whereby only the real part of the last operation is regarded. Solving the same problem using, for example, the first form without complex notation would be much more awkward.

An arbitrary *periodic* load with period T (which repeats itself infinitely many times) is addressed. It is plotted as solid and dashed lines in Figure B.1. The load is expressed in a *Fourier series* form where each term j in the series represents a *discrete* harmonic load with a specific frequency ω_j (Equation (B.2)). This leads to

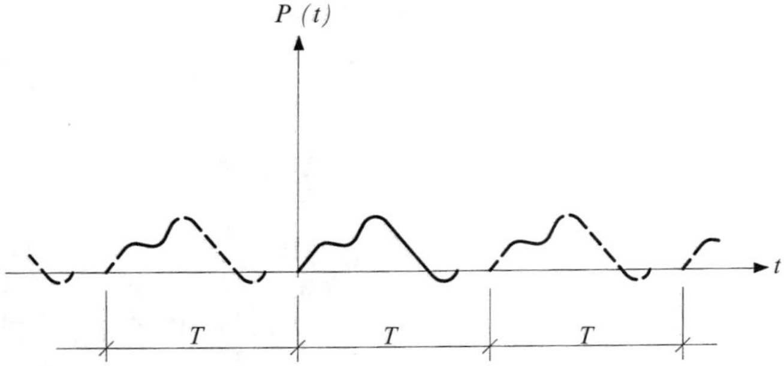

Figure B.1 Arbitrary periodic load represented by Fourier series

$$P(t) = a_0 + \sum_{j=1}^{\infty} a_j \cos \omega_j t + \sum_{j=1}^{\infty} b_j \sin \omega_j t \tag{B.22}$$

where $\omega_j = j\omega_1$ and with $\omega_1 = 2\pi/T$. Multiplying Equation (B.22) by $\cos \omega_k t$ $(k = 0,\ 1,\ \ldots,\infty)$ and integrating over T and with

$$\int_0^T \cos \omega_j t \cos \omega_k t \mathrm{d}t = \begin{Bmatrix} 0 & k \neq j \\ \frac{T}{2} & k = j \end{Bmatrix} \tag{B.23}$$

results in the coefficients

$$\begin{aligned} a_0 &= \frac{1}{T}\int_0^T P(t)\mathrm{d}t \\ a_j &= \frac{2}{T}\int_0^T P(t) \cos \omega_j \mathrm{d}t \end{aligned} \tag{B.24a}$$

Proceeding analogously with $\sin \omega_k t$ yields

$$b_j = \frac{2}{T}\int_0^T P(t) \sin \omega_j \mathrm{d}t \tag{B.24b}$$

The form with complex amplitudes and exponential functions, corresponding for a specific term to Equation (B.11), is derived by substituting for the trigonometric functions the exponential terms in Equation (B.22)

$$\begin{aligned} \sin \omega_j t &= -\ \frac{1}{2}\,\mathrm{i}(\mathrm{e}^{\mathrm{i}\omega_j t} - \mathrm{e}^{-\mathrm{i}\omega_j t}) \\ \cos \omega_j t &= \frac{1}{2}\,(\mathrm{e}^{\mathrm{i}\omega_j t} + \mathrm{e}^{-\mathrm{i}\omega_j t}) \end{aligned} \tag{B.25}$$

This results in

$$P(t) = \sum_{j=-\infty}^{\infty} c_j e^{i\omega_j t} \tag{B.26}$$

with analogous to Equation (B.24)

$$c_j = \frac{1}{T}\int_0^T P(t)e^{-i\omega_j t}dt = \frac{1}{T}\int_{-\frac{T}{2}}^{\frac{T}{2}} P(t)e^{-i\omega_j t}dt \tag{B.27}$$

The limits of integration have been changed to $-T/2$ and $T/2$, which spans the total load period.

This procedure can be generalised to a load which is *non-periodic*. It is sufficient to extend the period T towards infinity. In this case the series expansion, modified appropriately, will represent the load plotted as a solid line in Figure B.1.

Taking this limit $T \to \infty$, the frequency increment $\omega_1 = 2\pi/T$ separating any two distinct frequencies becomes infinitesimal $d\omega$ and the distinct frequencies ω_j are replaced by a *continuous* function ω. The subscript j in Equation (B.26) is dropped, and the summation approaches an integration. The definition

$$P(\omega) = \int_{-\infty}^{\infty} P(t)e^{-i\omega t}dt \tag{B.28}$$

is introduced. Substituting Equation (B.27) with $T = 2\pi/d\omega$ in Equation (B.26) and using Equation (B.28) yields

$$P(t) = \frac{1}{2\pi}\int_{-\infty}^{\infty} P(\omega)e^{i\omega t}d\omega \tag{B.29}$$

Thus, the Fourier series becomes a *Fourier integral*. The two integrals $P(\omega)$ (Equation (B.28)) and $P(t)$ (Equation (B.29)) are known as a *Fourier transform pair*, as the frequency function can be calculated from the time function, and vice versa by analogous operations. $P(\omega)$ is the Fourier transform of $P(t)$, and $P(t)$ is the inverse Fourier transform of $P(\omega)$.

The relationship between the *unit-impulse response coefficient* and the dynamic-stiffness coefficient of an unbounded domain is investigated. By definition, the unit-impulse response coefficient in time $S^{\infty}(t)$ is equal to the force as a function of time $P(t)$ that produces a displacement $u(t)$ of unit-impulse $\delta(t)$ (Dirac delta function). Transforming this displacement into the frequency domain (Equation (B.28)) yields

$$u(\omega) = \int_{-\infty}^{\infty} u(t)e^{-i\omega t}\,dt = \int_{-\infty}^{\infty} \delta(t)e^{-i\omega t}dt = 1 \tag{B.30}$$

The force-displacement relationship in the frequency domain with Equation (B.30) equals

$$P(\omega) = S^{\infty}(\omega)u(\omega) = S^{\infty}(\omega) \tag{B.31}$$

with the dynamic-stiffness coefficient $S^{\infty}(\omega)$. Applying the inverse transformation (Equation (B.29)) and substituting Equation (B.31) leads to

$$P(t) = \frac{1}{2\pi}\int_{-\infty}^{\infty} P(\omega)e^{i\omega t}d\omega = \frac{1}{2\pi}\int_{-\infty}^{\infty} S^{\infty}(\omega)e^{i\omega t}d\omega \tag{B.32}$$

As $S^{\infty}(t) = P(t)$, Equation (B.32) results in

$$S^{\infty}(t) = \frac{1}{2\pi}\int_{-\infty}^{\infty} S^{\infty}(\omega)e^{i\omega t}d\omega \tag{B.33}$$

that is, the *unit-impulse response coefficient* $S^{\infty}(t)$ and the *dynamic-stiffness coefficient* $S^{\infty}(\omega)$ form a *Fourier transform pair.*

Based on the so-called Fourier integral theorem, $S^{\infty}(t)$ exists when $S^{\infty}(\omega)$ appearing in Equation (B.33) is absolutely integrable over the ω-axis, that is

$$\int_{-\infty}^{\infty} | S^{\infty}(\omega)|d\omega < \infty \tag{B.34}$$

Due to the presence of radiation damping occurring in an unbounded domain, the values of the peaks of $|S^{\infty}(\omega)|$ multiplied by $d\omega$ remain finite. However, for $\omega \to \infty$, the dynamic-stiffness coefficient converges to a constant term, denoted as K_{∞}, and a linear term in $i\omega, i\omega C_{\infty}$. This limit is denoted as the singular part $S_s^{\infty}(\omega)$ (subscript s for singular).

$$\lim_{\omega \to \infty} S^{\infty}(\omega) = S_s^{\infty}(\omega) = K_{\infty} + i\omega C_{\infty} \tag{B.35}$$

Multiplying Equation (B.35) by $u(\omega)$, which leads to the corresponding force amplitude, it is apparent that K_{∞} represents a spring coefficient and with Equation (B.15) C_{∞} is a dashpot coefficient.

The inverse transformation of $S_s^{\infty}(\omega)$ follows from Equation (B.33) as

$$S_s^{\infty}(t) = \frac{1}{2\pi}\int_{-\infty}^{\infty} (K_{\infty} + i\omega C_{\infty})e^{i\omega t}d\omega = K_{\infty}\delta(t) + C_{\infty}\dot{\delta}(t) \tag{B.36}$$

The second term on the right-hand side is easily verified by calculating its Fourier transform. The first term is specified essentially in Equation (B.30), and the second equals after integration by parts

$$C_\infty \int_{-\infty}^{+\infty} \dot{\delta}(t)e^{-i\omega t}dt = C_\infty(\delta(t)e^{-i\omega t}|_{-\infty}^{+\infty} + i\omega \int_{-\infty}^{+\infty} \delta(t)e^{-i\omega t}dt) = i\omega C_\infty \tag{B.37}$$

$S^\infty(\omega)$ is decomposed into $S_s^\infty(\omega)$ and the remaining regular part $S_r^\infty(\omega)$ (subscript r for regular)

$$S^\infty(\omega) = K_\infty + i\omega C_\infty + S_r^\infty(\omega) \tag{B.38}$$

Substituted in Equation (B.33) and using Equation (B.36) leads to the unit-impulse response coefficient in time

$$S^\infty(t) = K_\infty\delta(t) + C_\infty\dot{\delta}(t) + S_r^\infty(t) \tag{B.39}$$

with

$$S_r^\infty(t) = \frac{1}{2\pi}\int_{-\infty}^{\infty} S_r^\infty(\omega)e^{i\omega t}d\omega \tag{B.40}$$

$S_r^\infty(t)$ exists, as for $\omega \to \infty$ $S_r^\infty(\omega)$ converges to zero.

The following inverse Fourier transformations apply:

$$F(\omega)\ G(\omega) \quad \to \quad \int_0^t F(t-\tau)G(\tau)d\tau \tag{B.41}$$

$$\frac{F(\omega)}{i\omega} \quad \to \int_0^t F(\tau)d\tau \tag{B.42}$$

$$F(\omega)_{,i\omega} \quad \to -tF(t) \tag{B.43}$$

$$\frac{1}{i\omega} \quad \to \mathrm{H}(t) \tag{B.44}$$

$$\frac{1}{(i\omega)^2} \quad \to t\mathrm{H}(t) \tag{B.45}$$

with the Heavyside step function $\mathrm{H}(t)$ (for $t < 0$ $\mathrm{H}(t) = 0$ and for $t \geq 0$ $\mathrm{H}(t) = 1$). Note that the inverse transform of a product of two functions in the frequency domain is a convolution integral in the time domain (Equation (B.41)).

Applying Equation (B.41) to the force-displacement relationship in the frequency domain

$$P(\omega) = S^\infty(\omega)u(\omega) \tag{B.46}$$

yields in the time domain

$$P(t) = \int_0^t S^\infty(t-\tau)u(\tau)\mathrm{d}\tau \tag{B.47}$$

Substituting Equation (B.39) leads to

$$P(t) = \int_0^t K_\infty \delta(t-\tau)u(\tau)\mathrm{d}\tau + \int_0^t C_\infty \dot{\delta}(t-\tau)u(\tau)\mathrm{d}\tau + \int_0^t S_r^\infty(t-\tau)u(\tau)\mathrm{d}\tau \tag{B.48}$$

or

$$P(t) = K_\infty u(t) + C_\infty \dot{u}(t) + \int_0^t S_r^\infty(t-\tau)u(\tau)\mathrm{d}\tau \tag{B.49}$$

The first two terms on the right-hand side represent the *instantaneous response*, and the third term the *lingering response*. It is obvious from this force-displacement relationship in the time domain that $K_\infty u(t)$ is a spring force and $C_\infty \dot{u}(t)$ a dashpot force.

The important result of Equation (B.41) that a product of the complex variables in the frequency domain corresponds to a convolution integral of the corresponding variables in the time domain can be derived using the force-displacement relationship. The principle of superposition and certain physically explainable properties of the unit-impulse response function are used. The unit-impulse response $S^\infty(t)$ is by definition equal to the force at time t required to produce a unit-impulse displacement at time zero, that is a delta function $\delta(t)$, and zero displacement at all other times $0 < \tau \leq t$. Due to causality, the force will vanish for all times before the displacement is enforced, that is, $S^\infty(t) = 0$ for $t < 0$. To determine the force $P(t)$ at time t which corresponds to the displacement $u(\tau)$, $0 \leq \tau \leq t$, the contribution to the force $P(t)$ at time t of the infinitesimal displacement pulse $u(\tau)\mathrm{d}\tau$ at τ, $0 \leq \tau \leq t$, and zero displacement for all other times from zero up to t is first addressed.

$$\mathrm{d}P(t) = S^\infty(t-\tau)u(\tau)\mathrm{d}\tau \tag{B.50}$$

For a unit-impulse displacement pulse $\delta(\tau)$ enforced at time τ, the force at time t will be equal to $S^\infty(t-\tau)$, that is, the time difference $t-\tau$ occurs (shifting property). To determine the contribution of all pulses $u(\tau)\mathrm{d}\tau$ for $0 \leq \tau \leq t$, Equation (B.50) is then integrated, yielding

$$P(t) = \int_0^t S^\infty(t-\tau)u(\tau)\mathrm{d}\tau \tag{B.51}$$

This convolution integral of the unit-impulse response and the displacement is similar to the well-known Duhamel integral introduced in elementary structural dynamics in the context of a flexibility formulation.

Appendix C – Dynamic Unbounded Medium-structure Interaction

To analyse dynamic unbounded medium-structure interaction, two *substructures* are defined, the structure and the unbounded medium, which are in contact at the structure-medium interface. As an example, the seismic interaction of a dam with its adjacent soil (Figure 1.1) can be addressed, with the dam of finite dimensions representing the (bounded) structure and the semi- infinite soil the unbounded medium. With respect to the degrees of freedom on the interface, properties describing the dynamic behaviour of the two substructures are established. These are summarised in Table 13.1 of Chapter 13. It is the aim of this appendix to demonstrate how the equations of motion of dynamic unbounded medium-structure interaction are formulated using these properties. Further elementary explanations can be found in Chapter 6 of the book [W2].

The Model Problem of Part I is first discussed. The dynamic interaction of the wedge, the structure, with the truncated semi- infinite wedge, the unbounded medium, is addressed. The equations of motion are formulated in the time and frequency domains. The procedure is then generalised from the scalar interaction force-displacement relationship on the interface of the Model Problem to the corresponding matrix relationship, permitting, for instance, a realistic dam interacting with its adjacent soil to be analysed. Dynamic loads applied to the structure and introduced into the coupled system via the unbounded medium, such as seismic excitation, are examined.

The coupled system corresponding to the model problem is shown in Figure C.1. It describes the two-dimensional out-of-plane (antiplane) motion with the displacement u and body load p perpendicular to the plane. The dynamic interaction of the wedge 0-1-2-0, the structure, with the truncated semi-infinite wedge ∞-1-2-∞, the unbounded medium, is addressed. The side 0-1-∞ is fixed ($u = 0$) and the side 0-2-∞ is free (surface traction $\tau_n = 0$). It is assumed that the coupled system is subjected to dynamic loads acting on the wedge. The line 1–2 represents the structure-medium interface.

As in Chapter 7, a single line element 1-2 is selected, resulting in one degree of freedom u_2 in node 2, denoted as u_b in the following (subscript b for base). The wedge is modelled with finite elements, as described in section 1.4. Two degrees of freedom in nodes 3 and 4 denoted as $\{u_s(t)\}$ are introduced. The subscript s for structure is used for nodes of the structure which are not in contact with the unbounded medium. It is assumed that concentrated nodal loads $\{P_s(t)\}$ act in these nodes and $P_b(t)$ in node 2.

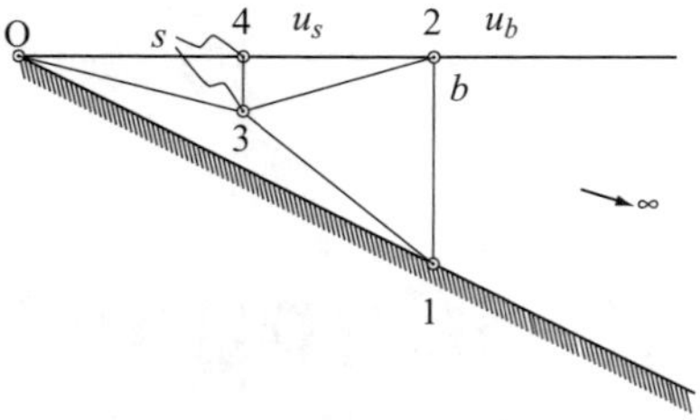

Figure C.1 Wedge (structure) interacting with truncated semi-infinite wedge (unbounded medium)

The truncated semi-infinite wedge is modelled with the scaled boundary finite element method as described for the Model Problem in Part I.

The differential equations of motion of the wedge, the structure, are formulated in the time domain as

$$\begin{bmatrix} [M_{ss}] & \{M_{sb}\} \\ [M_{bs}] & M_{bb} \end{bmatrix} \begin{Bmatrix} \{\ddot{u}_s(t)\} \\ \ddot{u}_b(t) \end{Bmatrix} + \begin{bmatrix} [K_{ss}] & \{K_{sb}\} \\ [K_{bs}] & K_{bb} \end{bmatrix} \begin{Bmatrix} \{u_s(t)\} \\ u_b(t) \end{Bmatrix} = \begin{Bmatrix} \{P_s(t)\} \\ P_b(t) - R(t) \end{Bmatrix} \quad \text{(C.1)}$$

The decomposition of the displacements $\{u(t)\}$ into the subvectors $\{u_s(t)\}$ and $u_b(t)$ is also performed with respect to the mass matrix $[M]$ and the static-stiffness matrix $[K]$. The displacement boundary conditions $u_0 = u_1 = 0$ are enforced. $[M]$ is specified for one finite element in Equation (1.90) and $[K]$ in Equation (1.63) $R(t)$ represents the interaction force of the truncated semi-infinite wedge, the unbounded medium.

The interaction force-displacement relationship in the time domain of the truncated semi-infinite wedge is specified in Equation (11.5). It is also derived in Equation (B.49). Adapting the nomenclature slightly yields

$$R(t) = K_\infty u_b(t) + C_\infty \dot{u}_b(t) + \int_0^t S_r^\infty(t-\tau) u_b(\tau) d\tau \quad \text{(C.2)}$$

with the spring coefficient K_∞ (Equation (10.4)) and the dashpot coefficient C_∞ (Equation (10.3)). The regular part of the unit-impulse response coefficient $S_r^\infty(t)$ follows numerically as described in section 11.2. Substituting Equation (C.2) into Equation (C.1) results in the *differential equations of motion in the time domain of unbounded medium-structure interaction for prescribed dynamic loads acting on the structure*

$$\begin{bmatrix} [M_{ss}] & \{M_{sb}\} \\ [M_{bs}] & M_{bb} \end{bmatrix} \begin{Bmatrix} \{\ddot{u}_s(t)\} \\ \ddot{u}_b(t) \end{Bmatrix} + \begin{bmatrix} [0] & \{0\} \\ [0] & C_\infty \end{bmatrix} \begin{Bmatrix} \{\dot{u}_s(t)\} \\ \dot{u}_b(t) \end{Bmatrix} + \begin{bmatrix} [K_{ss}] & \{K_{sb}\} \\ [K_{bs}] & K_{bb} + K_\infty \end{bmatrix} \begin{Bmatrix} \{u_s(t)\} \\ u_b(t) \end{Bmatrix}$$
$$+ \begin{Bmatrix} \{0\} \\ \int_0^t S_r^\infty(t-\tau) u_b(\tau) d\tau \end{Bmatrix} = \begin{Bmatrix} \{P_s(t)\} \\ P_b(t) \end{Bmatrix}$$
$$\text{(C.3)}$$

If the structure exhibits damping, the corresponding dashpot matrix can be incorporated into the second coefficient matrix on the left-hand side. When the structure is nonlinear, Equation (C.3) still applies in principle. The linear internal forces $[K]\{u(t)\}$ described in the third contribution on the left-hand side are replaced by the structure's nonlinear internal forces $\{F(t)\}$.

To derive the corresponding equations in the frequency domain, i.e. for harmonic excitation, complex amplitudes for the displacements and forces are introduced (Appendix B, Equations (B.13), (B.16) and (B.11)). The interaction force-displacement relationship in the frequency domain of the truncated semi-infinite wedge as specified in Equation (11.2) is used, which is also formulated in Equation (B.46). Substituting

$$\{u(t)\} = \{u(\omega)\}e^{i\omega t} \tag{C.4a}$$

$$\{\ddot{u}(t)\} = -\omega^2\{u(\omega)\}e^{i\omega t} \tag{C.4b}$$

$$\{P(t)\} = \{P(\omega)\}e^{i\omega t} \tag{C.4c}$$

$$R(t) = R(\omega)e^{i\omega t} \tag{C.4d}$$

and

$$R(\omega) = S^{\infty}(\omega)u_b(\omega) \tag{C.5}$$

in Equation (C.1) yields the *algebraic equations of motion in the frequency domain of unbounded medium-structure interaction for prescribed dynamic loads acting on the structure*

$$\begin{bmatrix} [S_{ss}(\omega)] & \{S_{sb}(\omega)\} \\ [S_{bs}(\omega)] & S_{bb}(\omega) + S^{\infty}(\omega) \end{bmatrix} \begin{Bmatrix} \{u_s(\omega)\} \\ u_b(\omega) \end{Bmatrix} = \begin{Bmatrix} \{P_s(\omega)\} \\ P_b(\omega) \end{Bmatrix} \tag{C.6}$$

The dynamic-stiffness matrix of the structure

$$[S(\omega)] = [K] - \omega^2[M] \tag{C.7}$$

follows from the corresponding static-stiffness matrix $[K]$ and mass matrix $[M]$ (Equation (1.91)). The dynamic-stiffness coefficient $S^{\infty}(\omega)$ of the unbounded medium is determined analytically in Equation (12.50).

To be able to formulate the contribution of the unbounded medium modelled with the scaled boundary finite element method to the coupled equations in the time domain (Equation (C.3)) and frequency domain (Equation (C.6)), K_{∞}, C_{∞}, $S_r^{\infty}(t)$ and $S^{\infty}(\omega)$, respectively, are required. The contribution of the structure modelled with the finite-element method can be constructed using $[K]$ and $[M]$ (in addition to nodal loads).

If the structure is also modelled with the scaled boundary finite element method leading to a super-element with degrees of freedom on the structure-medium interface only, the static-stiffness coefficient K^{b} (Equation (8.14)) and mass M^{b} (Equation (9.4)) are calculated. For vanishing dynamic loads in the interior of the structure $\{P_s(t)\} = 0$, the equations of motion of unbounded medium-structure interaction in the time and frequency domains equal

$$M^{\mathrm{b}}\ddot{u}_b(t) + C_\infty \dot{u}_b(t) + (K^{\mathrm{b}} + K_\infty)u_b(t) + \int_0^t S_r^\infty(t-\tau)u_b(\tau)\mathrm{d}\tau = P_b(t) \tag{C.8}$$

and

$$\left(S^{\mathrm{b}}(\omega) + S^\infty(\omega)\right)u_b(\omega) = P_b(\omega) \tag{C.9}$$

respectively. The dynamic-stiffness coefficient of the super-element (structure) equals

$$S^{\mathrm{b}}(\omega) = K^{\mathrm{b}} - \omega^2 M^{\mathrm{b}} \tag{C.10}$$

A more general case of dynamic unbounded medium-structure interaction is shown in Figure C.2. A (two-dimensional) dam A-B-C-D-E-F-A interacts with the adjacent soil G-B-C-D-E-I-H-G along the interface B-C-D-E. A seismic excitation introduced via the soil into the coupled system is assumed.

Subscripts are used to denote the nodes of the discretised system. The nodes on the dam-foundation interface are denoted by b (b for base), the remaining nodes of the dam by s (s for structure). The dam is discretised with finite elements introducing in each node two displacements. The vector of all these displacements $\{u\}$ is decomposed into $\{u_s\}$ and $\{u_b\}$. The dynamic behaviour of the dam is described by the static-stiffness matrix $[K]$ and the mass matrix $[M]$, which can both be decomposed analogously. The equations of motion in total displacements in the time-domain of the dam, the structure, is formulated for vanishing loads acting on the dam as

$$\begin{bmatrix} [M_{ss}] & [M_{sb}] \\ [M_{bs}] & [M_{bb}] \end{bmatrix} \begin{Bmatrix} \{\ddot{u}_s^t(t)\} \\ \{\ddot{u}_b^t(t)\} \end{Bmatrix} + \begin{bmatrix} [K_{ss}] & [K_{sb}] \\ [K_{bs}] & [K_{bb}] \end{bmatrix} \begin{Bmatrix} \{u_s^t(t)\} \\ \{u_b^t(t)\} \end{Bmatrix} = \begin{Bmatrix} \{0\} \\ -\{R(t)\} \end{Bmatrix} \tag{C.11}$$

The word total (superscript t) expresses that the motion is referred to an origin that does not move. $\{R(t)\}$ denotes the interaction forces of the unbounded soil acting in the nodes on the interface of the dam.

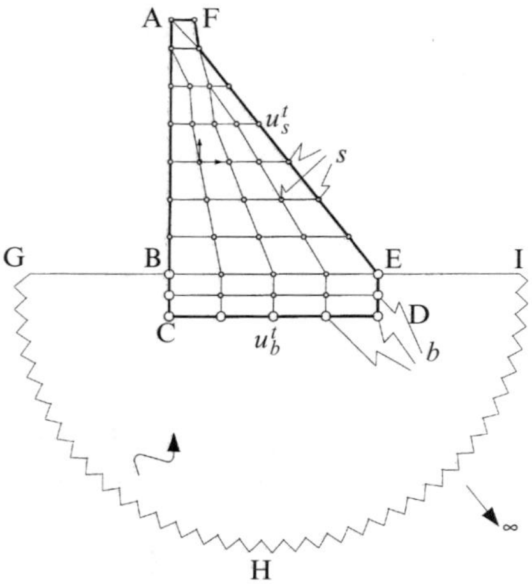

Figure C.2 Dam (structure) embedded in soil (unbounded medium)

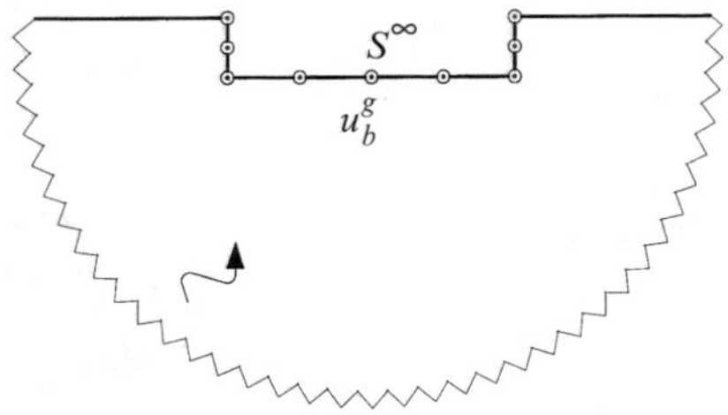

Figure C.3 Soil (with excavation) and corresponding effective seismic foundation input motion

The semi-infinite soil, the unbounded medium, is modelled with the scaled boundary finite element method using the displacements in the nodes on the dam-soil interface only $\{u_b\}$ as degrees of freedom (Figure C.3). In the interior of the soil and on its free surface extending to infinity (G-B, E-I) no discretisation is present. The dynamic behaviour of the soil is specified by its interaction force-displacement relationship on the interface, which is described by the unit-impulse response matrix $[S^\infty(t)]$ (i.e. $[K_\infty], [C_\infty], [S_r^\infty(t)]$) in the time domain and by the dynamic-stiffness matrix $[S^\infty(\omega)]$ in the frequency domain. For seismic excitation, the so-called *effective seismic foundation input motion* $\{u_b^g\}$ in the same nodes on the interface of the soil can be calculated. The superscript g denotes the system *ground*, i.e. the soil with excavation (Figure C.3). It can be shown that $\{u_b^g\}$ can be computed from the *free-field motion* of the soil (i.e. without the excavation) determined in the same nodes (see Chapter 6 of the book [W2]). For the motion $\{u_b^g\}$ the interaction forces acting on the interface vanish because for this loading state the interface in Figure C.3 is a free surface. The *interaction forces of the soil (ground) will thus depend upon the motion relative to the effective foundation input motion* $\{u_b^g\}$. In the time domain generalising Equation (B.47) to the matrix case

$$\{R(t)\} = \int_0^t [S^\infty(t-\tau)]\big(\{u_b^t(\tau)\} - \{u_b^g(\tau)\}\big)\mathrm{d}\tau \tag{C.12}$$

applies, or with Equation (B.49))

$$\begin{aligned}\{R(t)\} = {} & [K_\infty]\big(\{u_b^t(t)\} - \{u_b^g(t)\}\big) + [C_\infty]\big(\{\dot{u}_b^t(t)\} - \{\dot{u}_b^g(t)\}\big) \\ & + \int_0^t [S_r^\infty(t-\tau)]\big(\{u_b^t(\tau)\} - \{u_b^g(\tau)\}\big)\mathrm{d}\tau\end{aligned} \tag{C.13}$$

Analogously, in the frequency domain, using Equation (B.46)), the amplitudes of the interaction forces equal

$$\{R(\omega)\} = [S^\infty(\omega)](\{u_b^t(\omega)\} - \{u_b^g(\omega)\}) \tag{C.14}$$

Substituting Equation (C.13) into (C.11) leads to the *differential equations of motion in total displacements in the time domain of unbounded medium-structure interaction for seismic excitation*

$$\begin{bmatrix}[M_{ss}] & [M_{sb}]\\ [M_{bs}] & [M_{bb}]\end{bmatrix}\begin{Bmatrix}\{\ddot{u}_s^t(t)\}\\ \{\ddot{u}_b^t(t)\}\end{Bmatrix}+\begin{bmatrix}[0] & [0]\\ [0] & [C_\infty]\end{bmatrix}\begin{Bmatrix}\{\dot{u}_s^t(t)\}\\ \{\dot{u}_b^t(t)\}\end{Bmatrix}+\begin{bmatrix}[K_{ss}] & [K_{sb}]\\ [K_{bs}] & [K_{bb}]+[K_\infty]\end{bmatrix}\begin{Bmatrix}\{u_s^t(t)\}\\ \{u_b^t(t)\}\end{Bmatrix}$$
$$+\begin{Bmatrix}\{0\}\\ \int_0^t [S_r^\infty(t-\tau)]\{u_b^t(\tau)\}\mathrm{d}\tau\end{Bmatrix}=\begin{Bmatrix}\{0\}\\ [C_\infty]\{\dot{u}_b^g(t)\}+[K_\infty]\{u_b^g(t)\}+\int_0^t [S_r^\infty(t-\tau)]\{u_b^g(\tau)\}\mathrm{d}\tau\end{Bmatrix} \tag{C.15}$$

The *seismic motion* characterised by $\{u_b^g\}$ *is* thus *converted to an equivalent load* applied in the nodes on the structure-medium interface (subscript b) acting on the coupled system. This equivalent load, also called the *driving load*, is equal to the reaction forces when the structure is kept fixed. This is easily verified by setting $\{u_s^t(t)\}=0, \{u_b^t(t)\}=0$ in Equation (C.15). Note that in this formulation expressed in total displacements, the nodes in the interior of the structure (subscript s) are not loaded.

Using Equation (C.14) and proceeding analogously as in the derivation of Equation (C.6), the *algebraic equations of motion in total displacement amplitudes in the frequency domain of unbounded medium-structure interaction for seismic excitation* are formulated as

$$\begin{bmatrix}[S_{ss}(\omega)] & [S_{sb}(\omega)]\\ [S_{bs}(\omega)] & [S_{bb}(\omega)]+[S^\infty(\omega)]\end{bmatrix}\begin{Bmatrix}\{u_s^t(\omega)\}\\ \{u_b^t(\omega)\}\end{Bmatrix}=\begin{Bmatrix}\{0\}\\ [S^\infty(\omega)]\{u_b^g(\omega)\}\end{Bmatrix} \tag{C.16}$$

with the dynamic-stiffness matrix of the structure

$$[S(\omega)]=[K]-\omega^2[M] \tag{C.17}$$

Appendix D – Historical Note

It is appropriate to review the pioneering research which formed the basis of the development of the scaled boundary finite element method. Scientists in electrical engineering, mathematics and mechanics contributed to the novel procedure. Only key papers are mentioned, which in retrospect influenced decisively the development. It will become apparent that researchers worked on the same aspects at the same time without knowing about the other persons' activities, resulting in similar procedures. The relationship of the author's group to this early work is also addressed. A short description of the various phases of the development of the scaled boundary finite element method then follows.

Looking back, the roots of the scaled boundary finite element method are a combination of *infinite substructuring* and *finite elements*. An important quite well known algorithm was published in 1977 by Silvester's group to model an unbounded domain for Laplace's equation in electrostatics and magnetostatics [S1]. In this procedure, called *ballooning*, the outer boundary of a discretised region is moved outwards by a recursive technique using similarity with respect to a suitably chosen (similarity) centre. Rows of finite elements forming finite element cells are placed on the outside of the outer boundary, one at a time, and common nodes are eliminated from adjacent rows in such a way that a single cell surrounding the original region results with the outer nodes moving further outwards at each inclusion of a new row. The increase in dimensions of the boundary connecting the outer nodes follows a geometric series in this recursive technique. The prescribed boundary condition can then be enforced on the outer boundary.

A similar procedure using an infinite number of finite elements to represent Laplace's equation for an unbounded domain was proposed by Thatcher [T3]. The recursive technique is described by an eigenvalue problem involving a full matrix. Thatcher addressed even earlier in 1975 the representation of singularities in Laplace's equation in a bounded domain [T2]. Based on this *infinite grid refinement* technique, no *a priori* knowledge of the singularities is required and several singularities can be modelled in the same problem.

Independently, Ying developed a similar procedure, called the *infinite element* method to calculate stress singularities in elastostatics in a bounded domain [Y1], again resulting in an eigenvalue problem. The expression infinite element means that an infinite number of elements is present. As described in the foreword of his book [Y2], the paper was first presented at the National Conference on Fracture Mechanics in Nanning in 1974. The book [Y2] contains many references describing the work of the author's group (and others) including modelling of unbounded domains.

A the end of the 1970s the author was strongly influenced by Dasgupta who was in the process of developing the so-called *cloning* algorithm to model unbounded domains in a dynamic analysis. A bounded finite element cell, adjacent to the outer boundary with another similar boundary was introduced, analogous to ballooning in statics described above. Based on the static-stiffness and mass matrices of the cell, the dynamic-stiffness matrix of the unbounded domain followed from an eigenvalue problem. An average value of the two similar boundaries of the finite element cell and thus of the dimensionless frequency was introduced in the derivation. This means that the procedure is only valid when the dimensionless frequencies referred to the two similar boundaries of the cell are constant, which applies in statics and for a layer of constant depth in dynamics. In the latter case, the same equations as for the consistent boundary procedure, now called the thin-layer method, in two dimensions are obtained [L3], [W1], [K1]. In the general case of varying dimensionless frequency, incorrect results are calculated. Dasgupta published his work in 1982 [D2], with conference papers dating back to 1979 [D1]. The variation of the dimensionless frequency was correctly considered in 1982 in [W17], replacing the average value of the dimensionless frequency by two terms of a Taylor expansion. For a special case, the analytical limit of the infinitesimal cell width was also performed in the same article.

No additional work was performed by the authors group in the 1980s, until starting in 1990 Song developed multi-cell cloning to model unbounded domains [W6]. This rigorous procedure can today be regarded as outdated. As a major breakthrough, in the mechanically-based derivation of the scaled boundary finite element method the limit of the infinitesimal width of the finite element cell was performed analytically in elastodynamics of unbounded domains in the general case, which lead to the nonlinear first-order ordinary differential equations in the dynamic stiffness with the frequency as the independent variable. The procedure was denoted at the time as the *consistent infinitesimal finite element cell method* [W9]. Displacements in the interior were calculated for statics, incompressible elastodynamics was addressed, the diffusion equation was examined and bounded domains including singularities were examined. This material was incorporated in Part I of the book [W9]. The authors had no knowledge of the pioneering work of Thatcher and Ying described above when writing the book.

Further significant progress was achieved by Song when the fundamental equations were rederived based on the scaled boundary coordinate transformation. This led to the linear second-order ordinary differential equations in displacements with the radial coordinate as the independent variable [S7]. They can also be solved in dynamics and for body loads for unbounded and bounded domains analytically in the radial direction. The procedure is called the *scaled boundary finite element method.* The potential to solve all linear partial differential equations in engineering and physical science based on a straightforward coordinate transformation exists.

Finally, performing the integrations in the radial direction analytically, Deeks determined stress recovery and error estimations for bounded and unbounded domains based on the discretisation of the boundary only [D3]. An h-hierarchical adaptive procedure was derived for a prescribed error throughout the domain [D4].

Part II

Two- and Three-dimensional Elastodynamics, Statics and Diffusion

Part I addresses the Model Problem, which consists of the simplest case of wave propagation but still contains all essential features: the out-of-plane motion of a two-dimensional wedge and truncated semi-infinite wedge governed by the scalar Helmholz differential equation. Two derivations of the governing equations and the solution procedures are described in great detail. The features and significant advantages of the scaled boundary finite element method are clearly demonstrated.

Part II discusses the derivation of the fundamental equations and solution procedures from scratch for elastodynamics, statics and diffusion in two and three dimensions. In elastodynamics, the scalar and vector wave equations are addressed. Applications for bounded and unbounded media are presented. Extensions as well as error estimation and adaptivity are considered as additional topics.

The treatment is independent of Part I and quite general. However, when queries arise, the reader should consult the corresponding chapter of the Model Problem in Part I, which provides an example. To facilitate this, the same division and chapter headings are used in Parts I and II where feasible.

The derivation of the fundamental equations based on the scaled boundary transformation is described in Chapter 15, which thus corresponds to Chapter 5 for the Model Problem. The mechanically-based derivation presented in Chapter 6 is not addressed in Part II, but is available in the literature, where the procedure is called the consistent infinitesimal finite element cell method (Refs [S3]–[S6], [W7], [W8], [W10], [W11], and in the first part of [W9]). The solution procedures for statics and dynamics in the frequency and time domains, both numerically and analytically, for bounded and unbounded media are developed in Chapters 16–20, which correspond to Chapters 8–12. Various extensions are addressed in Chapter 21. Especially for bounded media, substructuring is a powerful tool which is examined in Chapter 22. Two- and three-dimensional examples in elastodynamics and diffusion for bounded and unbounded media are discussed in Chapters 23 and 24. Error estimation and adaptivity are addressed in Chapter 25, followed by concluding remarks in Chapter 26.

In contrast to Part I, where the theory and application to the Model Problem are intermingled, Part II first addresses all aspects of the general theory, and only then turns to examples. For the reader who would like to study a specific example in parallel to be able to master each step of the theory more efficiently, the corresponding aspect of Part I can be consulted.

15 Fundamental Equations

After a discussion of the salient concept of the scaled boundary finite element method (section 15.1), the fundamental equations are derived in great detail from scratch using the weighted-residual technique for the most general case, the three-dimensional vector-wave equation in elastodynamics (sections 15.2–15.4). This is followed by a summary of the relations for two-dimensional elastodynamics (section 15.5). The fundamental equations for the two- and three-dimensional diffusion equation are then addressed (section 15.6). Finally, an alternative derivation using the familiar virtual-work formulation is examined for elastodynamics (section 15.7).

Section 15.1 is based on Ref [W16], the key sections 15.2–15.5 on [S7], section 15.6 on [S9], and finally, section 15.7 on [D5].

15.1 SALIENT CONCEPT

For a brief characterisation of the scaled boundary finite element method, the reader is referred to Chapter 2, which also compares this novel computational procedure to the finite element and boundary element methods. In the scaled boundary finite element method, the geometry of the domain is described by a finite element discretisation with local coordinates on the boundary, and the so-called radial coordinate representing a scaling factor. This scaled boundary coordinate system is related to the Cartesian coordinate system by the so-called scaled boundary transformation, which actually describes similarity. The discretisation of the boundary permits arbitrary boundary geometries and discontinuous boundary conditions to be represented. This concept is described for three-dimensional bounded and unbounded media in Chapter 3. The reader should study the short Chapters 2 and 3 in depth. In particular, the significance of Figures 3.1 and 3.4 must be fully understood before continuing reading this section.

As another, but optional reference, the application to Laplace's equation in section 1.6 can serve as a first glimpse at the method.

To develop the salient concept of the scaled boundary finite element method, a limited overlap with the material in Chapters 2 and 3 will occur.

A Cartesian coordinate system $\hat{x}, \hat{y}, \hat{z}$ is, in general, not a good choice to base a numerical discretisation on (Figure 15.1). First, the boundary, where conditions are to be enforced, does not coincide with one of the Cartesian coordinate axes being constant. It is more appropriate to work in a local coordinate system ξ, η, ζ with the axes η, ζ lying in the circumferential directions (on the boundary or parallel to it). The third coordinate ξ measured from the so-called scaling centre O, the origin, is denoted as the

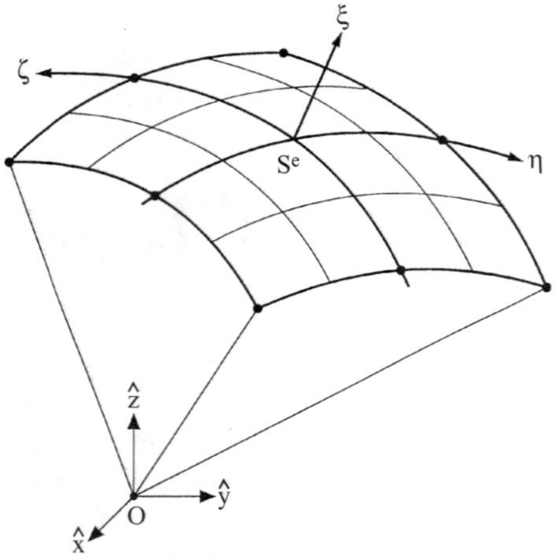

Figure 15.1 Scaled boundary transformation of geometry of surface finite element (Reprinted from Computer Methods in Applied Mechanics & Engineering, Vol. 180, Song and Wolf, 'Body loads in scaled boundary finite element method', pp. 117–135, © (1999), with permission from Elsevier Science)

radial coordinate. Secondly, this *scaled boundary coordinate system* ξ, η, ζ permits a numerical treatment in the circumferential directions η, ζ based on a weighted-residual technique, as in finite elements. This will transform the partial differential equations to ordinary differential equations in the radial coordinate ξ. The coefficients of the ordinary differential equations are determined by the finite element approximation in the circumferential directions. The ordinary differential equations, being much simpler, as only a single independent variable is present, than partial differential equations, are then solved analytically in the radial direction. Thus, a novel computational procedure, called the *scaled boundary finite element method* is developed, which *combines the advantages of the numerical and analytical approaches*. The method is a *semi-analytical* procedure to solve partial differential equations. In the circumferential directions (parallel to the boundary), where the behaviour is, in general smooth, the weighted-residual approximation of finite elements applies. For a bounded medium, the radial coordinate points from the boundary towards the interior, where the analytical solution permits stress concentrations and singularities to be captured directly without any prior knowledge of their behaviour. For an unbounded medium the radial coordinate points from the boundary towards infinity, where the boundary condition at infinity (radiation condition) can be incorporated exactly in the analytical solution.

In more detail, the origin of the new coordinate system, called the *scaling centre O*, is chosen in a zone from which the total boundary must be visible. (For the sake of simplicity, O coincides with the origin of the Cartesian coordinate system in Figure 15.1). For a bounded medium the scaling centre is located within the domain (Figure 3.1a.) As a special case, the scaling centre O can be chosen on the boundary (see Figure 3.1b). In this case, the total boundary is decomposed into two parts: that part of the boundary passing through the scaling centre denoted as the side-face A; and the remaining part S (Figure 3.1b). For an unbounded medium, the scaling centre is located outside the domain (Figure 3.4a). Again, as a special case the scaling centre O

can be chosen on the extension of the boundary (see Figure 3.4b), and the total boundary is decomposed into two parts: that part of the boundary with its extension passing through the scaling centre denoted as the side-face A; and the remaining part S (Figure 3.4b). Only the boundary S is discretised with (doubly-curved) surface finite elements, which could also be called scaled boundary elements. A typical finite element is shown in Figure 15.1. Denoting points on the boundary with x, y, z, the geometry is described in the local coordinate system η, ζ

$$\begin{aligned} x(\eta,\zeta) &= [N(\eta,\zeta)]\{x\} \\ y(\eta,\zeta) &= [N(\eta,\zeta)]\{y\} \\ z(\eta,\zeta) &= [N(\eta,\zeta)]\{z\} \end{aligned} \tag{15.1}$$

with the mapping functions $[N(\eta,\zeta)]$ and the coordinates of the nodes $\{x\}, \{y\}, \{z\}$. The three-dimensional medium is defined by scaling the boundary points with the dimensionless radial coordinate ξ measured from the scaling centre

$$\begin{aligned} \hat{x}(\xi,\eta,\zeta) &= \xi x(\eta,\zeta) \\ \hat{y}(\xi,\eta,\zeta) &= \xi y(\eta,\zeta) \\ \hat{z}(\xi,\eta,\zeta) &= \xi z(\eta,\zeta) \end{aligned} \tag{15.2}$$

with $\xi = 1$ on the boundary and $= 0$ at the scaling centre. For a bounded medium, $0 \leq \xi \leq 1$ applies and for an unbounded medium $1 \leq \xi < \infty$. The new coordinate system is defined by ξ and the two circumferential coordinates η, ζ. In the scaled boundary transformation $\hat{x}, \hat{y}, \hat{z}$ is replaced by ξ, η, ζ. Equation (15.2) describing scaling of the boundary has lead to the name of the method.

The components of the displacements, strains, stresses, etc. are still defined in the Cartesian coordinate system, but their position is specified in the scaled boundary coordinate system. The displacement amplitudes of the finite element on the boundary ($\xi = 1$) are interpolated using shape functions $[N(\eta,\zeta)]$, which for convenience are the same as the mapping functions. The discretisation is thus restricted to this boundary. It is postulated that *the same shape functions apply with the displacement amplitudes* $\{u(\xi)\}$ *for all surfaces with a constant* ξ

$$\{u(\xi,\eta,\zeta)\} = [N(\eta,\zeta)]\{u(\xi)\} \tag{15.3}$$

The displacement amplitudes along the line defined by the scaling centre and the node on the boundary $\{u(\xi)\}$ are analytical functions of the radial coordinate ξ only, which, as will be demonstrated, are calculated analytically from the corresponding ordinary differential equations. In the circumferential directions, the displacements are then interpolated using the same functions as used to describe the geometry of the boundary (Equation (15.1)) and the scaling equations (Equation (15.2)). The approximate solution determined from Equation (15.3) could be interpreted as applying in a generalised manner the procedure of separation of variables, interpolating in the circumferential directions η, ζ the discrete values of $\{u(\xi)\}$. Equation (15.3), together with the definition of the scaled boundary transformation (Equations (15.1) and (15.2)) forms the basis of the scaled boundary finite element method. The difference to the standard finite element method with the displacement amplitudes in distinct nodes $\{u\}$ located throughout the medium

$$\{u(\xi, \eta, \zeta)\} = [N(\xi, \eta, \zeta)]\{u\} \tag{15.4}$$

is clearly visible. Note that Equation (15.2) expresses similarity with the scaling centre coinciding with the similarity centre and ξ representing the similarity factor. The surfaces for a constant ξ are similar to the boundary. It is only thanks to this transformation that for a fixed η, ζ the displacements along a radial line are a function of ξ only, as expressed in Equation (15.3).

The material properties (elasticity matrix, mass density) can vary in the circumferential directions, as in the standard finite element method.

As only the boundary S is discretised, the novel computational procedure is a *boundary element method* based on finite element concepts. It combines the *advantages of the finite element and boundary element methods.* The scaled boundary finite element method has also got advantages of its own, such as no spatial discretisation of the side-faces A which form part of the boundary.

In a nutshell, the scaled boundary finite element method is a *semi-analytical fundamental solution-less boundary element method based on finite elements.*

15.2 GOVERNING EQUATIONS OF ELASTODYNAMICS IN SCALED BOUNDARY COORDINATES

As a first step in deriving the fundamental equations of the scaled boundary finite element method for three-dimensional elastodynamics, the governing equations with geometry in the transformed scaled boundary coordinate system are developed.

First, the governing equations of three-dimensional elastodynamics are formulated in Cartesian coordinates $\hat{x}, \hat{y}, \hat{z}$. A circumflex (^) is introduced to denote a point in space, as the coordinates x, y, z are reserved for the boundary, which is discretised. To simplify the nomenclature, the circumflex is omitted when used in a subscript to indicate a direction.

The bounded medium shown in Figure 3.1b is addressed. The boundary consisting of S_u and S_t is discretised, and the boundary passing through the scaling centre A_u and A_t is not discretised.

The differential equations of motion in the frequency domain expressed in displacement amplitudes $\{u\} = \{u(\hat{x}, \hat{y}, \hat{z})\} = [u_x u_y u_z]^T$ are formulated as

$$[L]^T\{\sigma\} + \{p\} + \omega^2\rho\{u\} = 0 \tag{15.5}$$

with the mass density ρ and the amplitudes of the body loads $\{p\}$. The stress amplitudes $\{\sigma\}$ follows from Hooke's law with the (anisotropic) elasticity matrix $[D]$ as

$$\{\sigma\} = [\sigma_x\ \sigma_y\ \sigma_z\ \tau_{yz}\ \tau_{xz}\ \tau_{xy}]^T = [D]\{\varepsilon\} \tag{15.6}$$

The strain amplitudes $\{\varepsilon\}$ are defined from the strain-displacement relationship

$$\{\varepsilon\} = [\varepsilon_x\ \varepsilon_y\ \varepsilon_z\ \gamma_{yz}\ \gamma_{xz}\ \gamma_{xy}]^T = [L]\{u\} \tag{15.7}$$

with the differential operator $[L]$

$$[L] = \begin{bmatrix} \frac{\partial}{\partial \hat{x}} & & \\ & \frac{\partial}{\partial \hat{y}} & \\ & & \frac{\partial}{\partial \hat{z}} \\ & \frac{\partial}{\partial \hat{z}} & \frac{\partial}{\partial \hat{y}} \\ \frac{\partial}{\partial \hat{z}} & & \frac{\partial}{\partial \hat{x}} \\ \frac{\partial}{\partial \hat{y}} & \frac{\partial}{\partial \hat{x}} & \end{bmatrix} \tag{15.8}$$

The boundary conditions are equal to

$$\{u\} = \{\bar{u}\} \text{ on } S_u \tag{15.9a}$$

$$\{t\} = \{\bar{t}\} \text{ on } S_t \tag{15.9b}$$

and

$$\{u\} = \{\bar{u}\} \text{ on } A_u \tag{15.10a}$$

$$\{t\} = \{\bar{t}\} \text{ on } A_t \tag{15.10b}$$

with the prescribed displacement amplitudes $\{\bar{u}\}$ and prescribed surface traction amplitudes $\{\bar{t}\}$. For simplicity, the right-hand side of Equation (15.10a) is assumed to be zero, $\{\bar{u}\} = 0$. The generalisation is addressed at the end of section 15.3.

Secondly, the transformation of the geometry corresponding to the scaled boundary coordinates discussed in section 15.1 is addressed. The coordinates on the (doubly-curved) boundary are denoted as x, y, z. A point on the boundary is described by its position vector

$$\boldsymbol{r} = x\mathbf{i} + y\mathbf{j} + z\mathbf{k} \tag{15.11}$$

The Cartesian coordinates $\hat{x}, \hat{y}, \hat{z}$ are transformed to the curvilinear coordinates ξ, η, ζ (Figure 15.2). The scaling centre O is selected in the interior of the domain (Figure 3.1a) or on the boundary (Figure 3.1b). It coincides with the origin of the coordinate system $\hat{x}, \hat{y}, \hat{z}$. The dimensionless radial coordinate ξ is measured from the scaling centre along the position vector

$$\hat{\boldsymbol{r}} = \hat{x}\mathbf{i} + \hat{y}\mathbf{j} + \hat{z}\mathbf{k} \tag{15.12}$$

(unit vectors **i, j, k**). ξ is constant (equal to 1) on the boundary. For the bounded medium (Figure 3.1), $0 \le \xi \le 1$ applies (and for an unbounded medium

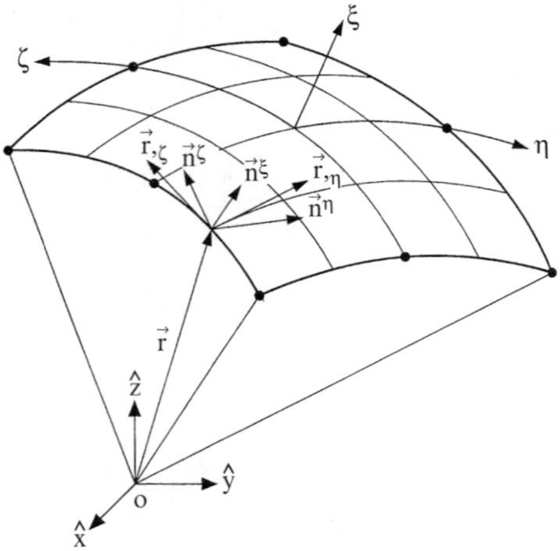

Figure 15.2 Scaled boundary transformation of geometry of surface finite element with derivatives of position vector and normals (Reprinted from Computer Methods in Applied Mechanics & Engineering, Vol. 147, Song and Wolf, 'The scaled boundary finite element method – alias consistent infinitesimal finite element cell method for elastodynamics, pp. 329–355, © (1997), with permission from Elsevier Science.)

$1 \leq \xi < \infty$). In a practical application, the geometry of the boundary is so general that only a piecewise description is feasible. (Doubly-curved) surface finite elements are used.

A specific finite element is addressed. The geometry of this finite element on the boundary is represented by interpolating its nodal coordinates $\{x\}, \{y\}$ and $\{z\}$ using the local coordinates $\eta,\ \zeta$

$$
\begin{aligned}
x(\eta, \zeta) &= [N(\eta, \zeta)]\{x\} \\
y(\eta, \zeta) &= [N(\eta, \zeta)]\{y\} \\
z(\eta, \zeta) &= [N(\eta, \zeta)]\{z\}
\end{aligned}
\tag{15.13}
$$

with the mapping functions

$$
[N(\eta, \zeta)] = [N_1(\eta, \zeta) N_2(\eta, \zeta) \ldots] \tag{15.14}
$$

A point in the domain is obtained by scaling that on the boundary

$$
\hat{\boldsymbol{r}} = \xi \boldsymbol{r} \tag{15.15}
$$

Expressed in coordinates

$$
\begin{aligned}
\hat{x}(\xi, \eta, \zeta) &= \xi x(\eta, \zeta) \\
\hat{y}(\xi, \eta, \zeta) &= \xi y(\eta, \zeta) \\
\hat{z}(\xi, \eta, \zeta) &= \xi z(\eta, \zeta)
\end{aligned}
\tag{15.16}
$$

applies with $\xi = 1$ on the boundary and $= 0$ at the scaling centre. The uniqueness of the transformation is guaranteed by the choice of the scaling centre, from which the total boundary must be visible.

To transform the differential operator $[L]$ in the $\hat{x}, \hat{y}, \hat{z}$ coordinate system (Equation (15.8)) to that in the ξ, η, ζ coordinate system, the Jacobian matrix is required

$$[\hat{J}(\xi, \eta, \zeta)] = \begin{bmatrix} \hat{x}_{,\xi} & \hat{y}_{,\xi} & \hat{z}_{,\xi} \\ \hat{x}_{,\eta} & \hat{y}_{,\eta} & \hat{z}_{,\eta} \\ \hat{x}_{,\zeta} & \hat{y}_{,\zeta} & \hat{z}_{,\zeta} \end{bmatrix} \tag{15.17}$$

The partial derivatives are calculated using Equations (15.16) and (15.13), leading, for instance, to

$$\hat{x}_{,\xi} = x = [N]\{x\} \tag{15.18a}$$

$$\hat{x}_{,\eta} = \xi x_{,\eta} = \xi[N_{,\eta}]\{x\} \tag{15.18b}$$

$$\hat{x}_{,\zeta} = \xi x_{,\zeta} = \xi[N_{,\zeta}]\{x\} \tag{15.18c}$$

Substituting into Equation (15.17) yields, with the diagonal matrix $\lceil 1\xi \ \xi \rfloor$,

$$[\hat{J}(\xi,\eta,\zeta)] = \lceil 1 \quad \xi \quad \xi \rfloor [J(\eta, \zeta)] \tag{15.19}$$

with the abbreviation

$$[J(\eta, \zeta)] = \begin{bmatrix} x & y & z \\ x_{,\eta} & y_{,\eta} & z_{,\eta} \\ x_{,\zeta} & y_{,\zeta} & z_{,\zeta} \end{bmatrix} \tag{15.20}$$

The arguments η and ζ are omitted for conciseness. $[J]$ depends upon the geometry of the boundary only. Its determinant is equal to

$$|J| = x(y_{,\eta}z_{,\zeta} - z_{,\eta}y_{,\zeta}) + y(z_{,\eta}x_{,\zeta} - x_{,\eta}z_{,\zeta}) + z(x_{,\eta}y_{,\zeta} - y_{,\eta}x_{,\zeta}) \tag{15.21}$$

$[J]$ and $|J|$ can be non-dimensionalised with a characteristic length r_0 with respect to the boundary as

$$[J] = r_0[\bar{J}] \tag{15.22a}$$

$$|J| = r_0^3|\bar{J}| \tag{15.22b}$$

where $[\bar{J}]$ and $|\bar{J}|$ are dimensionless. The inverse of the matrix $[J]$ is written as

$$[J]^{-1} = \frac{1}{|J|} \begin{bmatrix} y_{,\eta}z_{,\zeta} - z_{,\eta}y_{,\zeta} & zy_{,\zeta} - yz_{,\zeta} & yz_{,\eta} - zy_{,\eta} \\ z_{,\eta}x_{,\zeta} - x_{,\eta}z_{,\zeta} & xz_{,\zeta} - zx_{,\zeta} & zx_{,\eta} - xz_{,\eta} \\ x_{,\eta}y_{,\zeta} - y_{,\eta}x_{,\zeta} & yx_{,\zeta} - xy_{,\zeta} & xy_{,\eta} - yx_{,\eta} \end{bmatrix} \tag{15.23}$$

The coefficients in Equation (15.23) satisfy the identities

$$
\begin{aligned}
(zy_{,\zeta} - yz_{,\zeta})_{,\eta} + (yz_{,\eta} - zy_{,\eta})_{,\zeta} &= -2(y_{,\eta}z_{,\zeta} - z_{,\eta}y_{,\zeta}) \\
(xz_{,\zeta} - zx_{,\zeta})_{,\eta} + (zx_{,\eta} - xz_{,\eta})_{,\zeta} &= -2(z_{,\eta}x_{,\zeta} - x_{,\eta}z_{,\zeta}) \\
(yx_{,\zeta} - xy_{,\zeta})_{,\eta} + (xy_{,\eta} - yx_{,\eta})_{,\zeta} &= -2(x_{,\eta}y_{,\zeta} - y_{,\eta}x_{,\zeta})
\end{aligned} \tag{15.24}
$$

The derivatives with respect to $\hat{x}, \hat{y}, \hat{z}$ are transformed to those with respect to ξ, η, ζ using Equation (15.19)

$$
\begin{Bmatrix} \frac{\partial}{\partial \hat{x}} \\ \frac{\partial}{\partial \hat{y}} \\ \frac{\partial}{\partial \hat{z}} \end{Bmatrix} = [\hat{J}]^{-1} \begin{Bmatrix} \frac{\partial}{\partial \xi} \\ \frac{\partial}{\partial \eta} \\ \frac{\partial}{\partial \zeta} \end{Bmatrix} = [J]^{-1} \begin{Bmatrix} \frac{\partial}{\partial \xi} \\ \frac{1}{\xi}\frac{\partial}{\partial \eta} \\ \frac{1}{\xi}\frac{\partial}{\partial \zeta} \end{Bmatrix} \tag{15.25}
$$

The geometrical interpretation of the coordinate transformation is discussed (Figure 15.2). The derivatives of the position vector of a point on the boundary (Equation (15.11)) with respect to η and ζ represent two tangential vectors

$$\boldsymbol{r}_{,\eta} = x_{,\eta}\mathbf{i} + y_{,\eta}\mathbf{j} + z_{,\eta}\mathbf{k} \tag{15.26a}$$

$$\boldsymbol{r}_{,\zeta} = x_{,\zeta}\mathbf{i} + y_{,\zeta}\mathbf{j} + z_{,\zeta}\mathbf{k} \tag{15.26b}$$

with ζ and η, respectively, being constant. The three vectors $\mathbf{r}$, $\mathbf{r}_{,\eta}$, $\mathbf{r}_{,\zeta}$ form the base of the transformed coordinate system of the finite element. Equation (15.21) is equivalent to

$$|J| = \boldsymbol{r} \cdot (\boldsymbol{r}_{,\eta} \times \boldsymbol{r}_{,\zeta}) \tag{15.27}$$

The infinitesimal volume $\mathrm{d}V$ for any ξ is calculated as

$$\mathrm{d}V = \hat{\boldsymbol{r}}_{,\xi} \cdot (\hat{\boldsymbol{r}}_{,\eta} \times \hat{\boldsymbol{r}}_{,\zeta})\mathrm{d}\xi\, \mathrm{d}\eta\, \mathrm{d}\zeta \tag{15.28}$$

Using Equations (15.15) and (15.27) leads to

$$\mathrm{d}V = \xi^2 |J| \mathrm{d}\xi \mathrm{d}\eta \mathrm{d}\zeta \tag{15.29}$$

The outward normal vectors to the surfaces (η, ζ), (ζ, ξ) and (ξ, η), where the coordinates ξ, η and ζ, respectively, are constant, are equal on the boundary ($\xi = 1$) to

$$\boldsymbol{g}^{\xi} = \boldsymbol{r}_{,\eta} \times \boldsymbol{r}_{,\zeta} = (y_{,\eta}z_{,\zeta} - z_{,\eta}y_{,\zeta})\mathbf{i} + (z_{,\eta}x_{,\zeta} - x_{,\eta}z_{,\zeta})\mathbf{j} + (x_{,\eta}y_{,\zeta} - y_{,\eta}x_{,\zeta})\mathbf{k} \tag{15.30a}$$

$$\boldsymbol{g}^{\eta} = \boldsymbol{r}_{,\zeta} \times \boldsymbol{r} = (zy_{,\zeta} - yz_{,\zeta})\mathbf{i} + (xz_{,\zeta} - zx_{,\zeta})\mathbf{j} + (yx_{,\zeta} - xy_{,\zeta})\mathbf{k} \tag{15.30b}$$

$$\boldsymbol{g}^{\zeta} = \boldsymbol{r} \times \boldsymbol{r}_{,\eta} = (yz_{,\eta} - zy_{,\eta})\mathbf{i} + (zx_{,\eta} - xz_{,\eta})\mathbf{j} + (xy_{,\eta} - yx_{,\eta})\mathbf{k} \tag{15.30c}$$

The magnitude of a vector is denoted with the same symbol omitting the arrow sign (e.g. $g^\xi = |\boldsymbol{g}^\xi|$).

The transformation of infinitesimal surfaces for any ξ between the two coordinate systems is specified using Equations (15.15) and (15.30) as

$$\mathrm{d}S^\xi = |\hat{\boldsymbol{r}}_{,\eta} \times \hat{\boldsymbol{r}}_{,\zeta}|\mathrm{d}\eta\ \mathrm{d}\zeta = |\xi\boldsymbol{r}_{,\eta} \times \xi\boldsymbol{r}_{,\zeta}|\mathrm{d}\eta\ \mathrm{d}\zeta = \xi^2 g^\xi \mathrm{d}\eta\ \mathrm{d}\zeta \tag{15.31a}$$

$$\mathrm{d}S^\eta = |\hat{\boldsymbol{r}}_{,\zeta} \times \hat{\boldsymbol{r}}_{,\xi}|\mathrm{d}\zeta\ \mathrm{d}\xi = |\xi\boldsymbol{r}_{,\zeta} \times \boldsymbol{r}|\mathrm{d}\zeta\ \mathrm{d}\xi = \xi g^\eta \mathrm{d}\zeta\ \mathrm{d}\xi \tag{15.31b}$$

$$\mathrm{d}S^\zeta = |\hat{\boldsymbol{r}}_{,\xi} \times \hat{\boldsymbol{r}}_{,\eta}|\mathrm{d}\xi\ \mathrm{d}\eta = |\boldsymbol{r} \times \xi\boldsymbol{r}_{,\eta}|\mathrm{d}\xi\ \mathrm{d}\eta = \xi g^\zeta \mathrm{d}\xi\ \mathrm{d}\eta \tag{15.31c}$$

The unit normal vectors of the three surfaces (Figure 15.2) follow from Equation (15.30) as

$$\boldsymbol{n}^\xi = n_x^\xi \mathbf{i} + n_y^\xi \mathbf{j} + n_z^\xi \mathbf{k} = \frac{\boldsymbol{g}^\xi}{g^\xi} \tag{15.32a}$$

$$\boldsymbol{n}^\eta = n_x^\eta \mathbf{i} + n_y^\eta \mathbf{j} + n_z^\eta \mathbf{k} = \frac{\boldsymbol{g}^\eta}{g^\eta} \tag{15.32b}$$

$$\boldsymbol{n}^\zeta = n_x^\zeta \mathbf{i} + n_y^\zeta \mathbf{j} + n_z^\zeta \mathbf{k} = \frac{\boldsymbol{g}^\zeta}{g^\zeta} \tag{15.32c}$$

Equations (15.23), (15.30) and (15.32) yield

$$[J]^{-1} = \frac{1}{|J|}\begin{bmatrix} g^\xi n_x^\xi & g^\eta n_x^\eta & g^\zeta n_x^\zeta \\ g^\xi n_y^\xi & g^\eta n_y^\eta & g^\zeta n_y^\zeta \\ g^\xi n_z^\xi & g^\eta n_z^\eta & g^\zeta n_z^\zeta \end{bmatrix} \tag{15.33}$$

Substituting Equation (15.33) into Equation (15.25) results in

$$\begin{Bmatrix} \frac{\partial}{\partial \hat{x}} \\ \frac{\partial}{\partial \hat{y}} \\ \frac{\partial}{\partial \hat{z}} \end{Bmatrix} = \frac{g^\xi}{|J|}\{n^\xi\}\frac{\partial}{\partial \xi} + \frac{1}{\xi}\left(\frac{g^\eta}{|J|}\{n^\eta\}\frac{\partial}{\partial \eta} + \frac{g^\zeta}{|J|}\{n^\zeta\}\frac{\partial}{\partial \zeta}\right) \tag{15.34}$$

The partial derivatives with respect to $\hat{x}, \hat{y}, \hat{z}$ are expressed as a function of those with respect to ξ, η, ζ.

Thirdly, to derive the governing equations of elastodynamics with geometry in the transformed coordinate system, for the governing equations of elastodynamics in Cartesian coordinates, the scaled boundary transformation is applied to the geometry of the domain. Only the spatial coordinates are affected. The components of the displacements, strains and stresses are still defined in the original coordinate system $\hat{x}, \hat{y}, \hat{z}$. This is analogous to the procedure of mapping parent elements to curvilinear elements in the finite element method.

Thus, only the differential operator $[L]$ in Equation (15.8) needs to be addressed. Substituting Equation (15.34) into Equation (15.8) yields

$$[L] = [b^1]\frac{\partial}{\partial \xi} + \frac{1}{\xi}\left([b^2]\frac{\partial}{\partial \eta} + [b^3]\frac{\partial}{\partial \zeta}\right) \tag{15.35}$$

with

$$[b^1] = \frac{g^\xi}{|J|}\begin{bmatrix} n_x^\xi & 0 & 0 \\ 0 & n_y^\xi & 0 \\ 0 & 0 & n_z^\xi \\ 0 & n_z^\xi & n_y^\xi \\ n_z^\xi & 0 & n_x^\xi \\ n_y^\xi & n_x^\xi & 0 \end{bmatrix} \tag{15.36a}$$

$$[b^2] = \frac{g^\eta}{|J|}\begin{bmatrix} n_x^\eta & 0 & 0 \\ 0 & n_y^\eta & 0 \\ 0 & 0 & n_z^\eta \\ 0 & n_z^\eta & n_y^\eta \\ n_z^\eta & 0 & n_x^\eta \\ n_y^\eta & n_x^\eta & 0 \end{bmatrix} \tag{15.36b}$$

$$[b^3] = \frac{g^\zeta}{|J|}\begin{bmatrix} n_x^\zeta & 0 & 0 \\ 0 & n_y^\zeta & 0 \\ 0 & 0 & n_z^\zeta \\ 0 & n_z^\zeta & n_y^\zeta \\ n_z^\zeta & 0 & n_x^\zeta \\ n_y^\zeta & n_x^\zeta & 0 \end{bmatrix} \tag{15.36c}$$

Note that $[b^1]$, $[b^2]$ and $[b^3]$ are independent of ξ. The following identity can be verified by substituting Equations (15.36), (15.32), (15.30) and (15.24)

$$(|J|[b^2])_{,\eta} + (|J|[b^3])_{,\zeta} = -2|J|[b^1] \tag{15.37}$$

The amplitudes of the surface tractions $\{t\} = [t_x \ t_y \ t_z]^T$ on any boundary with the outward unit normal vector $\boldsymbol{n} = n_x\mathbf{i} + n_y\mathbf{j} + n_z\mathbf{k}$ equal

$$\{t\} = \begin{bmatrix} n_x & 0 & 0 & 0 & n_z & n_y \\ 0 & n_y & 0 & n_z & 0 & n_x \\ 0 & 0 & n_z & n_y & n_x & 0 \end{bmatrix}\{\sigma\} \tag{15.38}$$

By comparing Equations (15.38) and (15.36), the traction amplitudes on the surfaces $(\eta, \zeta), (\zeta, \xi)$ and (ξ, η), where the coordinates ξ, η and ζ, respectively, are constant, are written as

$$\{t^\xi\} = \frac{|J|}{g^\xi}[b^1]^T\{\sigma\} \tag{15.39a}$$

$$\{t^\eta\} = \frac{|J|}{g^\eta}[b^2]^T\{\sigma\} \tag{15.39b}$$

$$\{t^\zeta\} = \frac{|J|}{g^\zeta}[b^3]^T\{\sigma\} \tag{15.39c}$$

15.3 METHOD OF WEIGHTED RESIDUALS

The second step in deriving the fundamental equations consists in deriving the weak form of the differential equations of motion in the scaled boundary coordinate system starting from the strong form. This will lead to the so-called scaled boundary finite element equation in displacement.

The differential equations of motion in the frequency domain (equilibrium equations) with the geometry in the scaled boundary coordinate system ξ, η, ζ are specified in Equation (15.5) with $[L]$ in Equation (15.35). These partial differential equations are exact, representing the strong form. In the two circumferential directions η, ζ the method of weighted residuals is applied, leading to the weak form. As explained in section 1.3 for Laplace's equation from Equation (1.21) onwards, the equilibrium equations without any boundary conditions are multiplied by the weighting functions. After performing integration by parts, the boundary conditions can then be introduced.

The pyramid with one finite element on the boundary forming its base (Figure 3.1) is addressed. The superscript e is dropped for conciseness. To derive a finite element approximation, Equation (15.5), after substituting Equation (15.35) is multiplied by a weighting function $\{w\} = \{w(\xi, \eta, \zeta)\}$ and integrated over the domain of the pyramid

$$\int_V \{w\}^T[b^1]^T\{\sigma_{,\xi}\}\,dV + \int_V \{w\}^T\frac{1}{\xi}([b^2]^T\{\sigma_{,\eta}\}+[b^3]^T\{\sigma_{,\zeta}\})\,dV + \int_V \{w\}^T\{p\}\,dV + \omega^2\int_V \{w\}^T\rho\{u\}\,dV = 0 \tag{15.40}$$

The second term of Equation (15.40) denoted as I is examined. Substituting Equation (15.29) yields

$$I = \int_0^1 \xi\left(\int_{S^\xi} \{w\}^T|J|([b^2]^T\{\sigma_{,\eta}\} + [b^3]^T\{\sigma_{,\zeta}\})d\eta d\zeta\right) d\xi \tag{15.41}$$

where S^ξ denotes the surface with a constant ξ. Applying the integration-by-parts formula to the surface integral results in

$$I = \int_0^1 \xi\left(\oint_{\Gamma^\xi} \{w\}^T(|J|[b^3]^T\{\sigma\}d\eta + |J|[b^2]^T\{\sigma\}d\zeta) - \int_{S^\xi} ((\{w\}^T|J|[b^2]^T)_{,\eta} + (\{w\}^T|J|[b^3]^T)_{,\zeta})\{\sigma\}d\eta d\zeta\right) d\xi \tag{15.42}$$

with the contour Γ^{ξ} (resulting from scaling the edge of the surface finite element) of S^{ξ}. Substituting Equations (15.39b) and (15.39c) into Equation (15.42) and using Equation (15.37) leads to

$$I = \int_0^1 \xi \left(\oint_{\Gamma^{\xi}} \{w\}^T(\{t^{\zeta}\}g^{\zeta}\mathrm{d}\eta + \{t^{\eta}\}g^{\eta}\mathrm{d}\zeta) \right.$$
$$\left. - \int_{S^{\xi}} (-2\{w\}^T[b^1]^T + \{w_{,\eta}\}^T[b^2]^T + \{w_{,\zeta}\}^T[b^3]^T)\{\sigma\}|\mathrm{J}|\mathrm{d}\eta\mathrm{d}\zeta \right) \mathrm{d}\xi \qquad (15.43)$$

Substituting Equation (15.29) in the first, third and fourth terms in Equation (15.40), and using Equation (15.43), yields

$$\int_0^1 \left(\xi^2 \int_{S^{\xi}} \{w\}^T[b^1]^T\{\sigma_{,\xi}\}|J|\mathrm{d}\eta\mathrm{d}\zeta + \xi \oint_{\Gamma^{\xi}} \{w\}^T(\{t^{\zeta}\}g^{\zeta}\mathrm{d}\eta + \{t^{\eta}\}g^{\eta}\mathrm{d}\zeta) \right.$$
$$- \xi \int_{S^{\xi}} (-2\{w\}^T[b^1]^T + \{w_{,\eta}\}^T[b^2]^T + \{w_{,\zeta}\}^T[b^3]^T)\{\sigma\}|J|\mathrm{d}\eta\mathrm{d}\zeta$$
$$\left. +\xi^2 \int_{S^{\xi}} \{w\}^T\{p\}|J|\mathrm{d}\eta\mathrm{d}\zeta + \omega^2\xi^2 \int_{S^{\xi}} \{w\}^T\rho\{u\}|J|\mathrm{d}\eta\mathrm{d}\zeta \right) \mathrm{d}\xi = 0 \qquad (15.44)$$

Equation (15.44) is satisfied by setting the integrand of the integral over ξ equal to zero. This corresponds to *enforcing the equation of motion* Equation (15.5) with the differential operator $[L]$ (Equation (15.35)) *exactly in the* ξ *direction*

$$\xi^2 \int_{S^{\xi}} \{w\}^T[b^1]^T\{\sigma_{,\xi}\}|J|\mathrm{d}\eta\mathrm{d}\zeta + \xi \oint_{\Gamma^{\xi}} \{w\}^T(\{t^{\zeta}\}g^{\zeta}\mathrm{d}\eta + \{t^{\eta}\}g^{\eta}\mathrm{d}\zeta)$$
$$- \xi \int_{S^{\xi}} (-2\{w\}^T[b^1]^T + \{w_{,\eta}\}^T[b^2]^T + \{w_{,\zeta}\}^T[b^3]^T)\{\sigma\}|J|\mathrm{d}\eta\mathrm{d}\zeta$$
$$+\xi^2 \int_{S^{\xi}} \{w\}^T\{p\}|J|\mathrm{d}\eta\mathrm{d}\zeta + \omega^2\xi^2 \int_{S^{\xi}} \{w\}^T\rho\{u\}|J|\mathrm{d}\eta\mathrm{d}\zeta = 0 \qquad (15.45)$$

Note that no volume integrals are present in Equation (15.45).

The displacement amplitudes of the finite element on the boundary ($\xi = 1$) are interpolated using shape functions $[N(\eta, \zeta)]$, which for the sake of simplicity, are selected as equal to the mapping functions (Equation (15.14)), expanded and reordered correspondingly. It is postulated that the same shape functions apply with the displacement amplitudes $\{u(\xi)\}$ for all surfaces S^{ξ} with a constant ξ (see also Equation (15.3)):

$$\{u(\xi,\eta,\zeta)\} = [N(\eta,\zeta)]\{u(\xi)\} \tag{15.46}$$

The stress amplitudes are calculated substituting Equations (15.46) and (15.35) into Equation (15.7) using Equation (15.6):

$$\{\sigma\} = \{\sigma(\xi,\eta,\zeta)\} = [D]\left([B^1]\{u(\xi)\}_{,\xi} + \frac{1}{\xi}[B^2]\{u(\xi)\}\right) \tag{15.47}$$

with

$$[B^1] = [b^1][N(\eta,\zeta)] \tag{15.48a}$$

$$[B^2] = [b^2][N(\eta,\zeta)]_{,\eta} + [b^3][N(\eta,\zeta)]_{,\zeta} \tag{15.48b}$$

$[B^1]$ and $[B^2]$ are independent of ξ. The elasticity matrix $[D(\eta,\zeta)]$ can vary in the circumferential directions η,ζ. The displacement discretisation also applies to the weighting function $\{w\} = \{w(\xi,\eta,\zeta)\}$, as in standard finite elements

$$\{w(\xi,\eta,\zeta)\} = [N(\eta,\zeta)]\{w(\xi)\} \tag{15.49}$$

As in Equation (15.44) the integrand in the integral over ξ is set equal to zero, the weighting function only 'acts' in the circumferential directions η,ζ.

Substituting Equation (15.49) into Equation (15.45) for arbitrary $\{w(\xi)\}$ yields

$$\xi^2\int_{S^\xi}[B^1]^T\{\sigma_{,\xi}\}|J|\mathrm{d}\eta\mathrm{d}\zeta - \xi\int_{S^\xi}(-2[B^1]^T + [B^2]^T)\{\sigma\}|J|\mathrm{d}\eta\mathrm{d}\zeta$$
$$+\,\omega^2\xi^2\int_{S^\xi}[N(\eta,\zeta)]^T\rho\{u\}|J|\mathrm{d}\eta\mathrm{d}\zeta + \xi\{F^{\mathrm{t}}(\xi)\} + \xi^2\{F^{\mathrm{b}}(\xi)\} = 0 \tag{15.50}$$

with

$$\{F^{\mathrm{t}}(\xi)\} = \oint_{\Gamma^\xi}[N(\eta,\zeta)]^T(\{t^\zeta\}g^\zeta\mathrm{d}\eta + \{t^\eta\}g^\eta\mathrm{d}\zeta) \tag{15.51}$$

$$\{F^{\mathrm{b}}(\xi)\} = \int_{S^\xi}[N(\eta,\zeta)]^T\{p\}|J|\mathrm{d}\eta\mathrm{d}\zeta \tag{15.52}$$

$\{F^{\mathrm{t}}\}$ corresponds to the amplitudes of the nodal forces resulting from the surface tractions acting on those surfaces A which pass through the scaling centre (Figure 3.1a) and $\{F^{\mathrm{b}}\}$ from the body loads. Substituting the stress amplitudes of Equation (15.47) into Equation (15.50) results in the differential equations expressed in displacement amplitudes

$$\xi^2 \int_{S^\xi} [B^1]^T[D]\left([B^1]\{u(\xi)\}_{,\xi\xi} + \frac{1}{\xi}[B^2]\{u(\xi)\}_{,\xi} - \frac{1}{\xi^2}[B^2]\{u(\xi)\}\right)|J|\mathrm{d}\eta\mathrm{d}\zeta$$
$$-\xi \int_{S^\xi} (-2[B^1]^T + [B^2]^T)[D]\left([B^1]\{u(\xi)\}_{,\xi} + \frac{1}{\xi}[B^2]\{u(\xi)\}\right)|J|\mathrm{d}\eta\mathrm{d}\zeta$$
$$+\omega^2\xi^2 \int_{S^\xi} [N(\eta,\zeta)]^T \rho[N(\eta,\zeta)]\{u(\xi)\}|J|\mathrm{d}\eta\mathrm{d}\zeta + \xi\{F^{\mathrm{t}}\} + \xi^2\{F^{\mathrm{b}}\} = 0 \quad (15.53)$$

Introducing the coefficient matrices

$$[E^0] = \int_{S^\xi} [B^1]^T[D][B^1]|J|\mathrm{d}\eta\mathrm{d}\zeta \quad (15.54a)$$

$$[E^1] = \int_{S^\xi} [B^2]^T[D][B^1]|J|\mathrm{d}\eta\mathrm{d}\zeta \quad (15.54b)$$

$$[E^2] = \int_{S^\xi} [B^2]^T[D][B^2]|J|\mathrm{d}\eta\mathrm{d}\zeta \quad (15.54c)$$

$$[M^0] = \int_{S^\xi} [N(\eta,\zeta)]^T\rho[N(\eta,\zeta)]|J|\mathrm{d}\eta\mathrm{d}\zeta \quad (15.55)$$

and

$$\{F(\xi)\} = \xi\{F^{\mathrm{t}}\} + \xi^2\{F^{\mathrm{b}}\} \quad (15.56)$$

yields

$$[E^0]\xi^2\{u(\xi)\}_{,\xi\xi} + (2[E^0] - [E^1] + [E^1]^T)\xi\{u(\xi)\}_{,\xi}$$
$$+([E^1]^T - [E^2])\{u(\xi)\} + \omega^2[M^0]\xi^2\{u(\xi)\} + \{F(\xi)\} = 0 \quad (15.57)$$

Note that the coefficient matrices $[E^0], [E^1], [E^2], [M^0]$ are independent of ξ. Integrations over the surface finite element (at $\xi = 1$) only are involved. For later use, the coefficient matrices are non-dimensionalised

$$[E^0] = Gr_0[\bar{E}^0] \quad (15.58a)$$

$$[E^1] = Gr_0[\bar{E}^1] \quad (15.58b)$$

$$[E^2] = Gr_0[\bar{E}^2] \quad (15.58c)$$

and

$$[M^0] = \rho r_0^3[\bar{M}^0] \quad (15.59)$$

with the characteristic length r_0 with respect to the boundary and shear modulus G.

Equation (15.57) applies to the domain of the pyramid corresponding to one finite element on the boundary. To model the total domain, an assemblage as in the conventional finite element method is performed and the (homogeneous) displacement boundary conditions on A_u are enforced (Equation (15.10a), Figure 3.1a). To simplify the nomenclature, the same symbols are used for the assembled coefficient matrices and the assembled displacement amplitudes in the following. In the assemblage process the amplitudes of the nodal forces $\{F^t\}$ in Equation (15.51) will cancel with the exception of those on the part A_t of the total boundary passing through the scaling centre with prescribed surface tractions (Figure 3.1b). The assemblage process yields

$$\begin{aligned}&[E^0]\xi^2\{u(\xi)\}_{,\xi\xi} + (2[E^0] - [E^1] + [E^1]^T)\xi\{u(\xi)\}_{,\xi}\\&\quad + ([E^1]^T - [E^2])\{u(\xi)\} + \omega^2[M^0]\xi^2\{u(\xi)\} + \{F(\xi)\} = 0\end{aligned} \tag{15.60}$$

Equation (15.60) represents the *scaled boundary finite element equation in displacement* formulated in the frequency domain for elastodynamics in three dimensions. It is a system of linear second-order ordinary differential equations for the displacement amplitudes $\{u(\xi)\}$ with the dimensionless radial coordinate ξ as the independent variable. As already pointed out, the coefficient matrices $[E^0], [E^1], [E^2], [M^0]$ are independent of ξ. In the case of prescribed surface tractions on the side-face A_t and body loads, the system of differential equations is non-homogeneous.

Although the derivation is performed for a bounded medium ($0 \leq \xi \leq 1$) with the integrations over ξ in the weighted-residual statement running from 0 to 1 in Equation (15.44), it also applies for an unbounded medium ($1 \leq \xi < \infty$) with integrations from 1 towards ∞ (Figure 3.4). Thus, the scaled boundary finite element equation in displacement (Equation (15.60)) is also valid for an unbounded medium with the same definitions for the coefficient matrices $[E^0], [E^1], [E^2]$, (Equation 15.54), $[M^0]$ (Equation (15.55)).

It can be shown (section 5.2.1 of [W9]) that $[E^0]$ is positive definite. It follows from the construction of $[E^2]$ (Equation (15.54c) with a positive definite elasticity matrix $[D]$) that $[E^2]$ is semi-positive definite. Introducing the displacement boundary condition on A_u can result in a positive definite $[E^2]$.

As the last item in this section, the procedure to incorporate non-homogeneous displacement boundary conditions $\{\bar{u}\}$ on the side-faces A_u (Equation (15.10a)) is discussed. For homogeneous boundary conditions $\{u(\xi)\} = \{\bar{u}\} = 0$ for the degrees of freedom in the nodes on A_u is enforced when assembling the pyramids. The differential equations corresponding to the degrees of freedom on A_u are not present in the assembled system. (They could be used to calculate the corresponding nodal reaction forces.) For non-homogeneous boundary conditions $\{u(\xi)\} = \{\bar{u}(\xi)\}$ for the degrees of freedom on A_u the procedure is analogous. Formulating Equation (15.60) for all degrees of freedom which are not on A_u yields

$$\begin{aligned}&[E^0]\xi^2\{u(\xi)\}_{,\xi\xi} + [E^0_u]\xi^2\{\bar{u}(\xi)\}_{,\xi\xi}\\&+ (2[E^0] - [E^1] + [E^1]^T)\xi\{u(\xi)\}_{,\xi} + (2[E^0_u] - [E^1_u] + [E^1_u]^T)\xi\{\bar{u}(\xi)\}_{,\xi}\\&+ ([E^1]^T - [E^2])\{u(\xi)\} + ([E^1_u]^T - [E^2_u])\{\bar{u}(\xi)\}\\&+ \omega^2[M^0]\xi^2\{u(\xi)\} + \omega^2[M^0_u]\xi^2\{\bar{u}(\xi)\} + \{F(\xi)\} = 0\end{aligned} \tag{15.61}$$

$[E_u^0], [E_u^1], [E_u^2], [M_u^0]$ are the "off-diagonal" coefficient matrices providing the coupling to the degrees of freedom on A_u. Rearranging Equation (15.61) yields

$$\begin{aligned}&[E^0]\xi^2\{u(\xi)\}_{,\xi\xi} + (2[E^0] - [E^1] + [E^1]^T)\xi\{u(\xi)\}_{,\xi} \\ &+ ([E^1]^T - [E^2])\{u(\xi)\} + \omega^2[M^0]\xi^2\{u(\xi)\} + \{F(\xi)\} + \{F_u(\xi)\} = 0\end{aligned} \tag{15.62}$$

with the amplitudes of the nodal forces resulting from the prescribed displacements on the side-faces A_u

$$\begin{aligned}\{F_u(\xi)\} &= [E_u^0](\xi^2\{\bar{u}(\xi)\}_{,\xi\xi} + 2\xi\{\bar{u}(\xi)\}_{,\xi}) \\ &- [E_u^1]\xi\{\bar{u}(\xi)\}_{,\xi} + [E_u^1]^T(\xi\{\bar{u}(\xi)\}_{,\xi} + \{\bar{u}(\xi)\}) \\ &- [E_u^2]\{\bar{u}(\xi)\} + \omega^2[M_u^0]\xi^2\{\bar{u}(\xi)\}\end{aligned} \tag{15.63}$$

Obviously, this procedure is analogous to that used in the finite element method to incorporate prescribed displacements.

Initial strains and initial stresses can also be incorporated in the formulation straightforwardly. An efficient procedure addressing the initial stress state making use of the salient features of the scaled boundary finite element method is discussed in section 16.5.

This section addresses the three-dimensional vector-wave equation, i.e. in each node three displacement components occur. For the scalar wave equation with a single displacement in each node, simplifications in certain abbreviations arise. As far as the spatial discretisation is concerned, the equations are the same as for the diffusion equation (section 15.6). The reader should consult [W9] or [S10] for easy reference. References [S10] and [W9] also contain examples of the abbreviations and coefficient matrices for vector and scalar wave equations.

15.4 DYNAMIC-STIFFNESS MATRIX

As a third and final step in deriving the fundamental equations, the nodal force-displacement relationship is addressed. This will lead to the scaled boundary finite element equation in dynamic stiffness.

To determine the nodal force-(nodal) displacement relationship as in standard finite elements, the amplitudes of the internal nodal forces $\{Q(\xi)\}$, which are equal to the resultants of the surface tractions $\{t^\xi\}$ on a line with a constant ξ, are addressed. For any surface S^ξ with a constant ξ, the virtual work is formulated as

$$\{w(\xi)\}^T\{Q(\xi)\} = \int_{S^\xi} \{w\}^T\{t^\xi\}\mathrm{d}S^\xi \tag{15.64}$$

Substituting the weighting function $\{w\} = \{w(\xi, \eta, \zeta)\}$ (Equation (15.49)) and the infinitesimal surface $\mathrm{d}S^\xi$ (Equation (15.31a)) yields, for an arbitrary $\{w(\xi)\}$,

$$\{Q(\xi)\} = \int_{S^\xi} [N(\eta, \zeta)]^T\{t^\xi\}\xi^2 g^\xi \mathrm{d}\eta\mathrm{d}\zeta \tag{15.65}$$

Substituting Equation (15.47) into Equation (15.39a) and then in Equation (15.65) and using Equation (15.48) leads to

$$\{Q(\xi)\} = \int_{S^\xi} [B^1]^T [D]\left([B^1]\{u(\xi)\}_{,\xi} + \frac{1}{\xi}[B^2]\{u(\xi)\}\right)\xi^2 |J| \mathrm{d}\eta \mathrm{d}\zeta \tag{15.66}$$

Introducing $[E^0]$, $[E^1]$ (Equations (15.54a), (15.54b)) results in

$$\{Q(\xi)\} = [E^0]\xi^2\{u(\xi)\}_{,\xi} + [E^1]^T\xi\{u(\xi)\} \tag{15.67}$$

For a bounded medium (Figure 3.1), S^ξ is a positive force, i.e. $\{n^\xi\}$ (Figure 15.2) is the unit outward normal. For an unbounded medium (Figure 3.4), the opposite applies. The two cases can be unified as

$$\{R(\xi)\} = \pm\{Q(\xi)\} \tag{15.68}$$

with the upper and lower signs corresponding to the bounded and unbounded media, respectively. In the frequency domain, the amplitudes of the displacements $\{u(\xi)\}$ are related to those of the nodal forces $\{R(\xi)\}$ as

$$\{R(\xi)\} = [S(\omega, \xi)]\{u(\xi)\} - \{R^{\mathrm{F}}(\xi)\} \tag{15.69}$$

$[S(\omega, \xi)]$ denotes the dynamic-stiffness matrix on a line with a constant ξ and $\{R^{\mathrm{F}}(\xi)]$ represents the amplitudes of the nodal loads due to the body load and the surface traction. Substituting Equations (15.67) and (15.69) into Equation (15.68) yields

$$\pm[S(\omega, \xi)]\{u(\xi)\} \mp \{R^{\mathrm{F}}(\xi)\} = [E^0]\xi^2\{u(\xi)\}_{,\xi} + [E^1]^T\xi\{u(\xi)\} \tag{15.70}$$

From Equations (15.70) and (15.60), the equation for $[S(\omega, \xi)]$ can be derived. Differentiating Equation (15.70) with respect to ξ leads to

$$\begin{aligned} &\pm [S(\omega, \xi)]_{,\xi}\{u(\xi)\} \pm [S(\omega, \xi)]\{u(\xi)\}_{,\xi} \mp \{R^{\mathrm{F}}(\xi)\}_{,\xi} \\ &- [E^0]\xi^2\{u(\xi)\}_{,\xi\xi} - (2[E^0] + [E^1]^T)\xi\{u(\xi)\}_{,\xi} - [E^1]^T\{u(\xi)\} = 0 \end{aligned} \tag{15.71}$$

Adding Equation (15.60) and Equation (15.71) results in

$$\begin{aligned} &\pm [S(\omega, \xi)]_{,\xi}\{u(\xi)\} + (\pm[S(\omega, \xi)] - \xi[E^1])\{u(\xi)\}_{,\xi} \\ &\mp \{R^{\mathrm{F}}(\xi)\}_{,\xi} - [E^2]\{u(\xi)\} + \omega^2[M^0]\xi^2\{u(\xi)\} + \{F(\xi)\} = 0 \end{aligned} \tag{15.72}$$

Solving Equation (15.70) for $\{u(\xi)\}_{,\xi}$ and substituting in Equation (15.72) multiplied by ξ yields

$$\begin{aligned} &((\pm[S(\omega, \xi)] - \xi[E^1])(\xi[E^0])^{-1}(\pm[S(\omega, \xi)] - \xi[E^1]^T) \\ &\pm\xi[S(\omega, \xi)]_{,\xi} - \xi[E^2] + \omega^2\xi^3[M^0])\{u(\xi)\} \\ &\mp \xi\{R^{\mathrm{F}}(\xi)\}_{,\xi} \mp (\pm[S(\omega, \xi)] - \xi[E^1])(\xi[E^0])^{-1}\{R^{\mathrm{F}}(\xi)\} + \xi\{F(\xi)\} = 0 \end{aligned} \tag{15.73}$$

For arbitrary $\{u(\xi)\}$, its coefficient matrix must vanish, leading to

$$(\pm[S(\omega,\xi)] - \xi[E^1])(\xi[E^0])^{-1}(\pm[S(\omega,\xi)] - \xi[E^1]^T) \pm\xi[S(\omega,\xi)]_{,\xi} - \xi[E^2] + \omega^2\xi^3[M^0] = 0 \tag{15.74a}$$

which will permit $[S(\omega, \xi)]$ to be determined (see below). In addition, the remaining part of Equation (15.73) leads to

$$\xi\{R^F(\xi)\}_{,\xi} + (\pm[S(\omega,\xi)] - \xi[E^1])(\xi[E^0])^{-1}\{R^F(\xi)\} = \pm\xi\{F(\xi)\} \tag{15.74b}$$

with $[S(\omega, \xi)]$ determined from Equation (15.74a). This is a system of *linear ordinary differential equations of first order for the amplitudes of the nodal load* $\{R^F(\xi)\}$. The boundary condition to be able to integrate these differential equations is $\{R^F(\xi = 0)\} = 0$ for a bounded medium and $\{R^F(\xi \to \infty)\} = 0$ for an unbounded medium.

Introducing the dimensionless, dynamic-stiffness matrix denoted with a bar

$$[S(\omega,\xi)] = Gr_0\xi[\bar{S}(\omega,\xi)] \tag{15.75}$$

and substituting Equations (15.58) and (15.59) into Equation (15.74a) leads to

$$(\pm[\bar{S}(\omega,\xi)] - [\bar{E}^1])[\bar{E}^0]^{-1}(\pm[\bar{S}(\omega,\xi)] - [\bar{E}^1]^T) - [\bar{E}^2] \pm [\bar{S}(\omega,\xi)] \pm \xi[\bar{S}(\omega,\xi)]_{,\xi} + \left(\frac{\omega r_0}{c_s}\xi\right)^2[\bar{M}^0] = 0 \tag{15.76}$$

with the shear-wave velocity $c_s = \sqrt{G/\rho}$. The coefficient of the last term defines the dimensionless frequency for any ξ

$$a(\omega,\xi) = \frac{\omega r_0}{c_s}\xi \tag{15.77}$$

The second-last term is written as

$$\xi[\bar{S}(\omega,\xi)]_{,\xi} = a[\bar{S}(\omega,\xi)]_{,a} \tag{15.78}$$

After substituting Equations (15.77) and (15.78) into Equation (15.76), the only independent variable in the equation is a. Thus, the dimensionless dynamic-stiffness matrix is a function of a only

$$[\bar{S}(\omega,\xi)] = [\bar{S}(a)] \tag{15.79}$$

The term with the derivative $a[\bar{S}(a)]_{,a}$ can be interpreted either for varying ξ with fixed ω, or for varying ω with fixed ξ:

$$a[\bar{S}(a)]_{,a} = \xi[\bar{S}(a)]_{,\xi} = \omega[\bar{S}(a)]_{,\omega} \tag{15.80}$$

Differentiating Equation (15.75) with respect to ξ, and using Equations (15.79) and (15.80) yields

$$\xi[S(\omega, \xi)]_{,\xi} = [S(\omega, \xi)] + \omega[S(\omega, \xi)]_{,\omega} \tag{15.81}$$

The spatial derivative can thus be replaced by the frequency derivative. Substituting Equation (15.81) into Equation (15.74a) results in

$$\begin{aligned}(\pm[S(\omega, \xi)] - \xi[E^1])(\xi[E^0])^{-1}(\pm[S(\omega, \xi)] - \xi[E^1]^T) - \xi[E^2] \\ + [\pm S(\omega, \xi)] \pm \omega[S(\omega, \xi)]_{,\omega} + \omega^2\xi^3[M^0] = 0\end{aligned} \tag{15.82}$$

For the boundary ($\xi = 1$), the dynamic-stiffness matrix is expressed as

$$(\pm[S(\omega)] - [E^1])[E^0]^1(\pm[S(\omega)] - [E^1]^T) - [E^2] \pm [S(\omega)] \pm \omega[S(\omega)]_{,\omega} + \omega^2[M^0] = 0 \tag{15.83}$$

For a bounded medium, the dynamic-stiffness matrix $[S^b(\omega)]$ (superscript b for bounded) is expressed as

$$\begin{aligned}([S^b(\omega)] - [E^1])[E^0]^{-1}([S^b(\omega)] - [E^1]^T) + [S^b(\omega)] \\ + \omega[S^b(\omega)]_{,\omega} - [E^2] + \omega^2[M^0] = 0\end{aligned} \tag{15.84}$$

and for an unbounded medium $[S^\infty(\omega)]$ (superscript ∞ for unbounded) as

$$\begin{aligned}([S^\infty(\omega)] + [E^1])[E^0]^{-1}([S^\infty(\omega)] + [E^1]^T) - [S^\infty(\omega)] \\ - \omega[S^\infty(\omega)]_{,\omega} - [E^2] + \omega^2[M^0] = 0\end{aligned} \tag{15.85}$$

These represent the *scaled boundary finite element equations in dynamic stiffness* formulated in the frequency domain for elastodynamics in three dimensions. They are nonlinear first-order ordinary differential equations, with the frequency ω as the independent variable.

For a bounded medium, $[S^b(\omega)]$ is usually not calculated. The static-stiffness matrix $[K^b]$ and the mass matrix $[M^b]$ are used, which follow from the low-frequency expansion of $[S^b(\omega)]$

$$[S^b(\omega)] = [K^b] - \omega^2[M^b] + O(\omega^4) \tag{15.86}$$

15.5 EQUATIONS FOR TWO-DIMENSIONAL ELASTODYNAMICS

The derivation of the scaled boundary finite element equations for two-dimensional elastodynamics closely follows that of the three-dimensional case. For easy reference, the key equations used for the implementation of the method are listed.

The differential equations of motion (Equation (15.5)) still hold for the two-dimensional case with displacement amplitudes $\{u\} = [u_x \; u_y]^T$ and stress amplitudes $\{\sigma\} = [\sigma_x \; \sigma_y \; \tau_{xy}]^T$, where the differential operator equals

$$[L] = \begin{bmatrix} \frac{\partial}{\partial \hat{x}} & \\ & \frac{\partial}{\partial \hat{y}} \\ \frac{\partial}{\partial \hat{y}} & \frac{\partial}{\partial \hat{x}} \end{bmatrix} \tag{15.87}$$

The scaled boundary transformation of the geometry is defined with the dimensionless radial coordinate ξ by

$$\begin{aligned} \hat{x}(\xi, \eta) &= \xi x(\eta) \\ \hat{y}(\xi, \eta) &= \xi y(\eta) \end{aligned} \tag{15.88}$$

where the coordinates on the boundary

$$\begin{aligned} x(\eta) &= [N(\eta)]\{x\} \\ y(\eta) &= [N(\eta)]\{y\} \end{aligned} \tag{15.89}$$

are represented by interpolating the nodal coordinates $\{x\}, \{y\}$ using the mapping functions (shape functions) $[N(\eta)]$. The curvilinear coordinate η in $[N(\eta)]$ equals $+1$ and -1 at the two ends of the line element.

The derivatives equal

$$\begin{aligned} x_{,\eta} &= [N(\eta)]_{,\eta}\{x\} \\ y_{,\eta} &= [N(\eta)]_{,\eta}\{y\} \end{aligned} \tag{15.90}$$

The determinant of a matrix related to the Jacobian matrix equals

$$|J| = xy_{,\eta} - yx_{,\eta} \tag{15.91}$$

The infinitesimal area in the scaled boundary coordinates equals

$$dV = \xi |J| d\xi d\eta \tag{15.92}$$

For elastodynamics, the amplitudes of the displacements in a point ξ, η on the surface element $\{u(\xi, \eta)\} = [u_x(\xi, \eta) u_y(\xi, \eta)]^T$ are expressed as

$$\{u(\xi, \eta)\} = [N(\eta)]\{u(\xi)\} \tag{15.93}$$

with the shape functions expanded and reordered. The stress amplitudes equal

$$\{\sigma(\xi, \eta)\} = [D]\left([B^1]\{u(\xi)\}_{,\xi} + \frac{1}{\xi}[B^2]\{u(\xi)\}\right) \tag{15.94}$$

with the matrices related to the strain displacement relationship

$$[B^1] = [b^1][N(\eta)] \tag{15.95a}$$

$$[B^2] = [b^2][N(\eta)]_{,\eta} \tag{15.95b}$$

which are calculated with

$$[b^1] = \frac{1}{|J|}\begin{bmatrix} y_{,\eta} & 0 \\ 0 & -x_{,\eta} \\ -x_{,\eta} & y_{,\eta} \end{bmatrix}$$

$$[b^2] = \frac{1}{|J|}\begin{bmatrix} -y & 0 \\ 0 & x \\ x & -y \end{bmatrix} \tag{15.96}$$

$[B^1]$ and $[B^2]$ are independent of ξ. The (anisotropic) two-dimensional elasticity matrix is denoted as $[D]$. The weighting function equals

$$\{w(\xi, \eta)\} = [N(\eta)]\{w(\xi)\} \tag{15.97}$$

The coefficients for a (curved) line finite element in the scaled boundary finite element method are specified in the same form as those for the three-dimensional case (Equations (15.54) and (15.55))

$$[E^0] = \int_{-1}^{+1} [B^1]^T[D][B^1]|J|d\eta \tag{15.98a}$$

$$[E^1] = \int_{-1}^{+1} [B^2]^T[D][B^1]|J|d\eta \tag{15.98b}$$

$$[E^2] = \int_{-1}^{+1} [B^2]^T[D][B^2]|J|d\eta \tag{15.98c}$$

and

$$[M^0] = \int_{-1}^{+1} [N(\eta)]^T\rho[N(\eta)]|J|d\eta \tag{15.99}$$

The contribution of the body loads is determined as follows. With $\{p(\xi, \eta)\} = \{p(\xi)\} = [p_x(\xi)p_y(\xi)]^T$

$$\{F^{\mathrm{b}}(\xi)\} = \int_{-1}^{+1} [N(\eta)]^T \{p(\xi)\}|J|\mathrm{d}\eta \tag{15.100}$$

applies. The prescribed surface tractions $[t_x(\xi) t_y(\xi)]^T$ on the part of the boundary A_t (Figures 3.1b and 3.4b) equal

$$\{F^{\mathrm{t}}(\xi)\} = \left\{ \begin{matrix} \left(\sqrt{x^2+y^2}\{t(\xi)\}\right) & |_{\eta=-1} \\ \left(\sqrt{x^2+y^2}\{t(\xi)\}\right) & |_{\eta=+1} \end{matrix} \right\} \tag{15.101}$$

Again,

$$\{F(\xi)\} = \xi^2\{F^{\mathrm{b}}(\xi)\} + \xi\{F^{\mathrm{t}}(\xi)\} \tag{15.102}$$

applies.

The scaled boundary finite element equation in displacement in two dimensions equals

$$\begin{aligned} &[E^0]\xi^2\{u(\xi)\}_{,\xi\xi} + ([E^0] - [E^1] + [E^1]^T)\xi\{u(\xi)\}_{,\xi} \\ &\quad - [E^2]\{u(\xi)\} + \omega^2[M^0]\xi^2\{u(\xi)\} + \{F(\xi)\} = 0 \end{aligned} \tag{15.103}$$

which should be compared to Equation (15.60) in three dimensions. Prescribed displacement boundary conditions on the side-faces A_u will lead to an additional term, analogous to the three-dimensional case in Equations (15.62) and (15.63).

The same nodal force-displacement relationship

$$\{R(\xi)\} = [S(\omega, \xi)]\{u(\xi)\} - \{R^{\mathrm{F}}(\xi)\} \tag{15.104}$$

applies with the internal nodal forces

$$\{Q(\xi)\} = [E^0]\xi\{u(\xi)\}_{,\xi} + [E^1]^T\{u(\xi)\} \tag{15.105}$$

This yields

$$\xi\{R^{\mathrm{F}}(\xi)\}_{,\xi} + (\pm[S(\omega, \xi)] - [E^1])[E^0]^{-1}\{R^{\mathrm{F}}(\xi)\} = \pm\{F(\xi)\} \tag{15.106}$$

for the amplitudes of the nodal load $\{R^{\mathrm{F}}(\xi)\}$ with the upper and lower signs applying to the bounded and unbounded media, respectively. Using

$$[S(\omega, \xi)] = G[\bar{S}(a)] \tag{15.107}$$

yields

$$\xi[S(\omega, \xi)]_{,\xi} = \omega[S(\omega, \xi)]_{,\omega} \tag{15.108}$$

resulting in the scaled boundary finite element equations in dynamic stiffness for a bounded medium

$$([S^b(\omega)] - [E^1])[E^0]^{-1}([S^b(\omega)] - [E^1]^T) + \omega[S^b(\omega)]_{,\omega} - [E^2] + \omega^2[M^0] = 0 \tag{15.109}$$

and for an unbounded medium

$$([S^\infty(\omega)] + [E^1])[E^0]^{-1}([S^\infty(\omega)] + [E^1]^T) - \omega[S^\infty(\omega)]_{,\omega} - [E^2] + \omega^2[M^0] = 0 \tag{15.110}$$

Equations (15.109) and (15.110) should be compared to Equations (15.84) and (15.85), respectively, in three dimensions. Note that in two dimensions, the dynamic-stiffness matrix is a function of the dimensionless frequency a (Equation (15.77)) and not only the dimensionless dynamic-stiffness matrix as in three dimensions.

Finally, the three- and two-dimensional formulations can be unified introducing the spatial dimension $s(=2$ or $=3)$. In addition, the variable $\bar{\xi} = \xi\omega$ can be introduced as the independent variable. The *scaled boundary finite element equation in displacement* (Equations (15.60) and (15.103)) equals

$$[E^0]\xi^2\{u(\xi)\}_{,\xi\xi} + ((s-1)[E^0] - [E^1] + [E^1]^T)\xi\{u(\xi)\}_{,\xi} + ((s-2)[E^1]^T - [E^2])\{u(\xi)\} + \omega^2[M^0]\xi^2\{u(\xi)\} + \{F(\xi)\} = 0 \tag{15.111}$$

or

$$[E^0]\bar{\xi}^2\{u(\bar{\xi})\}_{,\bar{\xi}\bar{\xi}} + ((s-1)[E^0] - [E^1] + [E^1]^T)\bar{\xi}\{u(\bar{\xi})\}_{,\bar{\xi}} + ((s-2)[E^1]^T - [E^2])\{u(\bar{\xi})\} + [M^0]\bar{\xi}^2\{u(\bar{\xi})\} + \{F(\bar{\xi})\} = 0 \tag{15.112}$$

The equation could also be formulated with the same coefficients choosing the dimensionless frequency corresponding to the radial coordinate $r(=\xi r_0$, with the radial coordinate of the boundary r_0 measured from the scaling centre) as the independent variable

$$a = \frac{\omega\xi r_0}{c_s} \tag{15.113}$$

The *interaction force-displacement relationship* for vanishing loads at a surface with a constant ξ (Equations (15.70) and (15.105)) is formulated for the unbounded medium as

$$[S^\infty(\omega,\xi)]\{u(\xi)\} = -\xi^{s-2}([E^0]\xi\{u(\xi)\}_{,\xi} + [E^1]^T\{u(\xi)\}) \tag{15.114}$$

or

$$\xi^{2-s}[S^\infty(\omega,\xi)]\{u(\bar{\xi})\} = -[E^0]\bar{\xi}\{u(\bar{\xi})\}_{,\bar{\xi}} - [E^1]^T\{u(\bar{\xi})\} \tag{15.115}$$

It follows from Equation (15.115) that $\xi^{2-s}[S^\infty(\omega,\xi)]$ is a function of $\bar{\xi}$ (or alternatively of a). For $s=3$, the dynamic-stiffness matrix $[S^\infty(\omega,\xi)]$ is a function of ξ and $\bar{\xi}$, i.e. ξ and ω (or ξ and a). On the boundary $\xi=1$, $[S^\infty(\omega,\xi=1)]=[S^\infty(\omega)]$ applies.

The *scaled boundary finite element equation in dynamic stiffness* for a bounded medium (Equations (15.84) and (15.109)) equals

$$\begin{aligned}&(\xi^{2-s}[S^{\mathrm{b}}(\omega,\xi)]-[E^1])[E^0]^{-1}(\xi^{2-s}[S^{\mathrm{b}}(\omega,\xi)]-[E^1]^T)\\&+(s-2)\xi^{2-s}[S^{\mathrm{b}}(\omega,\xi)]+\bar{\xi}(\xi^{2-s}[S^{\mathrm{b}}(\omega,\xi)])_{,\bar{\xi}}-[E^2]+\bar{\xi}^2[M^0]=0\end{aligned} \tag{15.116}$$

and for an unbounded medium (Equations (15.85) and (15.110))

$$\begin{aligned}&(\xi^{2-s}[S^\infty(\omega,\xi)]+[E^1])[E^0]^{-1}(\xi^{2-s}[S^\infty(\omega,\xi)]+[E^1]^T)\\&-(s-2)\xi^{2-s}[S^\infty(\omega,\xi)]-\bar{\xi}(\xi^{2-s}[S^\infty(\omega,\xi)])_{,\bar{\xi}}-[E^2]+\bar{\xi}^2[M^0]=0\end{aligned} \tag{15.117}$$

These are nonlinear first-order ordinary differential equations in $\xi^{2-s}[S^\infty(\omega,\xi)]$, with $\bar{\xi}$ as the independent variable (or alternatively with a).

15.6 EQUATIONS FOR DIFFUSION

In the previous sections of this chapter, the wave propagation equation is addressed. In the time domain the governing partial differential equation is hyperbolic. In the frequency domain with statics as a special case, an elliptic partial differential equation is processed. This section examines the fundamental equations of the scaled boundary finite element method for the diffusion equation, which is a parabolic partial differential equation in the time domain. The diffusion equation occurs in many fields of engineering. Heat conduction is used for illustration.

In contrast to the wave equation with a second derivative of the function with respect to time, the diffusion equation exhibits only a first derivative of the function with respect to time, which leads to an instantaneous response over the entire medium. The spatial derivatives are in the same form as in the scalar wave equation, while the time discretisation will differ. To avoid unnecessary repetition in the derivation, only final results with some intermediate expressions are listed. The reader can also consult Reference [S9].

First, the three-dimensional diffusion equation is formulated in Cartesian coordinates $\hat{x},\hat{y},\hat{z}$. A circumflex ($\hat{\ }$) is introduced to denote a point in space, as the coordinates x,y,z are reserved for the boundary, which is discretised. To simplify the nomenclature, the circumflex is omitted when used in a subscript to indicate a direction.

The partial differential equation in the frequency domain expressed in the function amplitude $u=u(\hat{x},\hat{y},\hat{z})$ is formulated as

$$\{L\}^T\{q\}-p+\mathrm{i}\omega\mu u=0 \tag{15.118}$$

μ is a material constant. $\{q\}$ is the flux vector related to the function u as

$$\{q\}=-[\kappa]\{L\}u \tag{15.119}$$

with the positive-definite (anisotropic) material matrix $[\kappa]$. $\{L\}$ is the differential operator

$$\{L\} = \begin{Bmatrix} \frac{\partial}{\partial \hat{x}} \\ \frac{\partial}{\partial \hat{y}} \\ \frac{\partial}{\partial \hat{z}} \end{Bmatrix} \tag{15.120}$$

p is the amplitude of the source per unit volume. The boundary conditions are equal to

$$u = \bar{u} \quad \text{on} \quad S_u \tag{15.121a}$$

$$-q_n = -\{n\}^T\{q\} = \bar{q} \quad \text{on} \quad S_q \tag{15.121b}$$

and

$$u = 0 \ \text{ on } \ A_u \tag{15.122a}$$

$$q_n = \bar{q} \ \text{ on } \ A_q \tag{15.122b}$$

with the prescribed function amplitude $\bar{u}$ on S_u and the prescribed normal flux amplitude $\bar{q}$ on the boundary S_q with the outward normal vector $\{n\}$. For simplicity, the right-hand side of Equation (15.22a) is assumed to be zero.

For heat conduction, u corresponds to the temperature, $\{q\}$ to the heat flux, μ to the product of the mass density ρ and the specific heat of the material c, and $[\kappa]$ to the thermal conductivity. p is the heat source per unit volume.

Secondly, the transformation of the geometry corresponding to the scaled boundary coordinates discussed in section 15.2 is addressed. It applies fully to the diffusion equation (Equation (15.11)–(15.34)). No modifications occur. In particular, the differential operator in Equation (15.120) is formulated as (Equation (15.34))

$$\{L\} = \begin{Bmatrix} \frac{\partial}{\partial \hat{x}} \\ \frac{\partial}{\partial \hat{y}} \\ \frac{\partial}{\partial \hat{z}} \end{Bmatrix} = \frac{g^{\xi}}{|J|}\{n^{\xi}\}\frac{\partial}{\partial \xi} + \frac{1}{\xi}\left(\frac{g^{\eta}}{|J|}\{n^{\eta}\}\frac{\partial}{\partial \eta} + \frac{g^{\zeta}}{|J|}\{n^{\zeta}\}\frac{\partial}{\partial \zeta}\right) \tag{15.123}$$

Thirdly, the governing equations of diffusion with geometry in the transformed coordinate system are formulated. Only the spatial coordinates are affected. The function and its flux are still defined in the original coordinate system $\hat{x}, \hat{y}, \hat{z}$. This is analogous to the procedure of mapping parent elements to curvilinear elements in the finite element method.

Thus, only the differential operator $\{L\}$ in Equation (15.123) needs to be addressed. Equation (15.123) is formulated as

$$\{L\} = \{b^1\}\frac{\partial}{\partial \xi} + \frac{1}{\xi}\left(\{b^2\}\frac{\partial}{\partial \eta} + \{b^3\}\frac{\partial}{\partial \zeta}\right) \tag{15.124}$$

$$\{b^1\} = (g^{\xi}/|J|)\{n^{\xi}\} \tag{15.125a}$$

$$\{b^2\} = (g^{\eta}/|J|)\{n^{\eta}\} \tag{15.125b}$$

$$\{b^3\} = (g^{\zeta}/|J|)\{n^{\zeta}\} \tag{15.125c}$$

Note that $\{b^1\}, \{b^2\}$ and $\{b^3\}$ are independent of ξ. The following identity exists:

$$(|J|\{b^2\})_{,\eta} + (|J|\{b^3\})_{,\zeta} = -2|J|\{b^1\} \tag{15.126}$$

The amplitude of the normal flux on any boundary with the outward unit normal vector $\{n\}$ equals

$$q_n = \{n\}^T\{q\} \tag{15.127}$$

By comparing Equations (15.127) and (15.125), the amplitudes of the normal flux on the surfaces $(\eta, \zeta), (\zeta, \xi)$ and (ξ, η), where the coordinates ξ, η and ζ, respectively, are constant, are written as

$$q_n^{\xi} = \frac{|J|}{g^{\xi}}\{b^1\}^T\{q\} \tag{15.128a}$$

$$q_n^{\eta} = \frac{|J|}{g^{\eta}}\{b^2\}^T\{q\} \tag{15.128b}$$

$$q_n^{\zeta} = \frac{|J|}{g^{\zeta}}\{b^3\}^T\{q\} \tag{15.128c}$$

Fourthly, the weighted residual technique is applied to derive a finite element approximation, analogous to the procedure described in section 15.3. Multiplying Equation (15.118) by the weighting function $w = w(\xi, \eta, \zeta)$ and integrating yields

$$\int_V w^T\{b^1\}^T\{q_{,\xi}\}\mathrm{d}V + \int_V w^T\frac{1}{\xi}(\{b^2\}^T\{q_{,\eta}\} + \{b^3\}^T\{q_{,\zeta}\})\mathrm{d}V - \int_V w^T p\mathrm{d}V + \mathrm{i}\omega\int_V w^T\mu u\mathrm{d}V = 0 \tag{15.129}$$

Integration by parts in the circumferential η-, ζ-directions is performed, and the relation is enforced exactly in the radial ξ-direction. With the shape functions $[N(\eta, \zeta)]$, the function amplitudes are interpolated

$$u(\xi, \eta, \zeta) = [N(\eta, \zeta)]\{u(\xi)\} \tag{15.130}$$

and the flux amplitudes are equal to

$$\{q\} = -[\kappa]\left([B^1]\{u(\xi)\}_{,\xi} + \frac{1}{\xi}[B^2]\{u(\xi)\}\right) \tag{15.131}$$

with

$$[B^1] = \{b^1\}[N(\eta, \zeta)] \tag{15.131a}$$

$$[B^2] = \{b^2\}[N(\eta, \zeta)]_{,\eta} + \{b^3\}[N(\eta, \zeta)]_{,\zeta} \tag{15.131b}$$

$[B^1]$ and $[B^2]$ are independent of ξ.
With the amplitudes of the nodal forces from the normal fluxes and the sources

$$\{F^q(\xi)\} = -\oint_{\Gamma^\xi} [N(\eta, \zeta)]^T (q_n^\zeta g^\zeta \mathrm{d}\eta + q_n^\eta g^\eta \mathrm{d}\zeta) \tag{15.132a}$$

$$\{F^p(\xi)\} = \int_{S^\xi} [N(\eta, \zeta)]^T p|J| \mathrm{d}\eta \mathrm{d}\zeta \tag{15.132b}$$

combined as

$$\{F(\xi)\} = \xi\{F^q(\xi)\} + \xi^2\{F^p(\xi)\} \tag{15.133}$$

and introducing the coefficient matrices, which are independent of ξ

$$[E^0] = \int_{S^\xi} [B^1]^T[\kappa][B^1]|J| \mathrm{d}\eta \mathrm{d}\zeta \tag{15.134a}$$

$$[E^1] = \int_{S^\xi} [B^2]^T[\kappa][B^1]|J| \mathrm{d}\eta \mathrm{d}\zeta \tag{15.134b}$$

$$[E^2] = \int_{S^\xi} [B^2]^T[\kappa][B^2]|J| \mathrm{d}\eta \mathrm{d}\zeta \tag{15.134c}$$

and

$$[M^0] = \int_{S^\xi} [N(\eta, \zeta)]^T \mu[N(\eta, \zeta)]|J| \mathrm{d}\eta \mathrm{d}\zeta \tag{15.135}$$

yields

$$\begin{aligned} &[E^0]\xi^2\{u(\xi)\}_{,\xi\xi} + (2[E^0] - [E^1] + [E^1]^T)\xi\{u(\xi)\}_{,\xi} \\ &+ ([E^1]^T - [E^2])\{u(\xi)\} - \mathrm{i}\omega[M^0]\xi^2\{u(\xi)\} + \{F(\xi)\} = 0 \end{aligned} \tag{15.136}$$

Note that the coefficient matrices $[E^0]$, $[E^1]$, $[E^2]$, $[M^0]$ are independent of ξ. Integrations over the surface finite element (at $\xi = 1$) only are involved ($S^\xi = S$).

After assemblage of all finite elements on the boundary and enforcement of the (homogeneous) displacement boundary condition on the side-face A_u, and keeping the same nomenclature,

$$\begin{aligned}&[E^0]\xi^2\{u(\xi)\}_{,\xi\xi} + (2[E^0] - [E^1] + [E^1]^T)\xi\{u(\xi)\}_{,\xi}\\&+ ([E^1]^T - [E^2])\{u(\xi)\} - \mathrm{i}\omega[M^0]\xi^2\{u(\xi)\} + \{F(\xi)\} = 0\end{aligned} \qquad (15.137)$$

results. This equation represents the *scaled boundary finite element equation in function* formulated in the frequency domain in three dimensions. It is a system of linear second-order ordinary differential equations for the function amplitudes $\{u(\xi)\}$ in the dimensionless radial coordinate ξ as the independent variable.

Comparing Equation (15.137) with the corresponding equation in elastodynamics (Equation (15.60)), it follows that the relationship for diffusion follows from that in elastodynamics (scalar wave equation) by replacing ω^2 by -iω.

The scaled boundary finite element equation (Equation (15.137)) is valid for bounded and unbounded media. Following the procedure for two dimensions in section 15.5 leads to the corresponding relationship for diffusion, which again can be deduced directly from that of elastodynamics for the two-dimensional scalar wave equation (Equation (15.103)). Unifying the three- and two-dimensional cases with the spatial dimension s ($= 2$ or $= 3$) yields

$$\begin{aligned}&[E^0]\xi^2\{u(\xi)\}_{,\xi\xi} + ((s-1)[E^0] - [E^1] + [E^1]^T)\xi\{u(\xi)\}_{,\xi}\\&+ ((s-2)[E^1]^T - [E^2])\{u(\xi)\} - \mathrm{i}\omega[M^0]\xi^2\{u(\xi)\} + \{F(\xi)\} = 0\end{aligned} \qquad (15.138)$$

Fifthly, the dynamic-stiffness matrix can be derived, following the procedure in section 15.4. For any surface S^ξ with a constant ξ, the virtual work is formulated as

$$\{w(\xi)\}^T\{Q(\xi)\} = -\int_{S^\xi} w^T q_n^\xi \mathrm{d}S^\xi \qquad (15.139)$$

with the amplitudes of the internal nodal flux $\{Q(\xi)\}$ and of the normal flux q_n^ξ. This leads to

$$\{Q(\xi)\} = [E^0]\xi^2\{u(\xi)\}_{,\xi} + [E^1]^T\xi\{u(\xi)\} \qquad (15.140)$$

For vanishing nodal loads,

$$\pm[S(\omega,\xi)]\{u(\xi)\} = [E^0]\xi^2\{u(\xi)\}_{,\xi} + [E^1]^T\xi\{u(\xi)\} \qquad (15.141)$$

with the upper and lower signs applying to the bounded and unbounded media, respectively, is combined with Equation (15.137). After some algebra,

$$\begin{aligned}&(\pm[S(\omega,\xi)] - \xi[E^1])(\xi[E^0])^{-1}(\pm[S(\omega,\xi)] - \xi[E^1]^T) - \xi[E^2]\\&\pm \xi[S(\omega,\xi)]_{,\xi} - \mathrm{i}\omega\xi^3[M^0] = 0\end{aligned} \qquad (15.142)$$

results. Introducing the dimensionless dynamic-stiffness matrix in frequency domain denoted with a bar and a specific material constant κ

$$[S(\omega, \xi)] = \kappa r_0 \xi [\bar{S}(\omega, \xi)] \tag{15.143}$$

and

$$[E^0] = \kappa r_0 [\bar{E}^0], \ldots \tag{15.144}$$

$$[M^0] = \mu r_0^3 [\bar{M}^0] \tag{15.145}$$

leads to

$$\begin{aligned}(\pm[\bar{S}(\omega, \xi)] - [\bar{E}^1])([\bar{E}^0])^{-1}(\pm[\bar{S}(\omega, \xi)] - [\bar{E}^1]^T) - [\bar{E}^2] \\ \pm [\bar{S}(\omega, \xi)] \pm \xi[\bar{S}(\omega, \xi)]_{,\xi} - i\omega \frac{\mu}{\kappa} r_0^2 \xi^2 [\bar{M}^0] = 0\end{aligned} \tag{15.146}$$

The coefficient of the last term defines the dimensionless frequency for any ξ

$$a(\omega, \xi) = \frac{\omega \mu r_0^2}{\kappa} \xi^2 \tag{15.147}$$

Note the difference in the power of ξ to elastodynamics (Equation (15.77)). The second-last term is written as

$$\xi[\bar{S}(\omega, \xi)]_{,\xi} = 2a[\bar{S}(\omega, \xi)]_{,a} \tag{15.148}$$

After substituting Equations (15.147) and (15.148) into Equation (15.146), the only independent variable in the equation is a. Thus, the dimensionless stiffness matrix in frequency domain is a function of a only:

$$[\bar{S}(\omega, \xi)] = [\bar{S}(a)] \tag{15.149}$$

The term with the derivative $a[\bar{S}(a)]_{,a}$ can be interpreted either for varying ξ with fixed ω or for varying ω with fixed ξ:

$$a[\bar{S}(a)]_{,a} = 0.5\xi[\bar{S}(a)]_{,\xi} = \omega[\bar{S}(a)]_{,\omega} \tag{15.150}$$

Differentiating Equation (15.143) with respect to ξ, and using Equations (15.149) and (15.150), yields

$$\xi[S(\omega, \xi)]_{,\xi} = [S(\omega, \xi)] + 2\omega[S(\omega, \xi)]_{,\omega} \tag{15.151}$$

Substituting Equation (15.151) into Equation (15.142) results in

$$\begin{aligned}(\pm[S(\omega, \xi)] - \xi[E^1])(\xi[E^0])^{-1}(\pm[S(\omega, \xi)] - \xi[E^1]^T) - \xi[E^2] \\ + [\pm S(\omega, \xi)] \pm 2\omega[S(\omega, \xi)]_{,\omega} - i\omega\xi^3[M^0] = 0\end{aligned} \tag{15.152}$$

For the boundary ($\xi = 1$) the dynamic-stiffness matrix is expressed as

$$\begin{aligned}(\pm[S(\omega,\xi)] - [E^1])[E^0]^{-1}(\pm[S(\omega)] - [E^1]^T) - [E^2] \\ \pm [S(\omega)] \pm 2\omega[S(\omega)]_{,\omega} - \mathrm{i}\omega[M^0] = 0\end{aligned} \tag{15.153}$$

This represents the *scaled boundary finite element equation in dynamic stiffness* formulated in the frequency domain for diffusion in three dimensions. It is a system of nonlinear, first-order ordinary differential equations with the frequency as independent variable.

Note the differences with the scaled boundary finite element equation in dynamic stiffness for elastodynamics (Equation (15.83)). As already discussed, $\omega^2 = -(\mathrm{i}\omega)^2$ for elastodynamics corresponds to $-\mathrm{i}\omega$ for diffusion. This leads to the selection of a new variable Ω with $-\Omega^2 = \mathrm{i}\omega$ or

$$\omega = \mathrm{i}\Omega^2 \tag{15.154}$$

Substituting Equation (15.154) into Equation (15.153) yields

$$\begin{aligned}(\pm[S(\Omega)] - [E^1])[E^0]^{-1}(\pm[S(\Omega)] - [E^1]^T) - [E^2] \\ \pm [S(\Omega)] \pm \Omega[S(\Omega)]_{,\Omega} + \omega^2[M^0] = 0\end{aligned} \tag{15.155}$$

This equation is identical to the scaled boundary finite element equation in dynamic stiffness for elastodynamics (Equation (15.83)) with the independent variable ω replaced by Ω. The transformation between the relations valid for diffusion and elastodynamics described in Equation (15.154) can also be used when addressing the solution procedures in the following chapters.

Using the spatial dimension s, the scaled boundary finite element equation in dynamic stiffness can be written for a bounded medium as

$$\begin{aligned}([S^{\mathrm{b}}(\omega)] - [E^1])[E^0]^{-1}([S^{\mathrm{b}}(\omega)] - [E^1]^T) - [E^2] \\ + (s-2)[S^{\mathrm{b}}(\omega)] + 2\omega[S^{\mathrm{b}}(\omega)]_{,\omega} - \mathrm{i}\omega[M^0] = 0\end{aligned} \tag{15.156}$$

and for an unbounded medium as

$$\begin{aligned}([S^{\infty}(\omega)] + [E^1])[E^0]^{-1}([S^{\infty}(\omega)] + [E^1]^T) - [E^2] \\ - (s-2)[S^{\infty}(\omega)] - 2\omega[S^{\infty}(\omega)]_{,\omega} - \mathrm{i}\omega[M^0] = 0\end{aligned} \tag{15.157}$$

The nodal flux-function relationship for an unbounded medium equals (Equation (15.141))

$$\xi^{2-s}[S^{\infty}(\omega,\xi)]\{u(\bar{\xi})\} = -[E^0]\bar{\xi}\{u(\bar{\xi})\}_{,\bar{\xi}} - [E^1]^T\{u(\bar{\xi})\} \tag{15.158}$$

15.7 ALTERNATIVE VIRTUAL WORK DERIVATION

The first derivation of the fundamental equations is based on the assemblage process of finite elements, and on similarity. This mechanically-based derivation discussed in Part I for the Model Problem (see also [W9]) appeals to structural engineers, but is quite involved as an analytical limit in the radial direction is performed working with a discretised system. The second derivation after a transformation to the scaled boundary coordinates is based on the weighted-residual method over the whole domain, being approximate in the circumferential directions but exact in the radial direction. This so-called scaled boundary transformation-based derivation should appeal to modern numerical analysts, as many other numerical procedures are based on a weighted-residual statement. All previous sections of this chapter describe this derivation. Both derivations lead to the same final results, enforcing a finite element type approximation in the circumferential directions, and an exact formulation in the radial direction.

A third alternative derivation is described in this section. After a transformation to the scaled boundary coordinates, instead of starting with a weighted-residual statement, a virtual-work formulation is applied, developed along similar lines as in the classical finite element method. It will also be demonstrated that the equivalent nodal forces are obtained directly from the virtual-work statement. It is hoped that this formulation will make the method more approachable to a wider engineering science community, in particular to engineers working in structural mechanics. This section is based on Reference [D5].

This virtual work derivation is discussed for three- dimensional elastodynamics. As already mentioned, the differential equations of motion are formulated as far as the geometry is concerned in the scaled boundary coordinates ξ, η, ζ (Figure 15.2) A bounded medium as shown in Figure 3.1a is addressed. Body loads are present, but for the sake of simplicity, no non-zero prescribed surface tractions on the side-face A_t. The differential operator $[L]$ is specified in Equation (15.35). The displacement amplitudes are interpolated as (Equation (15.46) or Equation (15.3))

$$\{u(\xi, \eta, \zeta)\} = [N(\eta, \zeta)]\{u(\xi)\} \tag{15.159}$$

The stress amplitudes are equal to (Equation (15.47))

$$\{\sigma(\xi, \eta, \zeta)\} = [D]\left([B^1]\{u(\xi)\}_{,\xi} + \frac{1}{\xi}[B^2]\{u(\xi)\}\right) \tag{15.160}$$

The virtual displacement field is formed using the shape functions $[N(\eta, \zeta)]$ to interpolate between nodes in the circumferential directions. The amplitudes of the virtual displacements are of the form

$$\{\delta u(\xi, \eta, \zeta)\} = [N(\eta, \zeta)]\{\delta u(\xi)\} \tag{15.161}$$

which is analogous to Equation (15.159). As is customary in books addressing structural mechanics, the virtual state is denoted with the symbol δ. The amplitudes of the corresponding virtual strains are specified as is apparent from Equation (15.47) as

$$\{\delta\varepsilon(\xi, \eta, \zeta)\} = [B^1]\{\delta u(\xi)\}_{,\xi} + \frac{1}{\xi}[B^2]\{\delta u(\xi)\} \tag{15.162}$$

The virtual work statement becomes

$$\int_V \{\delta\varepsilon(\xi,\eta,\zeta)\}^T\{\sigma(\xi,\eta,\zeta)\}\mathrm{d}V - \int_V \{\delta u(\xi,\eta,\zeta)\}^T\rho\omega^2\{u(\xi,\eta,\zeta)\}\mathrm{d}V$$
$$-\int_V \{\delta u(\xi,\eta,\zeta)\}^T\{p(\xi,\eta,\zeta)\}\mathrm{d}V - \int_S \{\delta u(\xi=1,\eta,\zeta)\}^T\{t(\eta,\zeta)\}\mathrm{d}S = 0 \tag{15.163}$$

where the first term represents the internal work, the second the work of the inertial loads and the third and forth terms the external work of the body loads and surface tractions. Substituting Equations (15.162), (15.161) and $\mathrm{d}V$ from Equation (15.29) into the first term yields

$$\int_V \left([B^1]\{\delta u(\xi)\}_{,\xi} + \frac{1}{\xi}[B^2]\{\delta u(\xi)\}\right)^T [D]\left([B^1]\{u(\xi)\}_{,\xi} + \frac{1}{\xi}[B^2]\{u(\xi)\}\right)\xi^2|J|\mathrm{d}\xi\mathrm{d}\eta\mathrm{d}\zeta$$
$$= \int_0^1 \{\delta u(\xi)\}^T_{,\xi}\int_S [B^1]^T[D][B^1]|J|\mathrm{d}\eta\mathrm{d}\zeta\xi^2\{u(\xi)\}_{,\xi}\mathrm{d}\xi$$
$$+\int_0^1 \{\delta u(\xi)\}^T_{,\xi}\int_S [B^1]^T[D][B^2]|J|\mathrm{d}\eta\mathrm{d}\zeta\xi\{u(\xi)\}\mathrm{d}\xi$$
$$+\int_0^1 \{\delta u(\xi)\}^T\int_S [B^2]^T[D][B^1]|J|\mathrm{d}\eta\mathrm{d}\zeta\xi\{u(\xi)\}_{,\xi}\mathrm{d}\xi$$
$$+\int_0^1 \{\delta u(\xi)\}^T\int_S [B^2]^T[D][B^2]|J|\mathrm{d}\eta\mathrm{d}\zeta\{u(\xi)\}\mathrm{d}\xi \tag{15.164}$$

Integration by parts in the radial coordinate ξ is applied for all terms containing derivatives of the virtual displacements with respect to ξ, $\{\delta u(\xi)\}_{,\xi}$. This Green's theorem introduces integrals over the boundary S. Equation (15.164) is rewritten as

$$\{\delta u(\xi)\}^T\int_S [B^1]^T[D][B^1]|J|\mathrm{d}\eta\mathrm{d}\zeta\xi^2\{u(\xi)\}_{,\xi}|_{\xi=1}$$
$$-\int_0^1 \{\delta u(\xi)\}^T\int_S [B^1]^T[D][B^1]|J|\mathrm{d}\eta\mathrm{d}\zeta(2\xi\{u(\xi)\}_{,\xi} + \xi^2\{u(\xi)\}_{,\xi\xi})\mathrm{d}\xi$$
$$+\{\delta u(\xi)\}^T\int_S [B^1]^T[D][B^2]|J|\mathrm{d}\eta\mathrm{d}\zeta\xi\{u(\xi)\}|_{\xi=1}$$
$$-\int_0^1 \{\delta u(\xi)\}^T\int_S [B^1]^T[D][B^2]|J|\mathrm{d}\eta\mathrm{d}\zeta(\{u(\xi)\} + \xi\{u(\xi)\}_{,\xi})\mathrm{d}\xi$$
$$+\int_0^1 \{\delta u(\xi)\}^T\int_S [B^2]^T[D][B^1]|J|\mathrm{d}\eta\mathrm{d}\zeta\xi\{u(\xi)\}_{,\xi}\mathrm{d}\xi$$
$$+\int_0^1 \{\delta u(\xi)\}^T\int_S [B^2]^T[D][B^2]|J|\mathrm{d}\eta\mathrm{d}\zeta\{u(\xi)\}\mathrm{d}\xi \tag{15.165}$$

For convenience, the following coefficient matrices (Equation (15.54)) are introduced

$$[E^0] = \int_S [B^1]^T [D] [B^1] |J| \mathrm{d}\eta \mathrm{d}\zeta \tag{15.166a}$$

$$[E^1] = \int_S [B^2]^T [D] [B^1] |J| \mathrm{d}\eta \mathrm{d}\zeta \tag{15.166b}$$

$$[E^2] = \int_S [B^2]^T [D] [B^2] |J| \mathrm{d}\eta \mathrm{d}\zeta \tag{15.166c}$$

These integrals can be computed element by element over the boundary and assembled together for the entire boundary in the same manner as the static-stiffness matrix is determined for the entire domain in the standard finite element method.

Using $\{u\}$ to represent $\{u(\xi = 1)\}$, and so forth, Equation (15.165) is expressed as

$$\int_V \{\delta\varepsilon\}^T \{\sigma\} \mathrm{d}V = \{\delta u\}^T ([E^0]\{u\}_{,\xi} + [E^1]^T \{u\})$$
$$- \int_0^1 \{\delta u(\xi)\}^T ([E^0]\xi^2 \{u(\xi)\}_{,\xi\xi} + (2[E^0] + [E^1]^T - [E^1])\xi\{u(\xi)\}_{,\xi} + ([E^1]^T - [E^2])\{u(\xi)\})\mathrm{d}\xi \tag{15.167}$$

Substituting Equations (15.161), (15.159) and (15.29) into the second term of Equation (15.163) yields

$$-\omega^2 \int_0^1 \{\delta u(\xi)\}^T \int_S [N]^T \rho [N] |J| \mathrm{d}\eta \mathrm{d}\zeta \xi^2 \{u(\xi)\} \mathrm{d}\xi \tag{15.168}$$

With the coefficient matrix (Equation (15.55))

$$[M^0] = \int_S [N]^T \rho [N] |J| \mathrm{d}\eta \mathrm{d}\zeta \tag{15.169}$$

Equation (15.168) is equal to

$$-\omega^2 \int_0^1 \{\delta u(\xi)\}^T [M^0] \xi^2 \{u(\xi)\} \mathrm{d}\xi \tag{15.170}$$

Analogously, the third term of Equation (15.163) is formulated as

$$-\int_0^1 \{\delta u(\xi)\}^T \xi^2 \int_{S^\xi} [N(\eta,\zeta)]^T \{p(\xi,\eta,\zeta)\}|J|\mathrm{d}\eta\mathrm{d}\zeta\mathrm{d}\xi$$

$$= -\int_0^1 \{\delta u(\xi)\}^T \xi^2 \{F^\mathrm{b}(\xi)\}\mathrm{d}\xi \tag{15.171}$$

with (Equation (15.52))

$$\{F^\mathrm{b}(\xi)\} = \int_{S^\xi} [N(\eta,\zeta)]^T \{p(\xi,\eta,\zeta)\}|J|\mathrm{d}\eta\mathrm{d}\zeta \tag{15.172}$$

Finally, the fourth term of Equation (15.163) with Equation (15.161) equals

$$-\{\delta u(\xi=1)\}^T \int_S [N(\eta,\zeta)]^T \{t(\eta,\zeta)\}\mathrm{d}S =$$

$$-\{\delta u\}^T \int_S [N(\eta,\zeta)]^T \{t(\eta,\zeta)\}\mathrm{d}S \tag{15.173}$$

The integral on the right-hand side can be identified as the amplitudes of the equivalent nodal forces due to the boundary tractions, which are denoted as $\{R\}$.

Substituting Equations (15.167), (15.170), (15.171), (15.173) into Equation (15.163) leads to

$$\begin{aligned}&\{\delta u\}^T([E^0]\{u\}_{,\xi} + [E^1]^T\{u\} - \{R\})\\ &-\int_0^1 \{\delta u(\xi)\}^T([E^0]\xi^2\{u(\xi)\}_{,\xi\xi} + (2[E^0] - [E^1] + [E^1]^T)\xi\{u(\xi)\}_{,\xi}\\ &+([E^1]^T - [E^2])\{u(\xi)\} + \omega^2[M^0]\xi^2\{u(\xi)\} + \xi^2\{F^\mathrm{b}(\xi)\})\mathrm{d}\xi = 0\end{aligned} \tag{15.174}$$

To satisfy Equation (15.174) for all $\{\delta u(\xi)\}$ with $0 \leq \xi \leq 1$, implying that the equilibrium equations (differential equations of motion) are satisfied exactly in the radial direction, and in the finite element sense in the circumferential directions, both of the following conditions must be satisfied:

$$\{R\} = [E^0]\{u\}_{,\xi} + [E^1]^T\{u\} \tag{15.175}$$

$$\begin{aligned}&[E^0]\xi^2\{u(\xi)\}_{,\xi\xi} + (2[E^0] - [E^1] + [E^1]^T)\xi\{u(\xi)\}_{,\xi}\\ &+ ([E^1]^T - [E^2])\{u(\xi)\} + \omega^2[M^0]\xi^2\{u(\xi)\} + \xi^2\{F^\mathrm{b}(\xi)\} = 0\end{aligned} \tag{15.176}$$

Equation (15.175) represents the nodal force-displacement relationship formulated at the boundary. It coincides with Equation (15.67) for $\xi = 1$ using Equation (15.68) formulated with the positive sign for a bounded medium. (Actually, the derivation also leads to the expression for any ξ, as is apparent from Equation (15.165) – see also section 21.1.) Equation (15.176) is the scaled boundary finite element equation in displacement, as already presented in Equation (15.57) (using Equation (15.56)).

The derivation could also be performed for an unbounded medium (Figure 3.4a), leading to the same result. The integrations run from $\xi = 1$ to $\xi \to \infty$ and the boundary represents a negative face, which both lead to a change in sign, which cancel each other in the scaled boundary finite element equation in displacement (Equation (15.176)). This derivation for an unbounded medium (with material properties varying in the radial direction) is presented in section 21.1.

16 Statics

As discussed in previous chapters, the scaled boundary finite element method is a semi-analytical procedure. It transforms the governing partial differential equations into ordinary differential equations. Although a system of differential equations must be addressed, there is only a single independent variable, the radial coordinate. For quite general (linear) cases, even involving dynamics, an analytical solution is feasible which can, however, become quite involved. For instance, for the wave equation, an infinite series is used which has to be terminated (Chapter 20). The efficiency of the overall procedure depends strongly upon the ease with which these ordinary differential equations can be solved. For statics, they are the well-known Euler–Cauchy differential equations, which are the simplest linear differential equations with variable coefficients to solve. A quadratic eigenvalue problem of the order of the number of degrees of freedom on the boundary is addressed. The boundary conditions for bounded and unbounded media are straightforwardly enforced. Body loads and prescribed surface tractions are also processed using elementary procedures.

The same applies in diffusion for steady state with sources and prescribed normal fluxes, as the spatial discretisation and thus the coefficient matrices are the same as for the scalar wave equations specialised for statics (section 15.6). The scaled boundary finite element equations for $\omega = 0$ are also identical to those in statics. In the following, statics is addressed, but all results also apply to steady state in diffusion.

Static (and steady state) problems for bounded and unbounded media are thus solved very accurately and efficiently applying the scaled boundary finite element method. This is demonstrated addressing error estimates and running times for many examples in Chapter 25. The fact that the method involves the solution of a quadratic eigenvalue problem, the computational expense of which rises rapidly as the number of degrees of freedom increases, is not really of any concern. It is sufficient to apply substructuring, i.e. to break the domain into a few subdomains. Each of these superelements with its own scaling centre interacts only on its boundary, which will be discretised with a small number of degrees of freedom. Thus, a few quadratic eigenvalue problems of modest size are calculated. In addition, for each subdomain at its scaling centre, stress discontinuities and singularities can be modelled, and certain traction and displacement boundary conditions on its side-faces (straight segments of boundary that intersect at the scaling centre) can be satisfied exactly. Of course, the restriction that the entire boundary must be visible from the scaling centre is easily satisfied on the level of the subdomains. The substructuring approach which is especially effective for a bounded medium is discussed in Chapter 22. Many examples for statics are described in Chapter 23.

Summarising, it is fair to state that (linear) statics, which is very common, is ideally suited not only to explain the scaled boundary finite element method and to demonstrate its high efficiency, but also to actually perform challenging calculations in practice.

Thus, the linear analysis in elastostatics of a bounded medium and of an unbounded (infinite) medium is addressed (Figure 16.1). In the domain V the governing partial differential equations in displacements u apply. Body loads p are present. On the boundary S either the displacements u on S_u or the surface tractions t on S_t are prescribed. For an unbounded medium (Figure 16.1b), an additional boundary condition at infinity must be satisfied.

In the scaled boundary finite element method, surface finite elements are used to discretise the boundary. The fundamental equations of Chapter 15 apply. The nodal displacements on the boundary are denoted as $\{u\}$ and the corresponding nodal forces as $\{R\}$ (Figure 16.1). The nodal force-displacement relationship on the boundary of the medium is formulated as (Equation (15.69) for statics)

$$\{R\} = [K]\{u\} - \{R^F\} \tag{16.1}$$

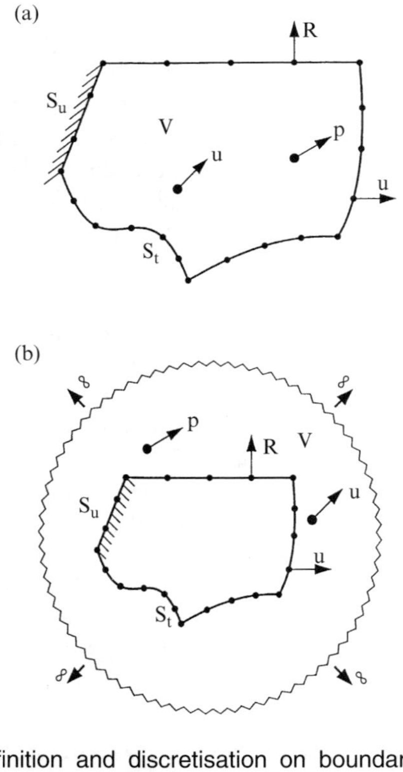

Figure 16.1 Problem definition and discretisation on boundary with surface finite elements (section): (a) bounded medium, (b) unbounded medium (Reprinted from Computer Methods in Applied Mechanics & Engineering, Vol. 180, Song and Wolf, 'Body loads in scaled boundary finite element method', pp. 117–135, © (1999), with permission from Elsevier Science)

with the static-stiffness matrix $[K]$ and the nodal loads due to body loads $\{R^{\mathrm{F}}\}$, determined from the scaled boundary finite element method. Non-zero prescribed surface tractions on that part of the boundary denoted as A_t, the side-face (Figures 3.1b, 3.4b), also contribute to $\{R^{\mathrm{F}}\}$. In this chapter the term body load also includes the surface tractions on A_t and prescribed displacements on A_u, when present. Enforcing the boundary conditions on S_u and S_t permits the displacements $\{u\}$ and nodal forces $\{R\}$ to be calculated in all nodes. The displacements, strains and stresses in the interior of the medium can then be calculated in the scaled boundary finite element method.

16.1 VANISHING BODY LOAD

For vanishing body loads $\{F(\xi)\} = 0$ and statics ($\omega = 0$), the homogeneous second-order ordinary differential equations, the scaled boundary finite element equation in displacement (Equation (15.111)) equals

$$[E^0]\xi^2\{u(\xi)\}_{,\xi\xi} + ((s-1)[E^0] - [E^1] + [E^1]^T)\xi\{u(\xi)\}_{,\xi} \\ +((s-2)[E^1]^T - [E^2])\{u(\xi)\} = 0 \tag{16.2}$$

They are applicable for two- and three-dimensional cases ($s = 2$ and $= 3$). These Euler–Cauchy equations can be solved analytically.

Equation (16.2) is transformed into first-order ordinary differential equations with twice the number of unknowns by introducing the variable

$$\{X(\xi)\} = \begin{Bmatrix} \xi^{+0.5(s-2)}\{u(\xi)\} \\ \xi^{-0.5(s-2)}\{Q(\xi)\} \end{Bmatrix} \tag{16.3}$$

with the internal nodal forces $\{Q(\xi)\}$ (Equations (15.67), (15.105))

$$\{Q(\xi)\} = \xi^{s-2}([E^0]\xi\{u(\xi)\}_{,\xi} + [E^1]^T\{u(\xi)\}) \tag{16.4}$$

This results in

$$\xi\{X(\xi)\}_{,\xi} = -[Z]\{X(\xi)\} \tag{16.5}$$

with a Hamiltonian matrix

$$[Z] = \begin{bmatrix} [E^0]^{-1}[E^1]^T - 0.5(s-2)[I] & -[E^0]^{-1} \\ -[E^2] + [E^1][E^0]^{-1}[E^1]^T & -([E^1][E^0]^{-1} - 0.5(s-2)[I]) \end{bmatrix} \tag{16.6}$$

The following eigenvalue problem is solved:

$$[Z][\Phi] = -[\Phi]\lceil\Lambda\rfloor \tag{16.7}$$

with the eigenvalues of the Hamiltonian matrix [P1]

$$\lceil \Lambda \rfloor = \begin{bmatrix} -\lceil \lambda_i \rfloor & \\ & \lceil \lambda_i \rfloor \end{bmatrix} \tag{16.8}$$

and the eigenvector matrix partitioned conformably

$$[\Phi] = \begin{bmatrix} [\Phi_{11}] & [\Phi_{12}] \\ [\Phi_{21}] & [\Phi_{22}] \end{bmatrix} \tag{16.9}$$

The real parts of all elements λ_i are negative [L1]. The solution of Equation (16.5) equals

$$\{X(\xi)\} = \begin{bmatrix} [\Phi_{11}] & [\Phi_{12}] \\ [\Phi_{21}] & [\Phi_{22}] \end{bmatrix} \begin{bmatrix} \lceil \xi^{-\lambda_i} \rfloor & \\ & \lceil \xi^{\lambda_i} \rfloor \end{bmatrix} \begin{Bmatrix} \{c_1\} \\ \{c_2\} \end{Bmatrix} \tag{16.10}$$

with the integration constants $\{c_1\}$, $\{c_2\}$. Substituting Equation (16.10) into (16.3) yields

$$\{u(\xi)\} = \xi^{-0.5(s-2)}([\Phi_{11}] \lceil \xi^{-\lambda_i} \rfloor \{c_1\} + [\Phi_{12}] \lceil \xi^{\lambda_i} \rfloor \{c_2\}) \tag{16.11a}$$

$$\{Q(\xi)\} = \xi^{+0.5(s-2)}([\Phi_{21}] \lceil \xi^{-\lambda_i} \rfloor \{c_1\} + [\Phi_{22}] \lceil \xi^{\lambda_i} \rfloor \{c_2\}) \tag{16.11b}$$

The integration constants are determined from the boundary conditions.

For a bounded medium, the displacements at $\xi = 0$ must remain finite. This condition $\{u(\xi = 0)\}$ leads to $\{c_2\} = 0$ in Equation (16.11a), as the real parts of all λ_i are negative:

$$\{u(\xi)\} = \xi^{-0.5(s-2)} [\Phi_{11}] \lceil \xi^{-\lambda_i} \rfloor \{c_1\} \tag{16.12a}$$

$$\{Q(\xi)\} = \xi^{+0.5(s-2)} [\Phi_{21}] \lceil \xi^{-\lambda_i} \rfloor \{c_1\} \tag{16.12b}$$

The displacements on the boundary $\{u(\xi = 1)\}$ determine $\{c_1\}$.

To calculate the static-stiffness matrix $[K^{\mathrm{b}}] = [K]$ (b for bounded) defined in Equation (16.1) with $\{R^{\mathrm{F}}\} = 0$, $\{c_1\}$ is eliminated from Equations (16.12a) and (16.12b). Substituting Equation (15.68) for a bounded medium $\{R\} = \{Q(\xi = 1)\}$ yields

$$\{R\} = [\Phi_{21}][\Phi_{11}]^{-1}\{u(\xi = 1)\} \tag{16.13}$$

Comparing with Equation (16.1) leads to

$$[K^{\mathrm{b}}] = [\Phi_{21}][\Phi_{11}]^{-1} \tag{16.14}$$

For an unbounded medium, the displacements for $\xi \to \infty$ must remain finite. This condition $\{u(\xi \to \infty)\}$ leads to $\{c_1\} = 0$ in Equation (16.11a):

$$\{u(\xi)\} = \xi^{-0.5(s-2)} [\Phi_{12}] \lceil \xi^{\lambda_i} \rfloor \{c_2\} \tag{16.15a}$$

$$\{Q(\xi)\} = \xi^{+0.5(s-2)}[\Phi_{22}]\lceil\xi^{\lambda_i}\rfloor\{c_2\} \tag{16.15b}$$

The displacements on the boundary $\{u(\xi = 1)\}$ determine $\{c_2\}$.

The static-stiffness matrix $[K^\infty] = [K]$ in Equation (16.1) follows from eliminating $\{c_2\}$ in Equations (16.15a) and (16.15b), and using Equation (15.68) for an unbounded medium $\{R\} = -\{Q(\xi = 1)\}$ as

$$[K^\infty] = -[\Phi_{22}][\Phi_{12}]^{-1} \tag{16.16}$$

A physical interpretation of the eigenvalue problem (Equation (16.7)) is discussed. Addressing the i-th column of Equation (16.11a) leads to

$$\{u_i(\xi)\} = \xi^{-0.5(s-2)}(\{\Phi_{11i}\}\xi^{-\lambda_i}c_{1i} + \{\Phi_{12i}\}\xi^{\lambda_i}c_{2i}) \tag{16.17}$$

Examining the first term on the right-hand side, the exponent $-0.5(s-2) - \lambda_i$ and the corresponding vector $\{\Phi_{11i}\}$ may be interpreted as an independent *mode of deformation*, which satisfies the scaled boundary finite element equation in displacement, and thus equilibrium in the radial ξ-direction. The integration constant c_{1i} represents the contribution of this mode to the solution, and is dependent on the boundary condition. The vector $\{\Phi_{11i}\}$ represents the modal displacements at the boundary nodes, while $-0.5(s-2) - \lambda_i$, and thus λ_i can be identified as the *modal scaling factor in the radial direction*. Analogously, the ith column of Equation (16.11b) equals

$$\{Q_i(\xi)\} = \xi^{+0.5(s-2)}(\{\Phi_{21i}\}\xi^{-\lambda_i}c_{1i} + \{\Phi_{22i}\}\xi^{\lambda_i}c_{2i}) \tag{16.18}$$

Examining the first term on the right-hand side, the internal nodal forces represent the corresponding modal forces. Again, $\{\Phi_{21i}\}$ represents the modal forces at the boundary nodes, and λ_i determines their variation in the radial direction.

Alternatively, the static-stiffness matrix of the bounded medium $[K^b]$ can also be determined from the scaled boundary finite element equation in dynamic stiffness (Equation (15.116)) formulated for statics on the boundary $\xi = 1$

$$([K^b] - [E^1])[E^0]^{-1}([K^b] - [E^1]^T) + (s-2)[K^b] - [E^2] = 0 \tag{16.19}$$

This can be reformulated as an algebraic Riccati equation:

$$\begin{aligned}[K^b][E^0]^{-1}[K^b] &- ([E^1][E^0]^{-1} - 0.5(s-2)[I])[K^b] \\ &- [K^b]([E^0]^{-1}[E^1]^T - 0.5(s-2)[I]) - [E^2] + [E^1][E^0]^{-1}[E^1]^T = 0\end{aligned} \tag{16.20}$$

In principle, an efficient solution is possible using Schur factorisation. The displacements in the interior can also be determined from combining Equations (16.1), (15.68) and (16.4) (or from Equation (15.114), but for a bounded medium):

$$[K^b(\xi)]\{u(\xi)\} = \xi^{s-2}([E^0]\xi\{u(\xi)\}_{,\xi} + [E^1]^T\{u(\xi)\}) \tag{16.21}$$

and solving for $\{u(\xi)\}$ analytically using the Schur factorisation. The same approach is also possible to address the static-stiffness matrix $[K^{\infty}]$ and the corresponding displacements in the interior $\{u(\xi)\}$ of the unbounded medium. It turns out that the Riccati equation has to be solved only once to determine $[K^{b}]$ and $[K^{\infty}]$. For details of this alternative procedure, References [S7] and [W8] should be consulted.

The first few eigenvalues and the corresponding variation of the displacements in the radial direction for the three-dimensional case are addressed. For the bounded medium, it follows from Equation (16.12a) with $s = 3$ that $\{u(\xi)\}$ is proportional to $\xi^{-\lambda_i-0.5}$. To be able to represent the three rigid body translations with $\{u(\xi)\} =$ constant, $-\lambda_i - 0.5 = 0$, resulting in $\lambda_1 = \lambda_2 = \lambda_3 = -0.5$. Analogously for the three rigid body rotations with $\{u(\xi)\} \propto \xi$, $-\lambda_i - 0.5 = 1$, yielding $\lambda_4 = \lambda_5 = \lambda_6 = -1.5$. These rigid body modes lead of course to a semi-positive definite static- stiffness matrix $[K^{b}]$ in Equation (16.14) with the rank reduced by 6. For the unbounded medium, the corresponding eigenvalues are known, as they are equal to those of the bounded medium, but with the sign reversed (Hamiltonian matrix; see Equation (16.8)). This is reflected in Equation (16.15a). Thus, with $s = 3$, $\{u(\xi)\}$ is proportional to $\xi^{\lambda_i-0.5}$. For $\lambda_1 = \lambda_2 = \lambda_3 = -0.5$, $\{u(\xi)\} \propto \xi^{-1}$ and for $\lambda_4 = \lambda_5 = \lambda_6 = -1.5\{u(\xi)\} \propto \xi^{-2}$ applies. Thus, the displacements diminish towards zero for $\xi \rightarrow \infty$. The corresponding static-stiffness matrix $[K^{\infty}]$ is positive definite.

Repeating this discussion for the two-dimensional case will lead to different conclusions for the unbounded case. For the bounded medium (Equation (16.12a) with $s = 2$), $\{u(\xi)\} \propto \xi^{-\lambda_i}$ results. The two rigid body translations $\{u(\xi)\} =$ constant lead to $\lambda_1 = \lambda_2 = 0$ and the rigid body rotation $\{u(\xi)\} \propto \xi$ yields $\lambda_3 = -1$. (The eigenvalue problem (Equation (16.7)) can still be solved for vanishing eigenvalues using, for example, a shift.) $[K^{b}]$ in Equation (16.14) will be semi-positive definite with the rank reduced by 3. For the unbounded medium, Equation (16.15a) leads to $\{u(\xi)\} \propto \xi^{\lambda_i}$. For $\lambda_1 = \lambda_2 = 0$, $\{u(\xi)\} =$ constant results, which corresponds to rigid body translational motions. For $\lambda_3 = -1\{u(\xi)\} \propto \xi^{-1}$ which diminishes towards zero for $\xi \rightarrow \infty$. The two translational modes will yield a semi-positive definite static-stiffness matrix $[K^{\infty}]$ with the rank reduced by 2. This means that a two-dimensional unbounded medium will undergo infinite displacements when subjected to static loads, in contrast to the three-dimensional case. This fact leads to an extension of the solution procedure of the scaled boundary finite element method for two-dimensional unbounded media, which is discussed in section 16.2.

16.2 UNBOUNDED TWO-DIMENSIONAL MEDIUM

In contrast to unbounded three-dimensional and axisymmetric media, unbounded two-dimensional media such as plane stress and plane strain domains undergo infinite displacement when subjected to static loads, even when the zero displacement boundary condition at infinity is enforced. This property creates a significant difficulty for displacement-based methods, such as the finite element method. However, the stress and strain fields for such problems are well behaved, and are of practical interest. The application of conventional displacement-based methods to these problems requires subtle modification of the problem to restrict the displacements to finite quantities.

Typically the domain is truncated at finite extent, small distributed springs are added over the domain, or the material stiffness is increased slightly but progressively away from the region of load application. Such procedures are unsatisfactory in several respects: the true problem is not solved, and aspects of the solution may be lost; the effect of the modification on the solution must be gauged; and the addition of very small parameters to stabilise the solution may result in the inversion of ill-conditioned matrices, with consequential numerical problems.

For the scaled boundary finite element method, two- dimensional cases for statics are calculated for practical cases by replacing the semi-positive definite coefficient matrix $[E^2]$ (Equation (15.98c)) by $[E^2] + \varepsilon G r_0^{s-2}$ (shear modulus G, characteristic length with respect to the boundary r_0), where ε is a very small positive dimensionless coefficient. Typically, ε equals 10^{-6} to 10^{-12}. This augmented $[E^2]$ becomes positive definite and the same also applies for the final result, the static-stiffness matrix $[K^\infty]$ [W9].

This section, which follows Reference [D6] closely, extends the solution procedure for unbounded two-dimensional media for statics without introducing any fudge: the original problem definition is not modified. Such an extension is possible because the approximate solution is obtained analytically in the radial direction. All aspects of the analytical solution are captured without numerical problems, and the complete analytical solution is obtained in the limit as the line finite element size on the boundary tends to zero.

Next, the static behaviour of a semi-infinite rod of varying cross-section is examined. This problem illustrates concisely the existence of situations in which displacements are infinite but well behaved. Then, the solution method of the scaled boundary finite element method is extended to include the unique deformation modes present in two-dimensional unbounded problems.

The behaviour of a rod (bar) of varying cross-sectional area is considered. The solution process for this problem reflects many aspects of the analytical part of the scaled boundary finite element solution process, and the problem itself illustrates the existence of displacement fields that are infinite, but otherwise well behaved. For these reasons the equations are developed in some detail. The situation is illustrated in Figure 16.2. For convenience, a coordinate axis x is defined from an origin O located so that the start of the rod is located at $x = x_0$. The x axis forms the central axis of the rod. The cross-sectional area is taken to vary as a power function of x, so that the cross-sectional area of the rod at any position x is

$$A(x) = A_0 \left(\frac{x}{x_0} \right)^n \tag{16.22}$$

where A_0 is the cross-sectional area of the rod at $x = x_0$, and n is defined by the geometry of the problem. (For example, $n = 2$ describes a semi-infinite truncated cone.) The displacement of the rod in the x-direction at any point is denoted as $u(x)$, while the force exerted on the end of the rod is P. Young's modulus of the rod is denoted as E. With these definitions the axial strain and stress in the rod can be expressed as

$$\varepsilon(x) = u(x)_{,x} \tag{16.23}$$

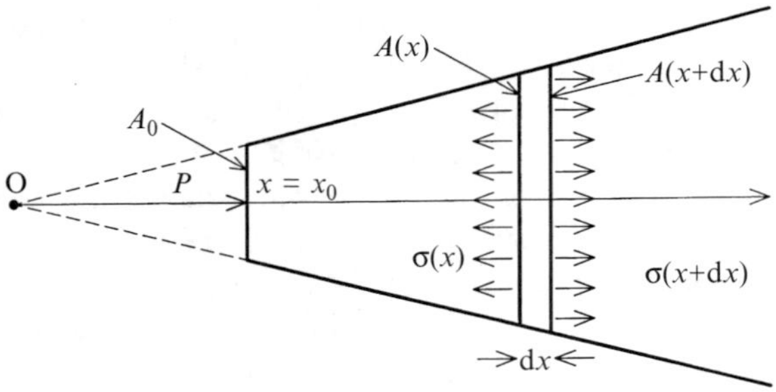

Figure 16.2 Semi-infinite axial rod with varying cross-section (Reproduced by permission of John Wiley & Sons Ltd)

and

$$\sigma(x) = Eu(x)_{,x} \tag{16.24}$$

respectively. Examining the equilibrium of a slice of the rod of infinitesimal thickness $\mathrm{d}x$ (illustrated in Figure 16.2):

$$\sigma(x + \mathrm{d}x)A(x + \mathrm{d}x) - \sigma(x)A(x) = 0 \tag{16.25}$$

With

$$\sigma(x + \mathrm{d}x) \approx \sigma(x) + \sigma(x)_{,x}\mathrm{d}x \tag{16.26}$$

and

$$A(x + \mathrm{d}x) \approx A(x) + A(x)_{,x}\mathrm{d}x \tag{16.27}$$

Substituting Equations (16.26) and (16.27) into (16.25) and simplifying, the equilibrium requirement becomes

$$\sigma(x)_{,x} A(x)\mathrm{d}x + \sigma(x)A(x)_{,x} \mathrm{d}x + \sigma(x)_{,x} A(x)_{,x} \mathrm{d}x^2 = 0 \tag{16.28}$$

Dividing through by $\mathrm{d}x$, and then taking the limit of each term as $\mathrm{d}x \to 0$, reduces this to

$$\sigma(x)_{,x} A(x) + \sigma(x)A(x)_{,x} = 0 \tag{16.29}$$

Substituting Equations (16.22) and (16.24) leads to the governing differential equation

$$EA_0\left(\frac{x}{x_0}\right)^n u(x)_{,xx} + \frac{nEA_0}{x_0}\left(\frac{x}{x_0}\right)^{n-1} u(x)_{,x} = 0 \tag{16.30}$$

or

$$x^2u(x)_{,xx} + n\,x\,u(x)_{,x} = 0 \tag{16.31}$$

For convenience, a dimensionless coordinate ξ is introduced, where

$$\xi = \frac{x}{x_0} \tag{16.32}$$

which implies

$$u(\xi) = u(x), \quad u(\xi)_{,\xi} = x_0 u(x)_{,x}, \quad u(\xi)_{,\xi\xi} = x_0^2 u(x)_{,xx} \tag{16.33}$$

Consequently, Equation (16.31) can be written as

$$\xi^2 u(\xi)_{,\xi\xi} + n\,\xi\, u(\xi)_{,\xi} = 0 \tag{16.34}$$

This equation is subject to two boundary conditions. The first is that the stress integrated over the area on the boundary at $x = x_0$ ($\xi = 1$) is in equilibrium with the applied load P, which requires

$$P = -\sigma(x_0)A(x_0) \tag{16.35}$$

or, on substitution of Equations (16.22), (16.24) and (16.33),

$$P = -EA_0 u(x_0)_{,x} = -\frac{EA_0}{x_0} u(\xi = 1)_{,\xi} \tag{16.36}$$

The second boundary condition can be constructed by requiring that the displacement approaches zero in the limit as ξ approaches infinity, or

$$\lim_{\xi \to \infty} u(\xi) = 0 \tag{16.37}$$

Inspection of the Euler–Cauchy equation (Equation (16.34)) reveals that the solutions may be found in the form

$$u(\xi) = \Phi\, \xi^{\lambda} + a \tag{16.38}$$

where Φ and a are constants. Substitution of Equation (16.38) into (16.34) yields the characteristic equation

$$(\lambda - 1)\lambda + n\lambda = 0 \tag{16.39}$$

This indicates the differential equation has two solutions of the assumed form, one with $\lambda = 0$, which corresponds to a rigid body translation of the rod, and one with $\lambda = 1 - n$. The rigid body translation cannot satisfy either boundary condition, and so need not be considered any further. If

$$u(\xi) = \Phi\, \xi^{1-n} + a \tag{16.40}$$

the first boundary condition (Equation (16.36)) requires that

$$\Phi = \frac{P x_0}{(n-1)EA_0} \tag{16.41}$$

If n is greater than 1

$$\lim_{\xi \to \infty} \xi^{1-n} \to 0 \tag{16.42}$$

and so the second boundary condition (Equation (16.37)) will be satisfied if $a = 0$. However, if n is less than one, this limit will be infinite, and so the second boundary condition will only be satisfied in the limit if

$$a = -\frac{P x_0}{(n-1)EA_0} \lim_{\xi \to \infty} \xi^{1-n} \tag{16.43}$$

implying that the displacement at any finite x is infinite. However, the stresses and strains throughout the rod can be computed, and the relative displacement between two finite points on the rod (ξ_1 and ξ_2) can be calculated as

$$u(\xi_2) - u(\xi_1) = \frac{P x_0}{(n-1)EA_0}(\xi_2^{1-n} - \xi_1^{1-n}) \tag{16.44}$$

If $n = 1$, corresponding to a wedge of uniform thickness but linearly increasing width, the above solution is not valid, as both roots of Equation (16.39) are zero, and so no solution in the form of Equation (16.38) exists that can satisfy the boundary conditions. In this case the original differential equation (Equation (16.34)) can be written as

$$\xi u(\xi)_{,\xi\xi} + u(\xi)_{,\xi} = 0 \tag{16.45}$$

By inspection, solutions can be found in the form

$$u(\xi) = \Psi \ln \xi + a \tag{16.46}$$

where Ψ and a are constants. Substitution of Equation (16.46) into (16.45) indicates that the differential equation is satisfied irrespective of the values of these constants. The constants are determined by the boundary conditions. Applying the first boundary condition (Equation (16.36)) yields

$$\Psi = -\frac{P x_0}{EA_0} \tag{16.47}$$

while the second boundary condition (Equation (16.37)) requires that

$$a = \frac{P x_0}{EA_0} \lim_{\xi \to \infty} \ln \xi \tag{16.48}$$

Again, the stress and strain fields are well behaved, and, although the displacements are infinite at every finite ξ, the relative displacement between any two finite points can be computed as

$$u(\xi_2) - u(\xi_1) = -\frac{P\,x_0}{EA_0}(\ln\xi_2 - \ln\xi_1) \tag{16.49}$$

Substituting Equations (16.46) with (16.47) into (16.24), the stress along the bar is

$$\sigma(\xi) = -\frac{P}{A_0}\frac{1}{\xi} \tag{16.50}$$

which is inversely proportional to the distance from the origin of the coordinate system.

It can be shown that the vertical displacements below a concentrated load acting on the surface of a half-plane exhibit the same behaviour as described above for the wedge (rod with $n = 1$), Equations (16.46), (16.49) and (16.50) with ξ denoting the distance from the free surface.

The special treatment which is necessary to solve unbounded two-dimensional media is discussed in detail. Denoting with n the number of degrees of freedom on the boundary, a total of n modes representing solutions to Equation (16.2), and satisfying the boundary conditions at infinity are required. The two rigid body translations corresponding to zero eigenvalues cannot be used as they do not satisfy the boundary conditions at infinity. The standard procedure described in section 16.1 thus obtains $n - 2$ of these in the form of Equation (16.15a). Two additional modes are required to complete the solution. The solution process cannot continue with only $n - 2$ modes, as the $[\Phi_{12}]$-matrix (which contains the modal displacements at the boundary in the columns) is not square, and Equation (16.16) cannot be solved. The remainder of this section will determine the two additional modes, permitting solution of unbounded two-dimensional domains. A parallel can be drawn with the axial behaviour of a semi-infinite rod of linearly varying cross-section discussed above.

Noting the parallel between the repeated rigid body solutions determined for the rod of linearly varying section and those determined from Equation (16.7), each additional mode of deformation satisfying the scaled boundary finite element equation in displacement (Equation (16.2)) is sought with unknown $\{\Psi\}$ and $\{\Phi\}$ in the form

$$\{u(\xi)\} = \{\Psi\}\ln\xi + \{\Phi\} \tag{16.51}$$

which is analogous to Equation (16.46). Here $\{\Phi\}$ represents a vector of constants. The symbol $\{\Phi\}$ is used for these constants because, as in Equation (16.15a), the vector contains the modal node displacements at the boundary $\xi = 1$. The vector $\{\Psi\}$ also contains constants. Both $\{\Psi\}$ and $\{\Phi\}$ are determined by the requirement that the deformation mode represented by Equation (16.51) satisfies Equation (16.2) for all values of ξ. Substitution of the mode into Equation (16.2) with $s = 2$ yields

$$-[E^0]\{\Psi\} + ([E^0] + [E^1]^T - [E^1])\{\Psi\} - [E^2](\{\Psi\}\ln\xi + \{\Phi\}) = 0 \tag{16.52}$$

or

$$([E^1]^T - [E^1])\{\Psi\} - [E^2]\{\Psi\}\ln\xi - [E^2]\{\Phi\} = 0 \tag{16.53}$$

This equation will be satisfied for all ξ if both

$$[E^2]\{\Psi\} = 0 \tag{16.54}$$

and

$$[E^2]\{\Phi\} = ([E^1]^T - [E^1])\{\Psi\} \tag{16.55}$$

Applying Equation (15.98c) with substitution of Equation (15.95b) indicates that if

$$[N(\eta)]_{,\eta}\{\Psi\} = 0 \tag{16.56}$$

then $\{\Psi\}$ satisfies Equation (16.54). Furthermore, the transpose of Equation (15.98b) with the substitution of Equations (15.95b) and (16.56) indicates that the same $\{\Psi\}$ must satisfy

$$[E^1]^T\{\Psi\} = 0 \tag{16.57}$$

Consequently, from Equation (16.55), for each of the logarithmic modes the vector $\{\Phi\}$ must satisfy the requirement that

$$[E^2]\{\Phi\} = -[E^1]\{\Psi\} \tag{16.58}$$

Note that satisfaction of Equation (16.56) does not imply

$$[N(\eta)]\{\Psi\} = 0 \tag{16.59}$$

and so, in general, Equations (15.98b) and (15.98c) imply

$$[E^1]\{\Psi\} \neq 0 \tag{16.60}$$

while Equations (15.98a) and (15.95a) imply

$$[E^0]\{\Psi\} \neq 0 \tag{16.61}$$

Returning to Equation (16.56), and as $[N(\eta)]\{\Psi\}$ denotes boundary displacements, this equation can be interpreted as requiring that the derivatives of the boundary displacements with respect to the boundary coordinate η are zero at each node. Consequently, vectors of nodal displacements representing rigid body translation of the boundary will be solutions of this equation. For convenience, the vector of boundary nodal displacements representing a rigid body translation of the boundary in the x-direction is designated $\{\Psi_x\}$, and the corresponding ridig body translation of the boundary in the y-direction is designated $\{\Psi_y\}$. In each of these vectors, the non-zero displacements are normalised to unity. Both of the vectors satisfy

Equation (16.56), and hence Equations (16.54) and (16.57). Since the two vectors are independent, two additional *logarithmic modes* can be obtained in the form of Equation (16.51). However, the corresponding vectors $\{\Phi_x\}$ and $\{\Phi_y\}$ must be determined for each to satisfy Equation (16.58), so that the modes, in turn, satisfy Equation (16.2).

However, $[E^2]$ is singular, having eigenvectors $\{\Psi_x\}$ and $\{\Psi_y\}$ corresponding to zero eigenvalues (as the vectors satisfy Equation (16.54)). Consequently, Equation (16.58) represents a linear system of n equations, only $n-2$ of which are independent. By introducing two constraint equations assigning zero to the displacements of a single node, the augmented set of equations can be solved for a vector $\{\Phi'\}$, which consists of the boundary node displacements relative to the selected node. Equation (16.58) is then satisfied for any vector $\{\Phi\}$ of the form

$$\{\Phi\} = \{\Phi'\} + a_x\{\Psi_x\} + a_y\{\Psi_y\} \tag{16.62}$$

where a_x and a_y are constants. The internal nodal forces at the boundary in equilibium with a logarithmic mode are found by substitution of Equation (16.51) into (16.4) (which is evaluated at $\xi = 1$) as

$$\{q\} = [E^0]\{\Psi\} + [E^1]^T\{\Phi\} \tag{16.63}$$

Introducing Equations (16.62) and (16.57), the modal node forces become

$$\{q\} = [E^0]\{\Psi\} + [E^1]^T\{\Phi'\} \tag{16.64}$$

These forces are independent of the node selected to have zero displacements, and do not depend on the constants a_x and a_y.

The two additional modes are now virtually complete. Taking $\{\Psi\}$ as $\{\Psi_x\}$ and substituting Equation (16.62) into (16.51), the nodal displacements and boundary nodal forces for the first mode can be expressed as

$$\{u(\xi)_x\} = \{\Psi_x\}(\ln\xi + a_{xx}) + \{\Phi'_x\} + a_{yx}\{\Psi_y\} \tag{16.65}$$

and

$$\{q_x\} = [E^0]\{\Psi_x\} + [E^1]^T\{\Phi'_x\} \tag{16.66}$$

where the additional x subscript indicates quantities for the logarithmic mode corresponding to rigid body motion of the boundary nodes in the x-direction.

Proceeding in the same manner for the second logarithmic mode corresponding to rigid body motion of the boundary nodes in the y-direction,

$$\{u(\xi)_y\} = \{\Psi_y\}(\ln\xi + a_{yy}) + \{\Phi'_y\} + a_{xy}\{\Psi_x\} \tag{16.67}$$

and

$$\{q_y\} = [E^0]\{\Psi_y\} + [E^1]^T\{\Phi'_y\} \tag{16.68}$$

follow.

There are four final constants to determine, a_{xx}, a_{yy}, a_{xy} and a_{yx}. To prevent the modal displacements from becoming infinite for $\xi \to \infty$, the first two constants should be

$$a_{xx} = a_{yy} = -\lim_{\xi\to\infty} \ln \xi \tag{16.69}$$

These terms represent the infinite part of displacement field expected over the unbounded domain. However, since only the relative displacements are of practical interest, the computation of which leads to the cancellation of the integration constants (as in Equation (16.49)), both a_{xx} and a_{yy} can be set to zero without any effect on the stress or the strain fields.

The final two constants are dictated by the Maxwell–Betti Reciprocal Theorem, which requires that the work calculated as the product of the nodal displacements of the first mode and the nodal forces of the second mode is equal to the work determined as the product of the nodal displacements of the second mode and the nodal forces of the first mode, both determined on the boundary, or

$$\{u(\xi = 1)_x\}^T\{q_y\} = \{u(\xi = 1)_y\}^T\{q_x\} \tag{16.70}$$

which on substitution of Equations (16.65) and (16.67) evaluated at the boundary and Equations (16.66) and (16.68) becomes

$$\begin{aligned} &a_{xx}\{\Psi_x\}^T\{q_y\} + (\{\Phi'_x\} + a_{yx}\{\Psi_y\})^T\{q_y\} \\ &\quad = a_{yy}\{\Psi_y\}^T\{q_x\} + (\{\Phi'_y\} + a_{xy}\{\Psi_x\})^T\{q_x\} \end{aligned} \tag{16.71}$$

The term $\{\Psi_x\}^T\{q_y\}$ expands to

$$\begin{aligned} \{\Psi_x\}^T\{q_y\} &= \{\Psi_x\}^T\{[E^0]\{\Psi_y\} + [E^1]^T\{\Phi'_y\}\} \\ &= \{\Psi_x\}^T[E^0]\{\Psi_y\} + \{\Psi_x\}^T[E^1]^T\{\Phi'_y\} \\ &= \{\Psi_x\}^T[E^0]\{\Psi_y\} - \{\Phi'_x\}^T[E^2]^T\{\Phi'_y\} \end{aligned} \tag{16.72}$$

on substitution of Equation (16.58). Similarly $\{\Psi_y\}^T\{q_x\}$ expands to

$$\{\Psi_y\}^T\{q_x\} = \{\Psi_y\}^T[E^0]\{\Psi_x\} - \{\Phi'_y\}^T[E^2]^T\{\Phi'_x\} \tag{16.73}$$

and so, since $[E^0]$ and $[E^2]$ are both symmetric,

$$\{\Psi_x\}^T\{q_y\} = \{\Psi_y\}^T\{q_x\} \tag{16.74}$$

Equating constants a_{xx} and a_{yy}, the reciprocal requirement is reduced to

$$(\{\Phi'_x\} + a_{yx}\{\Psi_y\})^T\{q_y\} = (\{\Phi'_y\} + a_{xy}\{\Psi_x\})^T\{q_x\} \tag{16.75}$$

which also follows directly from Equation (16.71) with $a_{xx} = a_{yy} = 0$.

This requirement is satisfied when both

$$(\{\Phi'_x\} + a_{yx}\{\Psi_y\})^T\{q_y\} = 0 \tag{16.76}$$

and

$$(\{\Phi'_y\} + a_{xy}\{\Psi_x\})^T\{q_x\} = 0 \tag{16.77}$$

are enforced, which requires that

$$a_{yx} = -\frac{\{\Phi'_x\}^T\{q_y\}}{\{\Psi_y\}^T\{q_y\}} \tag{16.78}$$

and

$$a_{xy} = -\frac{\{\Phi'_y\}^T\{q_x\}}{\{\Psi_x\}^T\{q_x\}} \tag{16.79}$$

The final representations of the two logarithmic modes used to complete the set of n independent displacement modes (satisfying the scaled boundary finite element equation in displacement) required to allow solution of a model with n degrees of freedom become

$$\{u(\xi)_x\} = \{\Psi_x\}\ln\xi + \{\Phi_x\} \tag{16.80}$$

with

$$\{\Phi_x\} = \{\Phi'_x\} - \frac{\{\Phi'_x\}^T\{q_y\}}{\{\Psi_y\}^T\{q_y\}}\{\Psi_y\} \tag{16.81}$$

where $\{\Phi'_x\}$ satisfies Equation (16.58) with $\{\Psi\} = \{\Psi_x\}$, and

$$\{u(\xi)_y\} = \{\Psi_y\}\ln\xi + \{\Phi_y\} \tag{16.82}$$

with

$$\{\Phi_y\} = \{\Phi'_y\} - \frac{\{\Phi'_y\}^T\{q_x\}}{\{\Psi_x\}^T\{q_x\}}\{\Psi_x\} \tag{16.83}$$

where $\{\Phi'_y\}$ satisfies Equation (16.58) with $\{\Psi\} = \{\Psi_y\}$. $\{q_x\}$ and $\{q_y\}$ are specified in Equations (16.66) and (16.68).

Equations (16.80) and (16.82), multiplied by the integration constants c_{2n-1} and c_{2n} respectively, are added to Equation (16.15a). The same also applies for Equations (16.66) and (16.68), but both formulated for any ξ, with respect to Equation (16.15b). n independent degrees of freedom are present. The analysis can now proceed analogously as in the three-dimensional case.

The displacements in the interior of the domain equal (Equations (15.93), (16.15a), (16.80), (16.82))

$$\{u(\xi,\eta)\} = [N(\eta)]\begin{pmatrix} \sum_{i=1}^{n-2} \{\Phi_{12i}\}\xi^{\lambda_i} c_{2i} + (\{\Psi_x\}\ln\xi + \{\Phi_x\})c_{2n-1} \\ +(\{\Psi_y\}\ln\xi + \{\Phi_y\})c_{2n} \end{pmatrix} \tag{16.84}$$

Displacement relative to a particular point can be determined by evaluating Equation (16.84) at the reference point, and then subtracting this displacement from the displacement computed at the point of interest. The absolute displacement computed by Equation (16.84) has no physical significance.

Substituting $\{u(\xi)\}$ of Equation (16.84) into Equation (15.94) and considering Equation (16.56) with Equation (15.95b) ($[B^2]\{\Psi_x\} = 0$, $[B^2]\{\Psi_y\} = 0$), the stress field is obtained as

$$\{\sigma(\xi,\eta)\} = [D]\begin{pmatrix} \sum_{i=1}^{n-2} (\lambda_i[B^1] + [B^2])\{\Phi_{12i}\}\xi^{\lambda_i - 1} c_{2i} \\ +([B^1]\{\Psi_x\} + [B^2]\{\Phi_x\})\frac{1}{\xi} c_{2n-1} \\ +([B^1]\{\Psi_y\} + [B^2]\{\Phi_y\})\frac{1}{\xi} c_{2n} \end{pmatrix} \tag{16.85}$$

For large ξ, the stresses associated with the logarithmic modes vary in the radial direction in proportion to $(1/\xi)$, which is consistent with the behaviour of the semi-infinite rod of a linearly varying cross-section (Equation (16.50)).

The unbounded two-dimensional domain without restraints on any side-faces present discussed in this section up to now is the standard case. It is illustrated in Figure 16.3a, where a semi-infinite wedge with two free side-faces is presented. Should either side-face be restrained in one direction (Figure 16.3b), n-1 modes with non-zero eigenvalues will be computed with Equation (16.7). Only the logarithmic mode corresponding to the unrestrained direction is then employed. The requirement of stiffness matrix symmetry is satisfied automatically, and so the constant a_{xy} (or a_{yx} as appropriate) is taken as zero. With the restrained direction denoted by the y-axis (Figure 16.3b), the logarithmic mode becomes

$$\{u(\xi)_x\} = \{\Psi_x\}\ln\xi + \{\Phi'_x\} \tag{16.86}$$

where $\{\Psi_x\}$ denotes the rigid body translation of the boundary nodes in the unrestrained direction x, and $\{\Phi'_x\}$ the solution of (Equation (16.58))

$$[E^2]\{\Phi'_x\} = -[E^1]\{\Psi_x\} \tag{16.87}$$

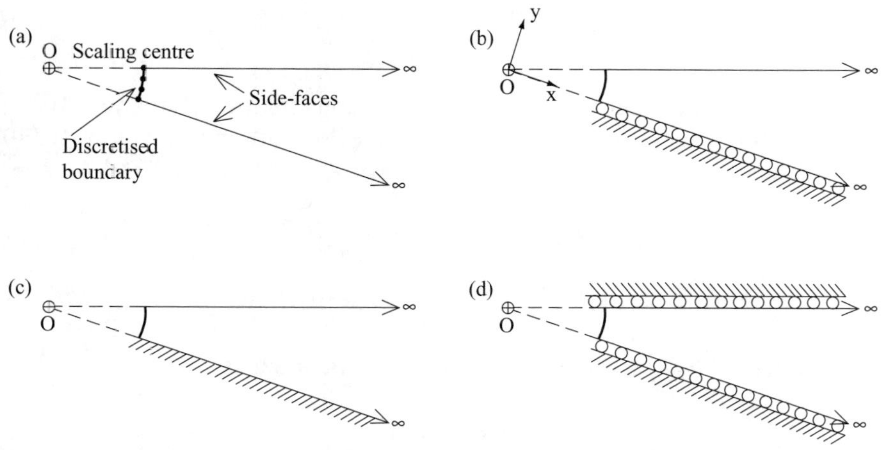

Figure 16.3 Semi-infinite wedge with (a) free side-faces, yielding two logarithmic modes, (b) one free side-face and one side-face restrained in one direction, yielding one logarithmic mode, (c) (d) side-faces restrained in two directions, yielding no logarithmic modes (Reproduced by permission of John Wiley & Sons Ltd)

with the displacement of the node in the x-direction in $\{\Phi'_x\}$ set to zero. Should restraint be provided in two coordinate directions on the side-faces, n modes with positive eigenvalues will be found, and neither of the logarithmic modes enters into the solution. Two such cases are shown in Figures 16.3c and 16.3d.

16.3 BODY LOAD

After determining the solution for the homogeneous case of the scaled boundary finite element equation in displacement (Equation (16.2)), that of the non-homogeneous case (Equation (15.111))

$$\begin{aligned}[E^0]\xi^2\{u(\xi)\}_{,\xi\xi} &+ ((s-1)[E^0]-[E^1]+[E^1]^T)\xi\{u(\xi)\}_{,\xi} \\ &+ ((s-2)[E^1]^T-[E^2])\{u(\xi)\}+\{F(\xi)\}=0\end{aligned} \tag{16.88}$$

follows from applying the technique of variation of parameters. $\{F(\xi)\}$ represents body loads, prescribed surface tractions on A_t and prescribed displacements on A_u, when applicable (see Equation (15.62) with $\{F(\xi)\}$ in Equation (16.88) including $\{F_u(\xi)\}$). Equation (16.88) is again transformed to first-order differential equations by introducing the variable $\{X(\xi)\}$ defined in Equation (16.3):

$$\xi\{X(\xi)\}_{,\xi} = -[Z]\{X(\xi)\} - \xi^{0.5(s-2)}\begin{Bmatrix}\{0\}\\ \{F(\xi)\}\end{Bmatrix} \tag{16.89}$$

with $[Z]$ specified in Equation (16.6). The solution of Equation (16.89) is postulated as

$$\{X(\xi)\} = \begin{bmatrix} [\Phi_{11}] & [\Phi_{12}] \\ [\Phi_{21}] & [\Phi_{22}] \end{bmatrix} \begin{bmatrix} \lceil \xi^{-\lambda_i} \rfloor & \\ & \lceil \xi^{\lambda_i} \rfloor \end{bmatrix} \begin{Bmatrix} \{c_1(\xi)\} \\ \{c_2(\xi)\} \end{Bmatrix} \tag{16.90}$$

Note that the integration constants of the homogeneous solution (Equation (16.10)) are replaced by $\{c_1(\xi)\}$, $\{c_2(\xi)\}$ which are functions of ξ. Substituting Equation (16.90) into (16.89) leads to

$$\{c_1(\xi)\}_{,\xi} = -\xi^{0.5s-2} \lceil \xi^{\lambda_i} \rfloor [A_{12}]\{F(\xi)\} \tag{16.91a}$$

$$\{c_2(\xi)\}_{,\xi} = -\xi^{0.5s-2} \lceil \xi^{-\lambda_i} \rfloor [A_{22}]\{F(\xi)\} \tag{16.91b}$$

where the inverse of $[\Phi]$ partitioned conformably is denoted as

$$[A] = \begin{bmatrix} [A_{11}] & [A_{12}] \\ [A_{21}] & [A_{22}] \end{bmatrix} = [\Phi]^{-1} \tag{16.92}$$

Its solution is formulated as

$$\{c_1(\xi)\} = \{c_1\} - \int_1^{\xi} \tau^{0.5s-2} \lceil \tau^{\lambda_i} \rfloor [A_{12}]\{F(\tau)\}\mathrm{d}\tau \tag{16.93a}$$

$$\{c_2(\xi)\} = \{c_2\} - \int_1^{\xi} \tau^{0.5s-2} \lceil \tau^{-\lambda_i} \rfloor [A_{22}]\{F(\tau)\}\mathrm{d}\tau \tag{16.93b}$$

$\{c_1\}$, $\{c_2\}$ are integration constants. Equations (16.90) and (16.3) yield

$$\{u(\xi)\} = \xi^{-0.5(s-2)}([\Phi_{11}] \lceil \xi^{-\lambda_i} \rfloor \{c_1(\xi)\} + [\Phi_{12}] \lceil \xi^{\lambda_i} \rfloor \{c_2(\xi)\}) \tag{16.94a}$$

$$\{Q(\xi)\} = \xi^{+0.5(s-2)}([\Phi_{21}] \lceil \xi^{-\lambda_i} \rfloor \{c_1(\xi)\} + [\Phi_{22}] \lceil \xi^{\lambda_i} \rfloor \{c_2(\xi)\}) \tag{16.94b}$$

The integration constants are determined from the boundary conditions.

For a bounded medium, the displacements at $\xi = 0$ must remain finite. This condition $\{u(\xi = 0)\}$ leads to $\{c_2(\xi = 0)\} = 0$ in Equation (16.94a), as the real parts of all λ_i are negative. Equation (16.93b) formulated at $\xi = 0$ yields

$$\{c_2\} = \int_1^{0} \tau^{0.5s-2} \lceil \tau^{-\lambda_i} \rfloor [A_{22}]\{F(\tau)\}\mathrm{d}\tau \tag{16.95}$$

Substituting into Equation (16.93b) leads to

$$\{c_2(\xi)\} = -\int_0^{\xi} \tau^{0.5s-2} \lceil \tau^{-\lambda_i} \rfloor [A_{22}]\{F(\tau)\}\mathrm{d}\tau \tag{16.96}$$

Substituting Equations (16.93a) and (16.96) into (16.94a) results in

$$\{u(\xi)\} = \xi^{-0.5(s-2)}([\Phi_{11}]\lceil\xi^{-\lambda_i}\rfloor(\{c_1\} + \int_{\xi}^{1}\tau^{0.5s-2}\lceil\tau^{\lambda_i}\rfloor[A_{12}]\{F(\tau)\}\mathrm{d}\tau) \\ -[\Phi_{12}]\lceil\xi^{\lambda_i}\rfloor\int_{0}^{\xi}\tau^{0.5s-2}\lceil\tau^{-\lambda_i}\rfloor[A_{22}]\{F(\tau)\}\mathrm{d}\tau)) \tag{16.97}$$

The displacements on the boundary $\{u(\xi = 1)\}$ determine $\{c_1\}$.

To determine the nodal loads $\{R^{\mathrm{F}}\}$ in Equation (16.1), Equation (16.96) is substituted into Equation (16.94b), which is formulated at the boundary $\xi = 1$. With Equation (15.68) $\{R\} = \{Q(\xi = 1)\}$ for the bounded medium

$$\{R\} = [\Phi_{21}]\{c_1\} - [\Phi_{22}]\int_{0}^{1}\tau^{0.5s-2}\lceil\tau^{-\lambda_i}\rfloor[A_{22}]\{F(\tau)\}\mathrm{d}\tau \tag{16.98a}$$

results with $\{c_1\} = \{c_1(\xi = 1)\}$ (Equation (16.93a)). Formulating Equation (16.97) for $\xi = 1$ yields

$$\{u(\xi = 1)\} = [\Phi_{11}]\{c_1\} - [\Phi_{12}]\int_{0}^{1}\tau^{0.5s-2}\lceil\tau^{-\lambda_i}\rfloor[A_{22}]\{F(\tau)\}\mathrm{d}\tau \tag{16.98b}$$

Eliminating $\{c_1\}$ from Equations (16.98a) and (16.98b) leads to

$$\{R\} = [\Phi_{21}][\Phi_{11}]^{-1}\{u(\xi = 1)\} \\ - ([\Phi_{22}] - [\Phi_{21}][\Phi_{11}]^{-1}[\Phi_{12}])\int_{0}^{1}\tau^{0.5s-2}\lceil\tau^{-\lambda_i}\rfloor[A_{22}]\{F(\tau)\}\mathrm{d}\tau \tag{16.99}$$

The coefficient matrix of $\{u(\xi = 1)\}$ is equal to $[K^{\mathrm{b}}]$ (Equation (16.14)). From the definition of $[A]$ (Equation (16.92))

$$[A_{22}]^{-1} = [\Phi_{22}] - [\Phi_{21}][\Phi_{11}]^{-1}[\Phi_{12}] \tag{16.100}$$

applies. Comparing Equation (16.99) with Equation (16.1) determines the nodal loads $\{R^{\mathrm{F}}\}$

$$\{R^{\mathrm{F}}\} = [A_{22}]^{-1}\int_{0}^{1}\tau^{0.5s-2}\lceil\tau^{-\lambda_i}\rfloor[A_{22}]\{F(\tau)\}\mathrm{d}\tau \tag{16.101}$$

For an unbounded medium, the displacements for $\xi \to \infty$ must remain finite. This condition $\{u(\xi \to \infty)\}$ leads to $\{c_1(\xi \to \infty)\} = 0$ in Equation (16.94a), as the real parts of all λ_i are negative. Equation (16.93a) formulated for $\xi \to \infty$ yields

$$\{c_1\} = \int_{1}^{\infty}\tau^{0.5s-2}\lceil\tau^{\lambda_i}\rfloor[A_{12}]\{F(\tau)\}\mathrm{d}\tau \tag{16.102}$$

Substituting into Equation (16.93a) leads to

$$\{c_1(\xi)\} = \int_{\xi}^{\infty} \tau^{0.5s-2} \lceil \tau^{\lambda_i} \rfloor [A_{12}]\{F(\tau)\}\mathrm{d}\tau \tag{16.103}$$

Substituting Equations (16.103) and (16.93b) into (16.94a) results in

$$\begin{aligned}\{u(\xi)\} = \xi^{-0.5(s-2)}([\Phi_{11}] \lceil \xi^{-\lambda_i} \rfloor \int_{\xi}^{\infty} \tau^{0.5s-2} \lceil \tau^{\lambda_i} \rfloor [A_{12}]\{F(\tau)\}\mathrm{d}\tau \\ +[\Phi_{12}] \lceil \xi^{\lambda_i} \rfloor (\{c_2\} - \int_{1}^{\xi} \tau^{0.5s-2} \lceil \tau^{-\lambda_i} \rfloor [A_{22}]\{F(\tau)\}\mathrm{d}\tau))\end{aligned} \tag{16.104}$$

The displacements on the boundary $\{u(\xi = 1)\}$ determine $\{c_2\}$.

To determine the nodal loads $\{R^{\mathrm{F}}\}$ in Equation (16.1), Equation (16.103) is substituted into Equation (16.94b), which is formulated at the boundary $\xi = 1$. With Equation (15.68) $\{R\} = -\{Q(\xi = 1)\}$ for the unbounded medium

$$\{R\} = -[\Phi_{21}] \int_{1}^{\infty} \tau^{0.5s-2} \lceil \tau^{\lambda_i} \rfloor [A_{12}]\{F(\tau)\}\mathrm{d}\tau - [\Phi_{22}]\{c_2\} \tag{16.105a}$$

results with $\{c_2\} = \{c_2(\xi = 1)\}$ (Equation (16.93b)). Formulating Equation (16.104) for $\xi = 1$ yields

$$\{u(\xi = 1)\} = [\Phi_{11}] \int_{1}^{\infty} \tau^{0.5s-2} \lceil \tau^{\lambda_i} \rfloor [A_{12}]\{F(\tau)\}\mathrm{d}\tau + [\Phi_{12}]\{c_2\} \tag{16.105b}$$

Eliminating $\{c_2\}$ from Equation (16.105) leads to

$$\begin{aligned}\{R\} &= -[\Phi_{22}][\Phi_{12}]^{-1}\{u(\xi = 1)\} \\ &- ([\Phi_{21}] - [\Phi_{22}][\Phi_{12}]^{-1}[\Phi_{11}]) \int_{1}^{\infty} \tau^{0.5s-2} \lceil \tau^{\lambda_i} \rfloor [A_{12}]\{F(\tau)\}\mathrm{d}\tau\end{aligned} \tag{16.106}$$

The coefficient matrix of $\{u(\xi = 1)\}$ is equal to $[K^{\infty}]$ (Equation (16.16)). From the definition of $[A]$ (Equation (16.92))

$$[A_{12}]^{-1} = [\Phi_{21}] - [\Phi_{22}][\Phi_{12}]^{-1}[\Phi_{11}] \tag{16.107}$$

applies. Comparing Equation (16.106) with (16.1) determines the nodal loads $\{R^{\mathrm{F}}\}$

$$\{R^{\mathrm{F}}\} = [A_{12}]^{-1} \int_{1}^{\infty} \tau^{0.5s-2} \lceil \tau^{\lambda_i} \rfloor [A_{12}]\{F(\tau)\}\mathrm{d}\tau \tag{16.108}$$

16.4 CERTAIN DISTRIBUTIONS OF BODY LOAD

For a general distribution of body loads (and prescribed surface tractions on A_t), a closed-form expression for $\{F^{\mathrm{b}}(\xi)\}$ and $\{F^{\mathrm{t}}(\xi)\}$ does not exist (e.g. see Equations (15.52) and (15.51)). This also applies to the non-homogeneous term $\{F(\xi)\}$ (Equation (15.56)) in the scaled boundary finite element equation in displacement (Equation (16.88)). A numerical integration over the domain is necessary to evaluate the displacement (e.g. Equation (16.97)) and the nodal loads (e.g. Equation (16.101)).

For certain distributions of the body loads, the integrations in the ξ-direction can be performed analytically independent of those in the other directions. No integration over the time domain occurs. Two cases are addressed: a concentrated load; and a body load varying as a power function of the radial coordinate ξ.

16.4.1 Concentrated Load

In two-dimensional elastostatics a concentrated load acting at ξ_0, η_0 is formulated as

$$\{p(\xi, \eta)\} = \delta(\xi - \xi_0)\,\delta\,(\eta - \eta_0)\{P\} \tag{16.109}$$

The components of the load are defined in the Cartesian coordinates $\{P\} = [P_x\ P_y]^T$. The Dirac-delta function is denoted as $\delta(.)$. As an example a three-node line element shown in Figure 16.4a is addressed. Substituting in Equation (15.100) and performing the integration in the η-direction for the corresponding finite element on the boundary yields

$$\{F^{\mathrm{b}}(\xi)\} = \delta(\xi - \xi_0)[N_0]\{P\} \tag{16.110}$$

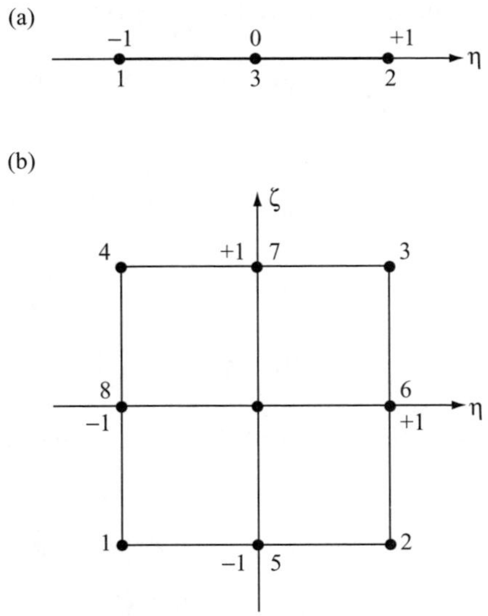

Figure 16.4 Parent element of curved finite element on boundary: (a) line element; (b) surface element (Reprinted from Computer Methods in Applied Mechanics and Engineering, Vol. 180, Song and Wolf, 'Body loads in scaled boundary finite element method', pp. 117–135, © (1999), with permission from Elsevier Science)

with

$$[N_0] = [N_1(\eta_0)[I] \quad N_2(\eta_0)[I] \quad N_3(\eta_0)[I]]^T |J(\eta_0)| \tag{16.111}$$

where $[I]$ is the 2×2 identity matrix. In three-dimensional elastostatics a concentrated load acting at ξ_0, η_0, ζ_0 is formulated as

$$\{p(\xi, \eta, \zeta)\} = \delta(\xi - \xi_0)\delta(\eta - \eta_0)\delta(\zeta - \zeta_0)\{P\} \tag{16.112}$$

with $\{P\} = [P_x \quad P_y \quad P_z]^T$. As an example the eight-node surface element shown in Figure 16.4b is examined. Substituting Equation (16.112) into (15.52) and performing the integration in the η- and ζ-directions yields

$$\{F^{\mathrm{b}}(\xi)\} = \delta(\xi - \xi_0)[N_0]\{P\} \tag{16.113}$$

with

$$[N_0] = [N_1(\eta_0, \zeta_0)[I] \quad N_2(\eta_0, \zeta_0)[I] \quad \cdots \quad N_8(\eta_0, \zeta_0)[I]]^T |J(\eta_0, \zeta_0)| \tag{16.114}$$

with the 3×3 identity matrix $[I]$. The contribution from the body loads (Equation (15.56)) to the scaled boundary finite element equation in displacement is unified for two and three-dimensions as

$$\{F(\xi)\} = \delta(\xi - \xi_0)\xi^2[N_0]\{P\} \tag{16.115}$$

For a bounded medium, the displacements follow from Equation (16.97) as

$$\{u(\xi)\} = \xi^{-0.5(s-2)}[\Phi_{11}]\lceil \xi^{-\lambda_i} \rfloor(\{c_1\} + \xi_0^{0.5s}\lceil \xi_0^{\lambda_i} \rfloor [A_{12}][N_0]\{P\}) \quad \text{for } \xi \leq \xi_0 \tag{16.116a}$$

$$\{u(\xi)\} = \xi^{-0.5(s-2)} \times ([\Phi_{11}]\lceil \xi^{-\lambda_i} \rfloor \{c_1\} - [\Phi_{12}]\lceil \xi^{\lambda_i} \rfloor \xi_0^{0.5s}\lceil \xi_0^{-\lambda_i} \rfloor [A_{22}][N_0]\{P\}) \quad \text{for } \xi \geq \xi_0 \tag{16.116b}$$

and the nodal loads from Equation (16.101)

$$\{R^{\mathrm{F}}\} = [A_{22}]^{-1}\xi_0^{0.5s}\lceil \xi_0^{-\lambda_i} \rfloor [A_{22}][N_0]\{P\} \tag{16.117}$$

For an unbounded medium, the displacements are calculated from Equation (16.104) as

$$\{u(\xi)\} = \xi^{-0.5(s-2)} \times ([\Phi_{11}]\lceil \xi^{-\lambda_i} \rfloor \xi_0^{0.5s}\lceil \xi_0^{\lambda_i} \rfloor [A_{12}][N_0]\{P\} + [\Phi_{12}]\lceil \xi^{\lambda_i} \rfloor \{c_2\}) \quad \text{for } \xi \leq \xi_0 \tag{16.118a}$$

$$\{u(\xi)\} = \xi^{-0.5(s-2)} \times [\Phi_{12}]\lceil \xi^{\lambda_i} \rfloor(\{c_2\} - \xi_0^{0.5s}\lceil \xi_0^{-\lambda_i} \rfloor [A_{22}][N_0]\{P\}) \quad \text{for } \xi \geq \xi_0 \tag{16.118b}$$

and the nodal loads from Equation (16.108)

$$\{R^{\mathrm{F}}\} = [A_{12}]^{-1}\xi_0^{0.5s}\lceil \xi_0^{\lambda_i} \rfloor [A_{12}][N_0]\{P\} \tag{16.119}$$

16.4.2 Load Varying as Power Function of Radial Coordinate

In two-dimensional elastostatics the distributed load equals

$$\{p(\xi,\eta)\} = \xi^d\{p_0(\eta)\} \tag{16.120}$$

The components of the load are defined in the Cartesian coordinates $\{p_0(\eta)\} = [p_{0x}(\eta) \quad p_{0y}(\eta)]^T$. The power d is a given constant. Substituting in Equation (15.100) and performing the integration in the η-direction yields

$$\{F^{\mathrm{b}}(\xi)\} = \xi^d\{P_0\} \tag{16.121}$$

with (Figure 16.4a)

$$\{P_0\} = \int_{-1}^{+1} [N_1[I] \quad N_2[I] \quad N_3[I]]^T\{p_0(\eta)\}|J|\mathrm{d}\eta \tag{16.122}$$

In three-dimensional elastostatics the distributed load equals

$$\{p(\xi,\eta,\zeta)\} = \xi^d\{p_0(\eta,\zeta)\} \tag{16.123}$$

with $\{p_0(\eta,\zeta)\} = [p_{0x}(\eta,\zeta) \quad p_{0y}(\eta,\zeta) \quad p_{0z}(\eta,\zeta)\,]^T$ and the constant power d. Substituting in Equation (15.52) and performing the integration in the η-, ζ-directions yields

$$\{F^{\mathrm{b}}(\xi)\} = \xi^d\{P_0\} \tag{16.124}$$

with (Figure 16.4b)

$$\{P_0\} = \int_{-1}^{+1}\int_{-1}^{+1} [N_1[I] \quad N_2[I] \quad \cdots \quad N_8[I]]^T\{p_0(\eta,\zeta)\}|J|\mathrm{d}\eta\mathrm{d}\zeta \tag{16.125}$$

The contribution from the body loads (Equation (15.56)) to the scaled boundary finite element equation in displacement is unified for two and three-dimensions as

$$\{F(\xi)\} = \xi^{d+2}\{P_0\} \tag{16.126}$$

For a bounded medium, the displacements follow from Equation (16.97) as

$$\begin{aligned}\{u(\xi)\} = {} & [\Phi_{11}](\xi^{-0.5(s-2)}\lceil \xi^{-\lambda_i} \rfloor\{c_1\} + \lceil d_i(\xi) \rfloor [A_{12}]\{P_0\}) \\ & - \xi^{d+2}[\Phi_{12}](-\lceil \lambda_i \rfloor + (0.5s + d + 1)[I])^{-1}[A_{22}]\{P_0\}\end{aligned} \tag{16.127}$$

with

$$d_i(\xi) = \begin{cases} \dfrac{\xi^{-\lambda_i-0.5s+1} - \xi^{d+2}}{\lambda_i + 0.5s + d + 1} & \text{for } \lambda_i + 0.5s + d \neq -1 \\ -\xi^{d+2} \ln \xi & \text{for } \lambda_i + 0.5s + d = -1 \end{cases} \tag{16.128}$$

and the nodal loads from Equation (16.101)

$$\{R^{\mathrm{F}}\} = [A_{22}]^{-1}(-\lceil \lambda_i \rfloor + (0.5s + 1 + d)[I])^{-1}[A_{22}]\{P_0\} \tag{16.129}$$

To achieve a finite displacement in the domain $\{u(\xi = 0)\}$, $d > -2$ must be satisfied.

For an unbounded medium, the displacements are calculated from Equation (16.104) as

$$\begin{aligned} \{u(\xi)\} = & -\xi^{d+2}[\Phi_{11}](\lceil \lambda_i \rfloor + (0.5s + d + 1)[I])^{-1}[A_{12}]\{P_0\} \\ & + [\Phi_{12}](\xi^{-0.5(s-2)}\lceil \xi^{\lambda_i} \rfloor \{c_2\} + \lceil d_i(\xi) \rfloor [A_{12}]\{P_0\}) \end{aligned} \tag{16.130}$$

with

$$d_i(\xi) = \begin{cases} \dfrac{\xi^{\lambda_i-0.5s+1} - \xi^{d+2}}{-\lambda_i + 0.5s + d + 1} & \text{for } -\lambda_i + 0.5s + d \neq -1 \\ -\xi^{d+2} \ln \xi & \text{for } -\lambda_i + 0.5s + d = -1 \end{cases} \tag{16.131}$$

and the nodal loads from Equation (16.108)

$$\{R^{\mathrm{F}}\} = [A_{12}]^{-1}(\lceil \lambda_i \rfloor + (0.5s + 1 + d)[I])^{-1}[A_{12}]\{P_0\} \tag{16.132}$$

To achieve a finite displacement for $\xi \to \infty$, $d < -2$ must be satisfied.

For the case of a load varying as a power function of the radial coordinate, a more straightforward approach is possible. Physical insight is gained expanding the modal interpretation of the solution process. This procedure, which is demonstrated for body loads, closely follows Reference [D5], where the case of prescribed surface tractions is also addressed.

For distributed loads with a variation ξ^d in the two- and three-dimensional cases specified in Equations (16.120) and (16.123), the non-homogeneous term $\{F(\xi)\}$ in the scaled boundary finite element equation in displacement (Equation (16.88)) varies as ξ^{d+2} (Equation (16.126)) with $\{P_0\}$ listed in Equations (16.122) and (16.125). A general solution to the non-homogeneous differential equation (Equation (16.88)) may be sought as a linear combination of the general solution of the homogeneous version (Equation (16.2)), and the particular solution of the same form as the terms $\xi^{d+2}\{P_0\}$. Since the general solution of Equation (16.2) is interpreted above as the combination of deformation modes, each of which satisfies internal equilibrium in the ξ-direction, the particular solution can also be interpreted as modes of deformation, which satisfy internal equilibrium but with the body loads. The modes representing the general solution of Equation (16.2) will be referred to as the 'homogeneous' modes, allowing differentiation between the mode types. The particular solution consisting of the body load mode displacements is of the form

$$\{u_b(\xi)\} = \xi^{d+2}\{\Phi_b\} \tag{16.133}$$

Substituting Equation (16.133) into (16.88) yields

$$((d+2)(s+d)[E^0] - (d+2)[E^1] + (s+d)[E^1]^T - [E^2])\{\Phi_b\} + \{P_0\} = 0 \tag{16.134}$$

Solving for $\{\Phi_b\}$, the nodal displacements for the body load mode are obtained as

$$\{\Phi_b\} = -((d+2)(s+d)[E^0] - (d+2)[E^1] + (s+d)[E^1]^T - [E^2])^{-1}\{P_0\} \tag{16.135}$$

(The case of a singular coefficient matrix is excluded. This corresponds to the eigenvalues $\pm\lambda_i = -1 - 0.5s - d$, which leads to logarithmic modes, as is visible from Equations (16.128) and (16.131) – see also section 16.2.) The equivalent internal nodal forces follow from substituting Equation (16.133) into (16.4):

$$\{Q_b(\xi)\} = \xi^{s+d}((d+2)[E^0] + [E^1]^T)\{\Phi_b\} \tag{16.136}$$

For a bounded medium, the solution to the inhomogeneous differential equations equals the sum of the homogeneous solution (Equation (16.12a)) and the particular solution (Equation (16.133)):

$$\{u(\xi)\} = \xi^{-0.5(s-2)}[\Phi_{11}]\lceil\xi^{-\lambda_i}\rfloor\{c_1\} + \xi^{d+2}\{\Phi_b\} \tag{16.137}$$

Analogously, for the internal nodal forces (Equations (16.12b), (16.136)):

$$\{Q(\xi)\} = \xi^{+0.5(s-2)}[\Phi_{21}]\lceil\xi^{-\lambda_i}\rfloor\{c_1\} + \xi^{s+d}((d+2)[E^0] + [E^1]^T)\{\Phi_b\} \tag{16.138}$$

results. Eliminating $\{c_1\}$ from Equations (16.137) and (16.138) and with $\{R\} = \{Q(\xi = 1)\}$ (Equation (15.68)) yields

$$\{R\} = [\Phi_{21}][\Phi_{11}]^{-1}\{u\} - ([\Phi_{21}][\Phi_{11}]^{-1} - (d+2)[E^0] + [E^1]^T)\{\Phi_b\} \tag{16.139}$$

In this nodal force displacement relationship on the boundary (Equation (16.1)), the coefficient matrix of $\{u\}$ is equal to the static-stiffness matrix $[K^{\mathrm{b}}]$ (Equation (16.14)), and the second term on the right-hand side represents $-\{R^{\mathrm{F}}\}$

$$\{R^{\mathrm{F}}\} = ([\Phi_{21}][\Phi_{11}]^{-1} - (d+2)[E^0] + [E^1]^T)\{\Phi_b\} \tag{16.140}$$

with $\{\Phi_b\}$ specified in Equation (16.135). Interpreting this equation physically, the first term denotes the nodal force required to counteract the displacement $\{\Phi_b\}$ on the boundary, and the second term is equal to the internal nodal force $\{Q_b\}$ on the boundary.

Analogously, for an unbounded medium (Equations (16.15a), (16.133)):

$$\{u(\xi)\} = \xi^{-0.5(s-2)}[\Phi_{12}]\lceil\xi^{\lambda_i}\rfloor\{c_2\} + \xi^{d+2}\{\Phi_b\} \tag{16.141}$$

and (Equations (16.15b), (16.136))

$$\{Q(\xi)\} = \xi^{+0.5(s-2)}[\Phi_{22}]\lceil\xi^{\lambda_i}\rfloor\{c_2\} + \xi^{s+d}((d+2)[E^0] + [E^1]^T)\{\Phi_b\} \tag{16.142}$$

apply. Eliminating $\{c_2\}$ from Equations (16.141) and (16.142), and with $\{R\} = -\{Q(\xi = 1)\}$ (Equation (15.68)) yields

$$\{R\} = -[\Phi_{22}][\Phi_{12}]^{-1}\{u\} - (-[\Phi_{22}][\Phi_{12}]^{-1} - (d+2)[E^0] + [E^1]^T)\{\Phi_b\} \tag{16.143}$$

Again (Equation (16.1)), the coefficient matrix of $\{u\}$ is equal to $[K^{\infty}]$ (Equation (16.16)), and the second term represents $-\{R^{\mathrm{F}}\}$

$$\{R^{\mathrm{F}}\} = (-[\Phi_{22}][\Phi_{12}]^{-1} - (d+2)[E^0] + [E^1]^T)\{\Phi_b\} \tag{16.144}$$

For the cases of prescribed surface tractions on A_t and prescribed displacements on A_u varying as power functions of the radial coordinate, analogous procedures for the particular solutions are possible.

16.5 INITIAL STRESS

Often it is appropriate to assume that at the outset of an analysis a state of residual stress exists. This initial stress must be known, determined for instance from measurements. As an example, tectonic stresses in rock can be mentioned. The corresponding displacements are assumed to vanish. This initial stress state applies to a simple structural system, called the free field, for instance a full plane or a half-plane with a horizontal free surface, for which an analytical solution, which is simple, exists. Obviously the initial stress is self-equilibrating. The initial stress $\{\sigma_0\}$ of the free field is simply added to the result $\{\sigma'\}$ of any static or dynamic analysis which as examples can address applied loads or a modification of the geometry by excavation.

The procedure is outlined for a simple example emphasising the advantages of the scaled boundary finite element method. The stress $\{\sigma\}$ in an irregular half-plane representing soil with an excavation and a horizontal and an inclined part of the free surface extending to infinity is to be determined (Figure 16.5a). The initial stress $\{\sigma_0\}$ consisting of constant normal stress in the horizontal direction acting in a full plane, representing the free field, is known (Figure 16.5b). As mentioned above, the displacements in the free field are set equal to zero and the surface tractions on a line in the free field corresponding to the free surface and the boundary of the excavation of the irregular half-plane can be determined easily. Applying those surface tractions together with the horizontal initial stress to the irregular half-plane leads to the same stress state as in the free field (Figure 16.5c). As the free surface and excavation boundary are stress free, the same surface tractions, but acting in the opposite direction, are applied to the irregular

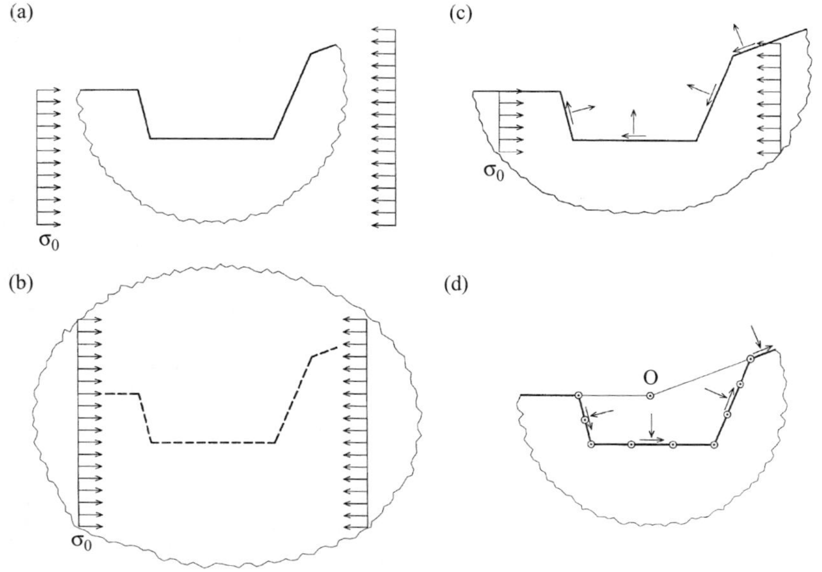

Figure 16.5 (a) Problem definition, (b) initial stress defined in full-plane, (c) irregular half-plane with surface tractions and initial stress of full-plane, (d) irregular half-plane with surface tractions acting in opposite direction

half-plane (Figure 16.5d). This analysis is efficiently performed with the scaled boundary finite element method. The scaling centre O is selected at the intersection of the extensions of the two parts of the free surface. The latter are thus side-faces with prescribed surface tractions and are not discretised. The spatial discretisation is limited to the boundary of the excavation. This part of the analysis will lead to stresses $\{\sigma'\}$ and displacements $\{u\}$. The final stress $\{\sigma\}$ satisfying the stress-free boundary conditions is equal to the superposition of $\{\sigma_0\}$ and $\{\sigma'\}$, determined from Figures 16.5b and 16.5d, respectively, and the final displacements are $\{u\}$.

The degrees of freedom are restricted to the boundary of the excavation which in a typical unbounded soil-structure-interaction analysis coincides with the structure-soil interface. The static-stiffness matrix with respect to these degrees of freedom and the corresponding loads due to the initial stress are determined. This procedure for statics is analogous to that in dynamics to analyse dynamic unbounded soil-structure interaction for seismic excitation, as outlined in Appendix C. The irregular half-plane corresponds to the soil system ground (Figure C.3). The free-field motion results in displacements (which are not set equal to zero as in the initial stress analysis) and in surface tractions on the two parts of the free surface and on the excavation boundary. The latter define the corresponding dynamic load. With the dynamic-stiffness matrix discretised in the degrees of freedom on the boundary of the excavation (structure-soil interface), the displacements in the nodes are calculated. Adding the free-field displacements yields the so-called effective seismic foundation input motion.

The significant advantages of the scaled boundary finite element method when calculating the (dynamic) stiffness also apply for the load corresponding to an initial stress state or to seismic excitation.

17 Mass Matrix of Bounded Medium

To perform a dynamic analysis of a bounded medium in the time domain, the mass matrix $[M^b]$ is required in addition to the static-stiffness matrix. The total domain or when substructuring is applied a subdomain is addressed. In both cases, degrees of freedom occur only on the boundary.

Two procedures leading to identical results exist to calculate $[M^b]$. In the first, the analytical solution of the displacements for statics can be used to derive shape functions in the radial direction, and the standard method of finite elements to calculate the mass matrix is followed. In the second, the low-frequency asymptotic expansion of the dynamic-stiffness matrix which involves the mass matrix is substituted into the scaled boundary finite element equation in dynamic stiffness, and addressing terms of order in $i\omega$ yields a (linear) equation in $[M^b]$.

Section 17.1 (based on Reference [S7]) examines the formulation for elastodynamics, and section 17.2 (based on Reference [S9]) outlines the slight modifications for diffusion.

17.1 ELASTODYNAMICS

To determine the mass matrix $[M^b]$ with respect to the degrees of freedom on the boundary, the analytical expression for the static displacements (Equation (16.12a) can be used to derive the shape functions in the ξ-direction. The standard procedure to calculate the mass matrix in finite elements can be followed.

The integration constants $\{c_1\}$ in Equation (16.12a) are expressed by the displacements on the boundary $\{u(\xi = 1)\} = \{u\}$

$$\{u\} = [\Phi_{11}]\{c_1\} \tag{17.1}$$

Solving for $\{c_1\}$ and substituting into Equation (16.12a) leads to

$$\{u(\xi)\} = \xi^{-0.5(s-2)}[\Phi_{11}] \lceil \xi^{-\lambda_i} \rfloor [\Phi_{11}]^{-1}\{u\} \tag{17.2}$$

The coefficient matrix of $\{u\}$ in Equation (17.2) represents the shape function in the ξ-direction. The shape functions $[N_V]$ in the domain V (subscript V for volume) follow for elastostatics in two dimensions (three-node line element in Figure 16.4a) where $s = 2$

$$\{u(\xi,\eta)\} = [N_1[I] \quad N_2[I] \quad N_3[I]]\{u(\xi)\} \tag{17.3}$$

with the 2×2 identity matrix $[I]$ and in three dimensions (eight-node surface element in Figure 16.4b)

$$\{u(\xi,\eta,\zeta)\} = [N_1[I] \quad N_2[I] \quad \ldots \quad N_8[I]]\{u(\xi)\} \tag{17.4}$$

with the 3×3 identity matrix $[I]$. For instance, for three-dimensional elastostatics

$$\{u(\xi,\eta,\zeta)\} = [N_V]\{u\} \tag{17.5}$$

where ($s = 3$)

$$[N_V] = [N_1[I] \quad N_2[I] \quad \ldots \quad N_8[I]]\xi^{-0.5}[\Phi_{11}]\lceil \xi^{-\lambda_i} \rfloor [\Phi_{11}]^{-1} \tag{17.6}$$

The mass matrix is calculated as

$$[M^b] = \int [N_V]^T \rho [N_V] \mathrm{d}V \tag{17.7}$$

with the mass density ρ. Substituting $\mathrm{d}V$ in Equation (15.92) or (15.29) and $[N_V]$ in Equation (17.7) yields

$$[M^b] = ([\Phi_{11}]^{-1})^T \int_0^1 \lceil \xi^{-\lambda_i} \rfloor [\Phi_{11}]^T [M^0][\Phi_{11}] \lceil \xi^{-\lambda_i} \rfloor \xi \mathrm{d}\xi \, [\Phi_{11}]^{-1} \tag{17.8}$$

where the coefficient matrix is defined as in two dimensions for elastostatics

$$[M^0] = \int_{-1}^{+1} [N_1[I] \quad N_2[I] \quad N_3[I]]^T \rho [N_1[I] \quad N_2[I] \quad N_3[I]] |J| \mathrm{d}\eta \tag{17.9}$$

and in three-dimensions for elastostatics

$$\begin{aligned} [M^0] = \int_{-1}^{+1}\int_{-1}^{+1} & [N_1[I] \quad N_2[I] \quad \ldots \quad N_8[I]]^T \\ & \times \rho [N_1[I] \quad N_2[I] \quad \ldots \quad N_8[I]] \, |J| \mathrm{d}\eta \mathrm{d}\zeta \end{aligned} \tag{17.10}$$

To be able to perform the integration in the ξ-direction analytically, the abbreviation

$$[m^0] = [\Phi_{11}]^T [M^0][\Phi_{11}] \tag{17.11}$$

is introduced. Each element of the matrix

$$[m^b] = \int_0^1 \lceil \xi^{-\lambda_i} \rfloor [m^0] \lceil \xi^{-\lambda_i} \rfloor \xi \mathrm{d}\xi \tag{17.12}$$

is evaluated analytically

$$m_{ij}^{b} = \int_{0}^{1} \xi^{-\lambda_i} m_{ij}^{0} \xi^{-\lambda_j} \xi \mathrm{d}\xi = \frac{m_{ij}^{0}}{-\lambda_i - \lambda_j + 2} \tag{17.13}$$

as the real parts of λ_i, λ_j are negative. The mass matrix follows from Equations (17.8) and (17.12) with Equation (17.11) as

$$[M^{b}] = ([\Phi_{11}]^{-1})^{T}[m^{b}][\Phi_{11}]^{-1} \tag{17.14}$$

Alternatively, $[M^{b}]$ can also be determined from the low-frequency asymptotic expansion of the dynamic-stiffness matrix $[S^{b}(\omega)]$. As in structural dynamics, $[S^{b}(\omega)]$ of the bounded medium is postulated as (Equation (15.86))

$$[S^{b}(\omega)] = [K^{b}] - \omega^{2}[M^{b}] \tag{17.15}$$

Terms of order in ω equal to or higher than four are neglected in this expansion. The static-stiffness matrix $[K^{b}]$ satisfies Equation (16.19).

Substituting Equation (17.15) into the scaled boundary finite element equation in dynamic stiffness (Equation (15.116)) leads to a constant term independent of ω, a term in ω^{2} and higher-order terms in ω, which are neglected. The constant term, which is equal to the left-hand side of Equation (16.19) vanishes. The coefficient matrix of the remaining term in ω^{2} is written as

$$\begin{aligned}&\left((-[K^{b}] + [E^{1}])[E^{0}]^{-1} - \frac{s}{2}[I]\right)[M^{b}] \\ &+ [M^{b}]\left([E^{0}]^{-1}(-[K^{b}] + [E^{1}]^{T}) - \frac{s}{2}[I]\right) + [M^{0}] = 0\end{aligned} \tag{17.16}$$

This is a *linear* equation to calculate the mass matrix $[M^{b}]$. Equation (17.16) is the Lyapunov equation

$$[A][X] + [X][A]^{T} = [C] \tag{17.17}$$

with the unknown $[X] = [M^{b}]$ and the other matrices defined straightforwardly. The solution procedure is described in Reference [B1]. Alternatively, intermediate results of the Schur factorisation [W9] can be used. $[M^{b}]$ is positive definite as $[M^{0}]$ has the same property.

Note that in the scaled boundary finite element method, the *mass matrix* follows from the low-frequency behaviour of the equation *without any additional assumptions*.

To be able to perform a dynamic analysis, a sufficient number of degrees of freedom must be present. For instance, at least 4–10 nodal values per wave length are necessary. To satisfy this requirement, substructuring may be necessary, which is discussed in Chapter 22.

The solution of the eigenvalue problem in statics (section 16.1) is computationally involved. In principle, this operation has to be performed for each subdomain.

Although only a small number of unknowns is processed for each subdomain, the total computational effort can be significant.

An attractive procedure in substructuring divides the domain into subdomains with boundaries which are similar to the boundary of the total domain (Figure 17.1). Or in other words, by scaling the boundary varying the scaling factor, a series of similar boundary finite element cells, which are the subdomains, is created. To calculate the static-stiffness and mass matrices of these subdomains, the same eigenvalues and eigenvectors as for the total domain apply. Thus, only one eigenvalue problem has to be solved.

The procedure is sketched as follows (Figure 17.1). To determine the static-stiffness matrix $[K^{\mathrm{b}}]$ and mass matrix $[M^{\mathrm{b}}]$ of the subdomain (boundary finite element cell) with degrees of freedom on the interior boundaries at ξ_i and on the exterior boundary at $\xi_e(\xi_i \leq \xi \leq \xi_e)$, the eigenvalue problem corresponding to the total domain with the boundary at $\xi = 1$ is solved. The solution is specified in Equation (16.11) as

$$\{u(\xi)\} = \xi^{-0.5(s-2)}([\Phi_{11}]\lceil \xi^{-\lambda_j} \rfloor\{c_1\} + [\Phi_{12}]\lceil \xi^{\lambda_j} \rfloor\{c_2\}) \tag{17.18a}$$

$$\{Q(\xi)\} = \xi^{+0.5(s-2)}([\Phi_{21}]\lceil \xi^{-\lambda_j} \rfloor\{c_1\} + [\Phi_{22}]\lceil \xi^{\lambda_j} \rfloor\{c_2\}) \tag{17.18b}$$

For the subdomain with $\xi_i \leq \xi \leq \xi_e$, for which Equation (17.18) applies, the scaling centre O lies outside the (bounded) subdomain. To determine the integration constants $\{c_1\}$, $\{c_2\}$, Equation (17.18a) is formulated for the degrees of freedom at the interior boundary $\{u_i\}$ with ξ_i and at the exterior boundary $\{u_e\}$ with ξ_e

$$\{u_i\} = \xi_i^{-0.5(s-2)}([\Phi_{11}]\lceil \xi_i^{-\lambda_j} \rfloor\{c_1\} + [\Phi_{12}]\lceil \xi_i^{\lambda_j} \rfloor\{c_2\}) \tag{17.19a}$$

$$\{u_e\} = \xi_e^{-0.5(s-2)}([\Phi_{21}]\lceil \xi_e^{-\lambda_j} \rfloor\{c_1\} + [\Phi_{22}]\lceil \xi_e^{\lambda_j} \rfloor\{c_2\}) \tag{17.19b}$$

Solving Equation (17.19) for $\{c_1\}$, $\{c_2\}$ and substituting in Equation (17.18a) yields the static displacements $\{u(\xi)\}$ as a function of the nodal values $\{u_i\}$, $\{u_e\}$ with ξ as the coordinate. This leads to the (analytical) shape functions in the ξ-direction for the

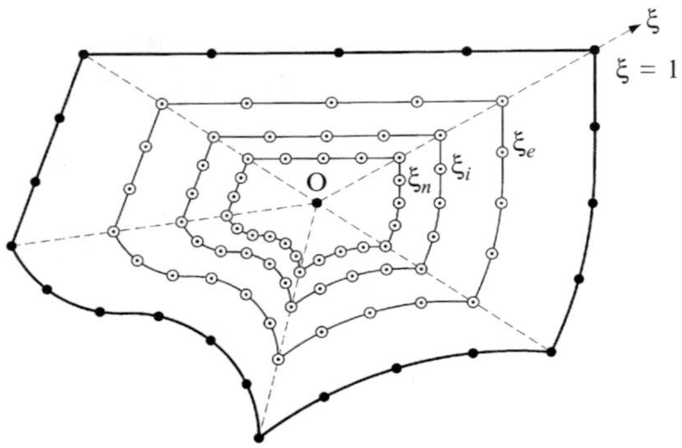

Figure 17.1 Substructuring into subdomains with similar boundaries (section)

subdomain. Multiplying $\{u(\xi)\}$ by the shape functions in the circumferential direction(s) (Equations (17.3) and (17.4) for the two- and three-dimensional cases) yields the final shape functions $[N_V]$. For instance, for the three-dimensional case

$$\{u(\xi,\eta,\zeta)\} = [N_V]\left\{\begin{array}{c}\{u_i\}\\ \{u_e\}\end{array}\right\} \tag{17.20}$$

applies. $[N_V]$ incorporates the polynomial variation in the η-, ζ-directions and the analytical variation in the ξ-direction. Equation (17.20) is in the same form as Equation (17.5). The remaining steps to calculate the mass matrix $[M^{\mathrm{b}}]$ specified in Equation (17.7) are the same as before. Again, the integration in the ξ-direction is performed analytically.

To determine the static-stiffness matrix $[K^{\mathrm{b}}]$ of the subdomain, the integration constants $\{c_1\}$, $\{c_2\}$ are substituted in Equation (17.18b), which is then evaluated for ξ_i and ξ_e. This yields the internal nodal forces $\{Q_i\}$ and $\{Q_e\}$ as a function of $\{u_i\}$, $\{u_e\}$. As the interior boundary and exterior boundary are negative and positive faces, respectively, of the (bounded) subdomain, the corresponding nodal forces are equal to (Equation (15.68))

$$\{R_i\} = -\{Q_i\} \tag{17.21a}$$

$$\{R_e\} = \{Q_e\} \tag{17.21b}$$

$[K^{\mathrm{b}}]$ then follows straightforwardly.

The property matrices of the innermost subdomain containing the scaling centre with its (only) boundary at ξ_n (Figure 17.1) follow from those of the total domain at $\xi = 1$ (Equations (16.14) and (17.14))

$$[K^{\mathrm{b}}(\xi = \xi_n)] = \xi_n^{s-2}[K^{\mathrm{b}}(\xi = 1)] \tag{17.22a}$$

$$[M^{\mathrm{b}}(\xi = \xi_n)] = \xi_n^{s}[M^{\mathrm{b}}(\xi = 1)] \tag{17.22b}$$

An even more efficient procedure results, when the radial coordinates of the boundaries of the subdomains (boundary finite element cells) are equal to the terms in a geometric series. This case is shown in Figure 17.1. Denoting the radial coordinate of the boundary of the total domain as $\xi_0 = 1$ and numbering the adjacent boundaries towards the scaling centre in ascending order $1, 2, \ldots k-1, k, \ldots n$, the ratio $\gamma = \xi_k/\xi_{k-1}$ is a constant < 1 for $k = 1, 2, \ldots, n$. The property matrices of the kth subdomain between the boundaries $\xi_i = \xi_k$ and $\xi_e = \xi_{k-1}$ equal

$$[K_k^{\mathrm{b}}] = \gamma^{(k-1)(s-2)}[K_1^{\mathrm{b}}] \tag{17.23a}$$

$$[M_k^{\mathrm{b}}] = \gamma^{(k-1)s}[M_1^{\mathrm{b}}] \tag{17.23b}$$

with $[K_1^b]$ and $[M_1^b]$ denoting the property matrices of the first subdomain ($k = 1$) between the boundaries $\xi_i = \xi_1$ and $\xi_e = \xi_0 = 1$. The static-stiffness and mass matrices of only one subdomain, the first one, are determined following the procedure outlined above.

17.2 DIFFUSION

As discussed in connection with Equation (15.137), ω^2 in elastodynamics is replaced by -iω to formulate the corresponding relationship in diffusion. Addressing the low-frequency asymptotic expansion of the dynamic-stiffness matrix $[S^b(\omega)]$, Equation (17.15) yields for diffusion

$$[S^b(\omega)] = [K^b] + i\omega[M^b] \tag{17.24}$$

Terms of order in iω equal to or higher than two are neglected in this expansion valid for low frequencies. The steady-state stiffness matrix $[K^b]$, determined using Equation (16.14), satisfies Equation (15.156) for $\omega = 0$.

Substituting Equation (17.24) into (15.156) leads to a constant term independent of iω, a term in iω and higher-order terms in iω which are neglected. The constant term, which is equal to Equation (15.156) formulated for $\omega = 0$, vanishes. The coefficient matrix of the remaining term in iω is written as

$$\begin{aligned}&\left((-[K^b]+[E^1])[E^0]^{-1}-\frac{s}{2}[I]\right)[M^b]\\&+[M^b]\left([E^0]^{-1}(-[K^b]+[E^1]^T)-\frac{s}{2}[I]\right)+[M^0]=0\end{aligned} \tag{17.25}$$

This is a linear equation to calculate the mass matrix $[M^b]$, which is identical to Equation (17.16) for elastodynamics. To further simplify the coefficient matrix, the eigenvalue problem (Equation (16.7)) is addressed. The first submatrix of the product in Equation (16.7) using Equations (16.6), (16.8) and (16.9), postmultiplied by $[\Phi_{11}]^{-1}$, leads to

$$[E^0]^{-1}(-[K^b]+[E^1]^T) - 0.5(s-2)[I] = [\Phi_{11}]\lceil\lambda_i\rfloor[\Phi_{11}]^{-1} \tag{17.26}$$

with $[K^b]$ defined in Equation (16.14). Substituting Equation (17.26) into (17.25) results in

$$-([\Phi_{11}]^{-1})^T\lceil\lambda_i\rfloor[\Phi_{11}]^T[M^b] - [M^b][\Phi_{11}]\lceil\lambda_i\rfloor[\Phi_{11}]^{-1} + 2[M^b] = [M^0] \tag{17.27}$$

Premultiplying Equation (17.27) by $[\Phi_{11}]^T$ and postmultiplying by $[\Phi_{11}]$ yields

$$([I]-\lceil\lambda_i\rfloor)[m^b] + [m^b]([I]-\lceil\lambda_i\rfloor) = [\Phi_{11}]^T[M^0][\Phi_{11}] \tag{17.28}$$

with the transformation

$$[m^b] = [\Phi_{11}]^T[M^b][\Phi_{11}] \tag{17.29}$$

As $\lceil \lambda_i \rfloor$ is a diagonal matrix, Equation (17.28) is solved by simple back substitution. As the real parts of the eigenvalues $\lceil \lambda_i \rfloor$ are negative, a solution of Equation (17.28) always exists. The mass matrix is obtained from Equation (17.29) as

$$[M^{\mathrm{b}}] = ([\Phi_{11}]^{-1})^T [m^{\mathrm{b}}][\Phi_{11}]^{-1} \tag{17.30}$$

The additional computational effort starting from $[K^{\mathrm{b}}]$ to calculate $[M^{\mathrm{b}}]$ is small.

18 High-frequency Asymptotic Expansion for Dynamic Stiffness of Unbounded Medium

For a bounded medium the mass matrix representing the inertial forces is the only additional matrix required to capture dynamic forces. The discretised equations of motion are formulated and solved as in the standard finite element method. The construction of the mass matrix in the scaled boundary finite element method is described in Chapter 17.

For an unbounded medium where a mass matrix does not exist the procedure is more complicated. To achieve a *unique solution*, boundary conditions at infinity must be formulated. This so-called *radiation condition* states for elastodynamics that no energy is radiated from infinity into the medium [S2].

In the scaled boundary finite element equations in displacement (Equation (15.112)) and in dynamic stiffness (Equation (15.117) for $\xi^{2-s}[S^{\infty}(\omega, \xi)]$) the independent variable is equal to $\bar{\xi} = \xi\omega$. Infinity corresponds to the dimensionless radial coordinate $\xi \to \infty$, which results in $\bar{\xi} \to \infty$. This limit $\bar{\xi} \to \infty$ can also be achieved with $\omega \to \infty$. Thus, to model the behaviour at infinity, the high-frequency limit $\omega \to \infty$ can be addressed.

This chapter examines the construction of a high-frequency asymptotic expansion for the dynamic-stiffness matrix, which permits the radiation condition to be satisfied exactly, and provides a starting value for the dynamic-stiffness matrix at a high but finite frequency for the numerical solution (Chapter 19), and analytical solution (Chapter 20) of the scaled boundary finite element equation in dynamic stiffness. This chapter is based on References [W9], [W11].

As already mentioned, the radiation condition addresses energy. The rate of energy transmission N through a boundary is defined as the product of the real part of the nodal forces $\{R\}e^{i\omega t}$, and of the real part of the velocity $\{\dot{u}\}e^{i\omega t}$, averaged over a period $2\pi/\omega$

$$N = \frac{\omega}{2\pi} \int_{0}^{2\pi/\omega} \mathrm{Re}(\{R\}^T e^{i\omega t})\mathrm{Re}(\{\dot{u}\}e^{i\omega t})\mathrm{d}t \tag{18.1}$$

The two vectors equal

$$\mathrm{Re}(\{R\}\mathrm{e}^{\mathrm{i}\omega t}) = \mathrm{Re}\{R\}\cos\omega t - \mathrm{Im}\{R\}\sin\omega t \tag{18.2a}$$

$$\mathrm{Re}(\{\dot{u}\}\mathrm{e}^{\mathrm{i}\omega t}) = \mathrm{Re}\{\dot{u}\}\cos\omega t - \mathrm{Im}\{\dot{u}\}\sin\omega t \tag{18.2b}$$

Substituting Equation (18.2) into Equation (18.1) and integrating over time leads to

$$N = \frac{1}{2}(\mathrm{Re}\{R\}^T\mathrm{Re}\{\dot{u}\} + \mathrm{Im}\{R\}^T\mathrm{Im}\{\dot{u}\}) \tag{18.3}$$

With

$$\{\dot{u}\} = \mathrm{i}\omega\{u\} = -\omega\mathrm{Im}\{u\} + \mathrm{i}\omega\mathrm{Re}\{u\} \tag{18.4}$$

$\mathrm{Re}\{\dot{u}\}$ and $\mathrm{Im}\{\dot{u}\}$ are identified.

The nodal force-displacement relationship for vanishing body loads equals, for an unbounded medium (section 15.4),

$$\{R\} = [S^\infty]\{u\} \tag{18.5}$$

Decomposing the dynamic-stiffness matrix into its real and imaginary parts

$$[S^\infty] = \mathrm{Re}[S^\infty] + \mathrm{i}\mathrm{Im}[S^\infty] \tag{18.6}$$

and substituting into Equation (18.5) results in

$$\mathrm{Re}\{R\} = \mathrm{Re}[S^\infty]\mathrm{Re}\{u\} - \mathrm{Im}[S^\infty]\mathrm{Im}\{u\} \tag{18.7a}$$

$$\mathrm{Im}\{R\} = \mathrm{Im}[S^\infty]\mathrm{Re}\{u\} + \mathrm{Re}[S^\infty]\mathrm{Im}\{u\} \tag{18.7b}$$

Substituting Equations (18.7) and (18.4) into Equation (18.3) yields

$$N = \frac{\omega}{2}(\mathrm{Re}\{u\}^T\mathrm{Im}[S^\infty]\mathrm{Re}\{u\} + \mathrm{Im}\{u\}^T\mathrm{Im}[S^\infty]\mathrm{Im}\{u\}^T) \tag{18.8}$$

Introducing the complex conjugate vector

$$\{u^*\} = \mathrm{Re}\{u\} - \mathrm{i}\mathrm{Im}\{u\} \tag{18.9}$$

Equation (18.8) is written as

$$N = \frac{\omega}{2}\{u\}^T\mathrm{Im}[S^\infty]\{u^*\} \tag{18.10}$$

Note that in the operation leading to Equation (18.8), the term with $\mathrm{Re}[S^\infty]$ cancels. The rate of energy transmission thus depends only upon $\mathrm{Im}[S^\infty]$. For a positive N the imaginary part of the dynamic-stiffness matrix (or the *damping matrix*) must be *positive definite* in the quadratic form.

As the radiation condition addresses the dynamic-stiffness matrix towards infinity, which can be achieved with $\omega \to \infty$, the asymptotic behaviour at high frequency of the scaled boundary finite element equation in dynamic stiffness (Equations (15.110) and Equation (15.117)) is examined.

$$(\xi^{2-s}[S^{\infty}(\omega,\xi)] + [E^1])[E^0]^{-1}(\xi^{2-s}[S^{\infty}(\omega,\xi)] + [E^1]^T) - (s-2)\xi^{2-s}[S^{\infty}(\omega,\xi)]$$
$$-\bar{\xi}(\xi^{2-s}[S^{\infty}(\omega,\xi)])_{,\bar{\xi}} - [E^2] + \bar{\xi}^2[M^0] = 0 \tag{18.11}$$

The dynamic-stiffness matrix $\xi^{2-s}[S^{\infty}(\omega,\ \xi)]$ at high frequency is expanded in a polynomial of $i\bar{\xi}$ in descending order starting at one

$$\xi^{2-s}[S^{\infty}(\omega,\xi)] \approx i\bar{\xi}[C_{\infty}] + [K_{\infty}] + \sum_{j=1}^{m} \frac{1}{(i\bar{\xi})^j}[A_j] \tag{18.12}$$

The first two terms on the right-hand side represent the constant dashpot matrix $[C_{\infty}]$ and the constant spring matrix $[K_{\infty}]$ (subscript ∞ for $\bar{\xi} \to \infty$ or with $\xi = 1,\ \omega \to \infty$). As discussed in Appendix B, they represent the singular part ($[S_s^{\infty}(\omega)]$ for $\xi = 1$ in Equation (B.35)) and the third term denotes the asymptotic expansion of the regular part ($[S_r^{\infty}(\omega)]$ in Equation (B.38)). All coefficient matrices $[C_{\infty}]$, $[K_{\infty}]$ and $[A_j]$ ($j = 1,\ 2, \ldots, m$) are unknown.

A concise formulation results when the transformation based on the following eigenvalue problem is introduced

$$[M^0][\Phi] = [E^0][\Phi]\lceil \Lambda^2 \rfloor \tag{18.13}$$

$[M^0]$ and $[E^0]$ are positive definite matrices resulting in positive eigenvalues $\lceil \Lambda^2 \rfloor$. The eigenvectors $[\Phi]$ are normalised as

$$[\Phi]^T[E^0][\Phi] = [I] \tag{18.14}$$

yielding

$$[\Phi]^T[M^0][\Phi] = \lceil \Lambda^2 \rfloor \tag{18.15}$$

and

$$[E^0]^{-1} = [\Phi][\Phi]^T \tag{18.16}$$

Premultiplying Equation (18.11) by $[\Phi]^T$ and post-multiplying by $[\Phi]$ results in

$$([s^{\infty}(\bar{\xi})] + [e^1])([s^{\infty}(\bar{\xi})] + [e^1]^T) - (s-2)[s^{\infty}(\bar{\xi})] - \bar{\xi}[s^{\infty}(\bar{\xi})]_{,\bar{\xi}} - [e^2] + \bar{\xi}^2\lceil \Lambda^2 \rfloor = 0 \tag{18.17}$$

where

$$[s^{\infty}(\bar{\xi})] = [\Phi]^T\xi^{2-s}[S^{\infty}(\omega,\xi)][\Phi] \tag{18.18}$$

and

$$[e^1] = [\Phi]^T [E^1] [\Phi] \tag{18.19a}$$

$$[e^2] = [\Phi]^T [E^2] [\Phi] \tag{18.19b}$$

Substituting Equation (18.12) into Equation (18.18) yields

$$[s^\infty(\bar{\xi})] \approx \mathrm{i}\bar{\xi}[c_\infty] + [k_\infty] + \sum_{j=1}^{m} \frac{1}{(\mathrm{i}\bar{\xi})^j}[a_j] \tag{18.20}$$

with

$$[c_\infty] = [\Phi]^T [C_\infty] [\Phi] \tag{18.21a}$$

$$[k_\infty] = [\Phi]^T [K_\infty] [\Phi] \tag{18.21b}$$

$$[a_j] = [\Phi]^T [A_j] [\Phi] \tag{18.21c}$$

Substituting Equation (18.20) into (18.17), and rearranging in descending order of the power series $\mathrm{i}\bar{\xi}$, leads to

$$\begin{aligned}
&(\mathrm{i}\bar{\xi})^2([c_\infty]^2 - \lceil \Lambda^2 \rfloor) + \mathrm{i}\bar{\xi}([c_\infty][k_\infty] + [k_\infty][c_\infty] + [c_\infty][e^1]^T + [e^1][c_\infty] - (s-1)[c_\infty]) \\
&\quad + [c_\infty][a_1] + [a_1][c_\infty] + ([k_\infty] + [e^1])([k_\infty] + [e^1]^T) - (s-2)[k_\infty] - [e^2] \\
&\quad + \frac{1}{\mathrm{i}\bar{\xi}}([c_\infty][a_2] + [a_2][c_\infty] + ([k_\infty] + [e^1])[a_1] + [a_1]([k_\infty] + [e^1]^T) \\
&\quad - (s-3)[a_1]) \approx 0
\end{aligned} \tag{18.22}$$

In Equation (18.22), $m = 2$ is selected. The generalisation to any m leading to additional terms is straightforward.

The coefficient matrix of each term of the power series in $\mathrm{i}\bar{\xi}$ is set equal to zero in descending order. The first yields

$$[c_\infty]^2 = \lceil \Lambda^2 \rfloor \tag{18.23}$$

Selecting the positive roots of each element on the diagonal of $\lceil \Lambda^2 \rfloor$ leads to

$$[c_\infty] = \lceil \Lambda \rfloor \tag{18.24}$$

The dashpot matrix $[C_\infty]$ follows from Equation (18.21a) as

$$[C_\infty] = ([\Phi]^{-1})^T \lceil \Lambda \rfloor [\Phi]^{-1} \tag{18.25}$$

As each coefficient of $\lceil \Lambda \rfloor$ is positive, $[C_\infty]$ will be positive definite. It follows from Equation (18.12) that in the limit $\xi \rightarrow \infty$, $[C_\infty]$ is proportional to the imaginary part of the dynamic-stiffness matrix, which is thus also positive definite. The rate of energy transmission N is positive (Equation (18.8)), satisfying exactly the radiation condition in the limit.

The second term in Equation (18.22) results, after substituting Equation (18.24) in

$$\lceil\Lambda\rfloor[k_\infty]+[k_\infty]\lceil\Lambda\rfloor=-\lceil\Lambda\rfloor[e^1]^T-[e^1]\lceil\Lambda\rfloor+(s-1)\lceil\Lambda\rfloor \tag{18.26}$$

This linear equation for $[k_\infty]$ is a Lyapunov equation similar to Equation (17.16), but with a diagonal coefficient matrix $\lceil\Lambda\rfloor$. Its solution for each element $k_{\infty kl}$ equals

$$k_{\infty kl}=\frac{1}{\Lambda_k+\Lambda_l}(-\Lambda_k e^1_{kl}-\Lambda_l e^1_{kl}+(s-1)\Lambda_k\delta_{kl}) \tag{18.27}$$

with the Kronecker delta $\delta_{kl}(=1$ for $k=l$; $=0$ for $k\neq l)$. The spring matrix $[K_\infty]$ is calculated from Equation (18.21b) as

$$[K_\infty]=([\Phi]^{-1})^T[k_\infty][\Phi]^{-1} \tag{18.28}$$

The third term in Equation (18.22) leads, after substituting Equation (18.24) with the known $[k_\infty]$, to

$$\lceil\Lambda\rfloor[a_1]+[a_1]\lceil\Lambda\rfloor=-([k_\infty]+[e^1])([k_\infty]+[e^1]^T)+(s-2)[k_\infty]+[e^2] \tag{18.29}$$

This equation for $[a_1]$ is in the same form as Equation (18.26). Analogously, $[a_2]$ is determined from the fourth term in Equation (18.22) with the known $[c_\infty]=\lceil\Lambda\rfloor$, $[k_\infty]$ and $[a_1]$

$$\lceil\Lambda\rfloor[a_2]+[a_2]\lceil\Lambda\rfloor=-([k_\infty]+[e^1])[a_1]-[a_1]([k_\infty]+[e^1]^T)+(s-3)[a_1] \tag{18.30}$$

The coefficient matrices $[A_j]$ result from Equation (18.21c) as

$$[A_j]=([\Phi]^{-1})^T[a_j][\Phi]^{-1} \tag{18.31}$$

After calculating $[C_\infty]$, $[K_\infty]$, $[A_1]$ and $[A_2]$, the asymptotic behaviour for $m=2$ follows from Equation (18.12).

For the boundary ($\xi=1$) the high-frequency asymptotic expansion (Equation (18.12)) equals

$$[S^\infty(\omega)]\approx i\omega[C_\infty]+[K_\infty]+\sum_{j=1}^{m}\frac{1}{(i\omega)^j}[A_j] \tag{18.32}$$

All equations discussed above apply to linear elastic material. They can straightforwardly be extended to viscoelastic material using the correspondence principle. For demonstration, the constant hysteretic material model is addressed with the same damping coefficient for shear and dilatational waves. The correspondence principle states that the solution for hysteretic material follows from the elastic results by replacing the real Lamé constants by the corresponding complex ones (multiplication by $1+2\mathrm{i}\zeta$ with the hysteretic damping ratio ζ). The wave velocities are also modified. The dimensionless frequency a (Equation (15.113)) is replaced by a^*

$$a^* = \frac{a}{\sqrt{1+2\mathrm{i}\zeta}} \tag{18.33}$$

Thus

$$[E^{0*}] = (1+2\mathrm{i}\zeta)[E^0] \tag{18.34a}$$

$$[E^{1*}] = (1+2\mathrm{i}\zeta)[E^1] \tag{18.34b}$$

$$[E^{2*}] = (1+2\mathrm{i}\zeta)[E^2] \tag{18.34c}$$

follow.

The high-frequency asymptotic expansion (Equation (18.12)) is written as

$$\xi^{2-s}[S^\infty(\omega,\xi)] \approx \mathrm{i}\bar{\xi}[C^*_\infty] + [K^*_\infty] + \sum_{j=1}^{m} \frac{1}{(\mathrm{i}\bar{\xi})^j}[A^*_j] \tag{18.35}$$

where

$$[C_\infty] = (1+2\mathrm{i}\zeta)^{1/2}[C_\infty] \tag{18.36a}$$

$$[K^*_\infty] = (1+2\mathrm{i}\zeta)[K_\infty] \tag{18.36b}$$

$$[A^*_j] = (1+2\mathrm{i}\zeta)^{1+\frac{1}{2}j}[A_j] \tag{18.36c}$$

The high-frequency asymptotic expansion of the dynamic-stiffness matrix $[S^\infty(\omega)]$ in the frequency domain corresponds to the *early-time asymptotic expansion* of the *unit-impulse response matrix* $[S^\infty(t)]$ in the time domain. This is a consequence of the initial value theorem. The inverse Fourier transform of Equation (18.32) equals

$$[S^\infty(t)] \approx [C_\infty]\dot{\delta}(t) + [K_\infty]\delta(t) + \sum_{j=1}^{m} \frac{1}{(j-1)!}[A_j]\,t^{j-1}\mathrm{H}(t) \tag{18.37}$$

with the Dirac delta function $\delta(t)$ and the Heaviside step function $\mathrm{H}(t)$. Appendix B can be consulted, in particular Equation (B.40), expressing that $[S^\infty(\omega)]$ and $[S^\infty(t)]$ form a Fourier transform pair. Thus, the asymptotic expansion of $[S^\infty(t)]$ for $t \to 0$ results in the same coefficient matrices $[C_\infty]$, $[K_\infty]$, and $[A_j]$ as the asymptotic expansion of $[S^\infty(\omega)]$ for $\omega \to \infty$.

It is interesting to note that the dashpot matrix $[C_\infty]$ can also be calculated without solving an eigenvalue problem, which is verified as follows.

The response of the unbounded medium at the initial time can be calculated directly. The direction perpendicular to the infinitesimal area $\mathrm{d}S$ of the boundary is addressed. The unbounded medium is initially at rest. After applying the load per unit area r during the first infinitesimal time dt the wave front is at the distance c_pdt (dilatational-wave velocity c_p) and the domain of influence equals c_pdtdS. The law of conservation of momentum is formulated for the first infinitesimal time dt. The initial momentum vanishes. The momentum (mass times velocity) at dt equals $\rho c_p \mathrm{d}t\mathrm{d}S\dot{u}$ (mass density ρ, velocity $\dot{u}$ at dt). The law is written as

$$r\mathrm{d}S\mathrm{d}t = \rho c_p \mathrm{d}t\mathrm{d}S\dot{u} \text{ for } t = 0 \tag{18.38}$$

which yields

$$r = \rho c_p \dot{u} \text{ for } t = 0 \tag{18.39}$$

The initial response perpendicular to the structure-soil interface of the unbounded soil is thus modelled by a dashpot with the coefficient per unit area ρc_p which is called the *impedance*. Analogously, the initial response in the tangential directions is described by dashpots with the coefficients ρc_s (shear-wave velocity c_s).

Applying virtual-work concepts, the interaction forces can be expressed by integrating the load per unit area r. The coefficient matrix of the interaction force-nodal velocity relationship is equal to $[C_\infty]$

$$\{R\} = [C_\infty]\{\dot{u}\} \text{ for } t = 0 \tag{18.40}$$

Substituting Equation (18.12) directly into Equation (18.11), and then addressing power series in $\bar{\xi}$ permits the coefficient matrices $[K_\infty]$ and $[A_j]$ to be calculated by solving Lyapunov equations, which are linear. For instance, the term in $\mathrm{i}\bar{\xi}$ yields the equation for $[K_\infty]$ with known $[C_\infty]$

$$\begin{aligned}[C_\infty][E^0]^{-1}[K_\infty] + [K_\infty][E^0]^{-1}[C_\infty] + [C_\infty][E^0]^{-1}[E^1]^T \\ +[E^1][E^0]^{-1}[C_\infty] - (s-1)[C_\infty] = 0\end{aligned} \tag{18.41}$$

Finally, the high-frequency asymptotic expansion of the dynamic-stiffness matrix for diffusion is addressed using the simple change of independent variable (Equation (15.154), which replaces $\mathrm{i}\omega$ in elastodynamics by $\sqrt{\mathrm{i}\omega}$ in diffusion, yielding from Equation (18.32)

$$[S^\infty(\omega)] \approx \sqrt{\mathrm{i}\omega}[C_\infty] + [K_\infty] + \sum_{j=1}^{m} \frac{1}{\sqrt{\mathrm{i}\omega}^{\,j}}[A_j] \tag{18.42}$$

19 Numerical Solution of Dynamic Stiffness, Unit-impulse Response and Displacement of Unbounded Medium

To determine the dynamic response of an unbounded medium, the governing ordinary differential equations can also be solved numerically in the $\bar{\xi}$-direction. Thus, a numerical procedure is applied in all directions. This is meaningful, as the transformation to the scaled boundary coordinates ξ, η, ζ does permit the radiation condition to be introduced rigorously for $\xi \to \infty$, which is a significant advantage.

To be able to enforce the boundary condition at infinity and at the boundary, the scaled boundary finite element equation in displacement (Equation (15.112)) is not solved directly. It is replaced by the two first-order differential equations, the scaled boundary finite element equation in dynamic stiffness (Equation (15.117)) and the nodal force-displacement relationship (Equation (15.115)).

The solution procedure of the scaled boundary finite element equation in dynamic stiffness, which permits the radiation condition to be enforced exactly, is described in section 19.1 [W9]. In this first step, the dynamic-stiffness matrix is calculated. The solution procedure of the nodal force-displacement relationship which satisfies the boundary condition at the boundary, is examined is section 19.3 [W3]. In this second step, the displacements are determined, whereby the dynamic-stiffness matrix is known from the first step. The scaled boundary finite element equation in dynamic stiffness can be transformed into the time domain, and then solved numerically in the time domain, which is discussed in section 19.2 [W12]. This results in the unit-impulse response function. The derivation is performed for elastodynamics.

19.1 DYNAMIC STIFFNESS

The scaled boundary finite element equation in dynamic stiffness for an unbounded medium (Equation (15.117)) is formulated as

$$(\xi^{2-s}[S^\infty(\omega,\xi)] + [E^1])[E^0]^{-1}(\xi^{2-s}[S^\infty(\omega,\xi)] + [E^1]^T) - (s-2)\xi^{2-s}[S^\infty(\omega,\xi)]$$
$$-\bar{\xi}(\xi^{2-s}[S^\infty(\omega,\xi)])_{,\bar{\xi}} - [E^2] + \bar{\xi}^2[M^0] = 0 \qquad (19.1)$$

These are nonlinear first-order ordinary differential equations in $\xi^{2-s}[S^\infty(\omega,\xi)]$ with $\bar{\xi} = \omega\xi$ as the independent variable (or alternatively with the dimensionless frequency a, Equation (15.113).

Equation (19.1) is solved numerically, incorporating the radiation condition formulated at infinity. The boundary condition (starting value) is calculated for a very large but finite $\bar{\xi}_h$ from the high-frequency asymptotic expansion of the dynamic-stiffness matrix (Equation 18.12)

$$\xi_h^{2-s}[S^\infty(\omega,\xi_h)] \approx \mathrm{i}\bar{\xi}_h[C_\infty] + [K_\infty] + \sum_{j=1}^{m} \frac{1}{(\mathrm{i}\bar{\xi}_h)^j}[A_j] \qquad (19.2)$$

The radiation condition is satisfied (either by constructing a positive definite $[C_\infty]$ from an eigenvalue problem (Equation (18.25)) or using the impedances (Equation (18.40)). The nonlinear first-order ordinary differential (Equation (19.1)) is then solved starting from $\bar{\xi}_h$ for decreasing $\bar{\xi}$ down to the boundary ($\xi_i = 1 \rightarrow \bar{\xi}_i = \omega$). A fourth-order Runge–Kutta scheme [P3] is applied. At the beginning for large $\bar{\xi}$, where the variables vary smoothly, an adaptive integration step size is determined. Later on, for smaller $\bar{\xi}$ a fixed integration step size is selected. This yields the dependent variable $\xi^{2-s}[S^\infty(\omega,\xi)]$ as a function of $\bar{\xi}$ (or of the dimensionless frequency a). The error introduced through the boundary condition (starting value) at $\bar{\xi}_h$ diminishes for decreasing $\bar{\xi}$.

For constant hysteretic material damping the scaled boundary finite element equation in dynamic stiffness (Equation (19.1)) equals

$$(\xi^{2-s}[S^\infty(\omega,\xi)] + [E^{1*}])[E^{0*}]^{-1}(\xi^{2-s}[S^\infty(\omega,\xi)] + [E^{1*}]^T) - (s-2)\xi^{2-s}[S^\infty(\omega,\xi)]$$
$$-\bar{\xi}(\xi^{2-s}[S^\infty(\omega,\xi)])_{,\bar{\xi}} - [E^{2*}] + \bar{\xi}^2[M^0] = 0 \qquad (19.3)$$

with the dynamic-stiffness matrix of the hysteretic material $[S^\infty(\omega,\xi)]$. The coefficient matrices are specified in Equation (18.34). $[M^0]$ as a function of the mass density ρ is not affected. The starting value corresponding to Equation (19.2) is listed in Equation (18.35) with $\bar{\xi} = \bar{\xi}_h$. No modification occurs in the numerical integration procedure for Equation (19.3). Even in the elastic case, the operations involve complex numbers!

The nonlinear ordinary differential equations of Equation (19.1) have to be solved numerically. A step size $\Delta\bar{\xi}$ must be chosen. At the discretised stations, the solutions can be interpreted physically as corresponding to fictitious boundaries, which are similar to the boundary for the unbounded medium. The medium between two subsequent boundaries represents a finite element cell of width $\Delta\bar{\xi}$. Assembling the dynamic stiffness of the unbounded medium with respect to a specific fictitious boundary and that of the cell yields the dynamic stiffness of the unbounded medium with respect to the neighbouring fictitious boundary. Thus, a procedure can be established which works only with the standard static-stiffness and mass matrices of the cell. This procedure called *dynamic condensation* is described in Reference [W9]. It is strongly related with the mechanically-based derivation which is not addressed in Part II.

The dynamic-stiffness matrix $[S^{\infty}(\omega)]$ is required to formulate the contribution of the unbounded medium to the equations of motion in the frequency domain of unbounded medium-structure interaction as outlined in Appendix C (Equation (C.16)).

For diffusion, the solution procedure to calculate the dynamic-stiffness for an unbounded medium matrix is analogous. The corresponding scaled boundary finite element equation in dynamic stiffness is specified in Equation (15.157) and the high-frequency asymptotic expansion in Equation (18.42), both formulated for $\xi = 1$, leading to $\bar{\xi} = \omega$.

19.2 UNIT-IMPULSE RESPONSE

To be able to formulate the nodal force-displacement relationship at the boundary in the time domain, the unit-impulse response functions are required. First, the scaled boundary finite element equation in dynamic stiffness is transformed from the frequency into the time domain using the inverse Fourier transformation. Then, a numerical discretisation in the time domain is applied to determine the unit-impulse response functions.

19.2.1 Inverse Fourier Transformation

The inverse Fourier transformation of the dynamic stiffness to the unit-impulse response is discussed for the scalar case in Appendix B.

In a frequency-domain analysis, the nodal force-displacement relationship at the boundary ($\xi = 1$) with the corresponding amplitudes is written for a specific frequency as

$$\{R(\omega)\} = [S^{\infty}(\omega)]\{u(\omega)\} \tag{19.4}$$

where $[S^{\infty}(\omega)]$ represents the dynamic-stiffness matrix of the unbounded medium. No body loads are present. In the time domain, the interaction force-displacement relationship is written for a specific time t as the convolution integral

$$\{R(t)\} = \int_{0}^{t} [S^{\infty}(t-\tau)]\{u(\tau)\}\mathrm{d}\tau \tag{19.5}$$

$[S^{\infty}(t)]$ is the response matrix of the unbounded medium to a unit impulse of displacements. At this stage of the derivation Equation (19.5) should be regarded as a formal expression. $[S^{\infty}(\omega)]$ and $[S^{\infty}(t)]$ form a Fourier transform pair (Equation (B.33)).

As already mentioned, it is the goal of this chapter to determine the corresponding equation for $[S^{\infty}(t)]$ in the time domain, and then calculate the values at discrete time stations.

As the inverse Fourier transform of $[S^{\infty}(\omega)]$ exists only in the sense of distributions, $[S^{\infty}(\omega)]$ is decomposed into the singular part and the remaining regular part $[S_r^{\infty}(\omega)]$

(subscript r for regular). The singular part is defined as the asymptotic value for $\omega \rightarrow \infty$ of $[S^{\infty}(\omega)]$, which equals $i\omega[C_{\infty}] + [K_{\infty}]$, the first two terms of the high-frequency asymptotic expansion at the boundary, (Equation (18.32)). The subscript ∞ represents the asymptotic value for $\omega \rightarrow \infty$. $[C_{\infty}]$ and $[K_{\infty}]$ are the constant dashpot matrix and constant spring matrix

$$[S^{\infty}(\omega)] = i\omega[C_{\infty}] + [K_{\infty}] + [S_r^{\infty}(\omega)] \tag{19.6}$$

The inverse Fourier transformation equals

$$[S^{\infty}(t)] = [C_{\infty}]\dot{\delta}(t) + [K_{\infty}]\delta(t) + [S_r^{\infty}(t)] \tag{19.7}$$

with the Dirac delta function $\delta(t)$. $[S_r^{\infty}(\omega)]$, which is square integrable, and $[S_r^{\infty}(t)]$ form a Fourier transform pair. Substituting Equation (19.7) into (19.5) yields the interaction force-displacement relationship

$$\{R(t)\} = [C_{\infty}]\{\dot{u}(t)\} + [K_{\infty}]\{u(t)\} + \int_0^t [S_r^{\infty}(t-\tau)]\{u(\tau)\}d\tau \tag{19.8}$$

The first two terms on the right-hand side of Equation (19.8) represent the instantaneous response and the convolution integral the lingering response.

Reference is made in the following to the high-frequency asymptotic expansion of the dynamic-stiffness matrix in Chapter 18 at the boundary $\xi = 1(\bar{\xi} = \omega)$.

To calculate the unit-impulse response matrix for all times, Equation (19.6) is substituted into Equation (18.18) yielding

$$[s^{\infty}(\omega)] = i\omega[c_{\infty}] + [k_{\infty}] + [s_r^{\infty}(\omega)] \tag{19.9}$$

with

$$[s_r^{\infty}(\omega)] = [\Phi]^T[S_r^{\infty}(\omega)][\Phi] \tag{19.10}$$

Note that $[s_r^{\infty}(\omega)]$ is of the order of $1/i\omega$ for $\omega \rightarrow \infty$ (Equation (18.20)). Substituting Equation (19.9) into (18.17) and rearranging in descending order of the power series of $i\omega$ leads to

$$\begin{aligned}[s_r^{\infty}(\omega)]^2 + i\omega[c_{\infty}][s_r^{\infty}(\omega)] + i\omega[s_r^{\infty}(\omega)][c_{\infty}] - \omega[s_r^{\infty}(\omega)]_{,\omega} + ([k_{\infty}] + [e^1])[s_r^{\infty}(\omega)] \\ +[s_r^{\infty}(\omega)]([k_{\infty}] + [e^1]^T) \\ -(s-2)[s_r^{\infty}(\omega)] + ([k_{\infty}] + [e^1])([k_{\infty}] + [e^1]^T) - [e^2] - (s-2)[k_{\infty}] = 0\end{aligned} \tag{19.11}$$

with the term in $(i\omega)^2$ and some terms in $i\omega$ vanishing, as $[c_{\infty}]$ and $[k_{\infty}]$ satisfy Equations (18.24) and (18.26), respectively.

To avoid the $\delta(t)$ function, and thus to simplify the numerical discretisation, the abbreviation

$$[v_r^\infty(\omega)] = \frac{1}{\mathrm{i}\omega}[s_r^\infty(\omega)] \tag{19.12}$$

is introduced. In the time domain Equation (19.12) corresponds to

$$[v_r^\infty(t)] = \int_0^t [s_r^\infty(\tau)]\mathrm{d}\tau \tag{19.13}$$

or

$$[s_r^\infty(t)] = [\dot{v}_r^\infty(t)] \tag{19.14}$$

The scaled boundary finite element equation in the time domain is developed for $[v_r^\infty(t)]$. The regular part of the transformed unit-impulse response matrix $[s_r^\infty(t)]$ is then calculated numerically as the time derivate of $[v_r^\infty(t)]$ (Equation (19.14)).

Dividing Equation (19.11) by $(\mathrm{i}\omega)^2$, and substituting Equation (19.12), results in

$$\begin{aligned}&[v_r^\infty(\omega)]^2 + [c_\infty][v_r^\infty(\omega)] + [v_r^\infty(\omega)][c_\infty] - [v_r^\infty(\omega)]_{,\mathrm{i}\omega} + \frac{1}{\mathrm{i}\omega}([k_\infty]\\&+[e^1])[v_r^\infty(\omega)] + \frac{1}{\mathrm{i}\omega}[v_r^\infty(\omega)]([k_\infty]+[e^1]^T) - (s-1)\frac{1}{\mathrm{i}\omega}[v_r^\infty(\omega)]\\&+\frac{1}{(\mathrm{i}\omega)^2}(([k_\infty]+[e^1])([k_\infty]+[e^1]^T) - [e^2] - (s-2)[k_\infty]) = 0\end{aligned} \tag{19.15}$$

Applying the inverse Fourier transformation to Equation (19.15) yields (Equations (B.41)–(B.45))

$$\begin{aligned}&\int_0^t [v_r^\infty(t-\tau)][v_r^\infty(\tau)]\mathrm{d}\tau + [c_\infty][v_r^\infty(t)] + [v_r^\infty(t)][c_\infty] + t[v_r^\infty(t)] + ([k_\infty]+[e^1])\int_0^t [v_r^\infty(\tau)]\mathrm{d}\tau\\&+\int_0^t [v_r^\infty(\tau)]\mathrm{d}\tau([k_\infty]+[e^1]^T) - (s-1)\int_0^t [v_r^\infty(\tau)]\mathrm{d}\tau\\&+t(([k_\infty]+[e^1])([k_\infty]+[e^1]^T) - [e^2] - (s-2)[k_\infty]) = 0\end{aligned} \tag{19.16}$$

This represents the *scaled boundary finite element equation in the time domain for* $[v_r^\infty(t)]$ *for* $t \geq 0$.

The inverse Fourier transformation of Equation (19.10) is written as

$$[s_r^\infty(t)] = [\Phi]^T[S_r^\infty(t)][\Phi] \tag{19.17}$$

19.2.2 Time Discretisation

The regular part of the transformed unit-impulse response matrix at $t = 0$ $[S_r^\infty(0)]$ is obtained by comparing Equations (19.7) and (18.37) formulated at $t = 0$ as

$$[S_r^\infty(0)] = [A_1] \tag{19.18}$$

The coefficient matrix $[A_1]$ follows from Equations (18.31) and (18.29). As $[S_r^\infty(0)]$ is finite (Equation (19.18)), $[S_r^\infty(0)]$ will also be finite (Equation (19.17)). Formulating Equation (19.13) for $t = 0$ yields

$$[v_r^\infty(0)] = 0 \tag{19.19}$$

After discretisation with respect to time, Equation (19.16) yields an equation for $[v_r^\infty]_n$ at each time station n ($n \geq 1$). It is assumed that $[v_r^\infty]_n$ is piecewise constant over the time $(n - 0.5)\Delta t \leq t \leq (n + 0.5)\Delta t$. At time station $n = 0$, $[v_r^\infty]_0 = 0$ (Equation (19.19)), which is assumed to apply for $0 \leq t < 0.5\Delta t$. The integral terms in Equation (19.16) are discretised as

$$\int_0^{n\Delta t} [v_r^\infty(\tau)]\mathrm{d}\tau = \Delta t \sum_{j=1}^{n-1} [v_r^\infty]_j + 0.5\Delta t [v_r^\infty]_n \tag{19.20}$$

$$\int_0^{n\Delta t} [v_r^\infty(n\Delta t - \tau)]\,[v_r^\infty(\tau)]\mathrm{d}\tau = \Delta t \sum_{j=1}^{n-1} [v_r^\infty]_j [v_r^\infty]_{n-j} \tag{19.21}$$

Substituting Equations (19.20) and (19.21) into Equation (19.16) formulated for $t = n\Delta t$ results in a *linear* equation for $[v_r^\infty]_n$

$$\begin{aligned}
&\left(\frac{1}{\Delta t}[c_\infty] + 0.5([k_\infty] + [e^1] + 0.5(1 - s)[I\,])\right)[v_r^\infty]_n \\
&+[v_r^\infty]_n\left(\frac{1}{\Delta t}[c_\infty] + 0.5([k_\infty] + [e^1]^T + 0.5(1 - s)[I\,])\right) + n[v_r^\infty]_n = -\sum_{j=1}^{n-1} [v_r^\infty]_j [v_r^\infty]_{n-j} \\
&-([k_\infty] + [e^1] + 0.5(1 - s)[I\,])\sum_{j=1}^{n-1} [v_r^\infty]_j + \sum_{j=1}^{n-1} [v_r^\infty]_j([k_\infty] + [e^1]^T + 0.5(1 - s)[I\,]) \\
&+n([e^2] + (s - 2)[k_\infty] - ([k_\infty] + [e^1])([k_\infty] + [e^1]^T))
\end{aligned} \tag{19.22}$$

Equation (19.22) is the Lyapunov equation in the form

$$[A][X] + [X][A]^T + n[X] = [C] \tag{19.23}$$

with $[X] = [v_r^\infty]_n$ and the other matrices defined straightforwardly.

The Lyapunov equation can be solved based on the Schur factorisation, as described in Reference [B1]. The Schur factorisation is performed only once, and for each time station a back substitution is required.

To prove that $[v_r^\infty]_n$ is symmetric, the transpose of Equation (19.22) is subtracted from Equation (19.22) with the symmetric matrices $[e^2]$, $[c_\infty]$, $[k_\infty]$ and $[v_r^\infty]_j$ ($j = 1, \ldots, n-1$) leading to

$$\left(\frac{1}{\Delta t}[c_\infty] + 0.5([k_\infty] + [e^1] + 0.5(1-s)[I])\right)([v_r^\infty]_n - [v_r^\infty]_n^T)$$
$$+([v_r^\infty]_n - [v_r^\infty]_n^T)\left(\frac{1}{\Delta t}[c_\infty] + 0.5([k_\infty] + [e^1]^T + 0.5(1-s)[I])\right) \tag{19.24}$$
$$+n\left([v_r^\infty]_n - [v_r^\infty]_n^T\right) = 0$$

This linear equation is only satisfied for $[v_r^\infty]_n = [v_r^\infty]_n^T$.

The matrix $[s_r^\infty]_n$ follows as (Equation (19.14))

$$[s_r^\infty]_n = \frac{1}{\Delta t}\left([v_r^\infty]_n - [v_r^\infty]_{n-1}\right) \tag{19.25}$$

The regular part of the unit-impulse response matrix at the nth time station is calculated from Equation (19.17) as

$$[S_r^\infty]_n = ([\Phi]^{-1})^T [s_r^\infty]_n [\Phi]^{-1} \tag{19.26}$$

As an alternative, the response functions in the time domain to unit impulse of acceleration can be determined [W9].

The unit-impulse response function matrix $[S^\infty(t)]$ consisting of $[C_\infty]$, $[K_\infty]$ and $[S_r^\infty(t)]$ is required to formulate the contribution of the unbounded medium to the differential equations of motion in the time domain of unbounded medium-structure interaction, as outlined in Appendix C (Equation (C.15)).

Summarising, starting from the scaled boundary finite element equation in frequency domain for the dynamic stiffness and performing the inverse Fourier transformation, the corresponding equation in time domain for the response to unit impulse of displacement is derived. The unit-impulse response consists of a singular and a regular part. To determine the singular part, an eigenvalue problem of positive definite matrices (or directly based on impedances) yields the dashpot matrix and a linear Lyapunov equation the spring matrix. To calculate the regular part, the decomposition of the coefficient matrix of the Lyapunov equation is performed once, and for each time station a back substitution is required. The early-time response which corresponds to the radiation condition at infinity is satisfied rigorously.

For diffusion, the general procedure is the same. For details, References [S4] and [W9] can be consulted.

19.3 DISPLACEMENT

To determine the displacements in the interior of the unbounded medium, the nodal force-displacement relationship at a constant ξ is addressed (Equation (15.115)

$$\xi^{2-s}[S^{\infty}(\omega,\xi)]\{u(\bar{\xi})\} = -[E^0]\bar{\xi}\{u(\bar{\xi})\}_{,\bar{\xi}} - [E^1]^T\{u(\bar{\xi})\} \tag{19.27}$$

The dynamic-stiffness matrix $\xi^{2-s}[S^{\infty}(\omega,\xi)]$, which is a function of $\bar{\xi} = \omega\xi$ (or a, Equation (15.113)), is known from solving the scaled boundary finite element equation in dynamic stiffness (Equation (19.1)). Thus, the linear first-order ordinary differential equations with known but varying coefficients (Equation (19.27)) are solved numerically. As boundary conditions formulated at $\xi = 1(\bar{\xi} = \omega)$, either the displacement amplitudes $\{u(\bar{\xi} = \omega)\}$ are enforced or the prescribed nodal force amplitudes $\{R(\bar{\xi} = \omega)\}$ are converted into displacement amplitudes at the boundary

$$\{u(\bar{\xi} = \omega)\} = [S(\omega)]^{-1}\{R(\bar{\xi} = \omega)\} \tag{19.28}$$

Equation (19.27) is then solved from $\bar{\xi} = \omega$ for increasing $\bar{\xi}$. For instance, an explicit forward Euler scheme is applied. The same integration step size as when solving Equation (19.1) is used, as $\xi^{2-s}[S^{\infty}(\omega,\xi)]$ must be known. This yields the dependent variable $\{u(\bar{\xi})\}$.

For constant hysteretic material damping, the nodal force-displacement relationship equals

$$\xi^{2-s}[S^{\infty}(\omega,\xi)]\{u(\bar{\xi})\} = -[E^{0*}]\bar{\xi}\{u(\bar{\xi})\}_{,\bar{\xi}} - [E^{1*}]^T\{u(\bar{\xi})\} \tag{19.29}$$

with the dynamic-stiffness matrix of the hysteretic material determined from Equation (19.3). The same numerical integration procedure with complex variables applies.

Thus, an efficient numerical procedure results for very accurately calculating the dynamic-stiffness matrix and displacements throughout the unbounded medium by solving numerically two first-order ordinary differential equations.

For diffusion, the procedure is analogous. Equation (15.158) is solved numerically with the dynamic-stiffness matrix known from addressing Equation (15.157).

20 Analytical Solution in Frequency Domain

Based on the transformation of the governing partial differential equations to the much simpler ordinary differential equations, depending on a single independent variable, an analytical solution is feasible. The analytical expressions in the radial coordinate with corresponding integration constants permit the boundary conditions, such as the radiation condition at infinity, to be enforced exactly, and yield as part of the solution process, in general, a closed-form expression, for instance describing the characteristics of the stress singularities. In addition, results can be calculated selectively. In contrast to the numerical methods addressed in Chapter 19, for instance, the dynamic-stiffness matrix for a specific frequency or the displacements at a specific location can directly be determined without having to calculate results for a large range of values.

As demonstrated for statics in Chapter 16, the analytical solution procedure is straightforward and very efficient in this case. The number of terms in the series is equal to the number of degrees of freedom on the boundary, and the form of each term is known: a power series with the powers calculated from an eigenvalue problem is constructed. In contrast, in dynamics the analytical solution is much more complicated, also from a programming point of view. As will become apparent, an infinite series is applied, of which the number of terms to be processed has to be selected by the analyst. The solution involves, in general, matrix functions with logarithmic terms. For an unbounded medium, the same high-frequency asymptotic expansion of the dynamic stiffness as in the numerical procedure has to be used. Thus, room for improvement in formulating an analytical solution procedure exists! For instance, a high-frequency asymptotic expansion for the displacements satisfying the radiation condition would be valuable. Of course, the final goal consists of constructing a series expansion for the displacements in the vector case which satisfies the radiation condition at infinity in the same way as the Hankel function does for the scalar case.

This chapter, which closely follows Reference [S8], addresses elastodynamics in the frequency domain. Section 20.1 examines the series solution for displacements, sections 20.2 and 20.3 describe analytical expressions for the dynamic stiffnesses of bounded and unbounded media, respectively, and finally, section 20.4 discusses the analysis of far-field displacements.

The reader is reminded that in Chapter 12 the Model Problem is solved analytically specifying all details. This mathematically challenging Chapter 20 can also be skipped without loss of continuity (with the exception of the example addressed in section 24.2).

20.1 DISPLACEMENT

For vanishing body load, the scaled boundary finite element equation in displacement equals (Equation (15.111))

$$\begin{aligned}[E^0]\xi^2\{u(\xi)\}_{,\xi\xi} &+ ((s-1)[E^0] - [E^1] + [E^1]^T)\xi\{u(\xi)\}_{,\xi} \\ &+ ((s-2)[E^1]^T - [E^2])\{u(\xi)\} + \omega^2[M^0]\xi^2\{u(\xi)\} = 0\end{aligned} \tag{20.1}$$

To solve these second-order ordinary differential equations of order n analytically, a transformation to first-order ordinary differential equations of order $2n$ is performed. Besides the displacement amplitudes $\{u(\xi)\}$, the corresponding internal nodal force amplitudes $\{Q(\xi)\}$ defined in Equations (15.67) and (15.105) are introduced. For a concise procedure, the dependent variables $\{X(\bar{\xi})\}$ of the first-order ordinary differential equations are defined as

$$\{X(\bar{\xi})\} = \begin{Bmatrix} \xi^{+0.5(s-2)}\{u(\xi)\} \\ \xi^{-0.5(s-2)}\{Q(\xi)\} \end{Bmatrix} \tag{20.2}$$

which is also introduced in statics (Equation (16.3)). Equation (20.1) is formulated as

$$\bar{\xi}\{X(\bar{\xi})\}_{,\bar{\xi}} = -[Z]\{X(\bar{\xi})\} - \bar{\xi}^2\begin{bmatrix} 0 & 0 \\ [M^0] & 0 \end{bmatrix}\{X(\bar{\xi})\} \tag{20.3}$$

with the coefficient matrix

$$[Z] = \begin{bmatrix} [E^0]^{-1}[E^1]^T - 0.5(s-2)[I] & -[E^0]^{-1} \\ -[E^2] + [E^1][E^0]^{-1}[E^1]^T & -([E^1][E^0]^{-1} - 0.5(s-2)[I]) \end{bmatrix} \tag{20.4}$$

and the independent variable $\bar{\xi} = \omega\xi$. Note that $\bar{\xi}$ is proportional to the dimensionless frequency $a(\omega,\ \xi)$ for any ξ (Equation (15.77)).

A derivation which is comparatively easy to understand is presented. It could be improved in computational efficiency.

An independent set of particular solutions which can be grouped in the matrix $[X(\bar{\xi})]$ is determined. Any other solution $\{X(\bar{\xi})\}$ can be expressed as

$$\{X(\bar{\xi})\} = [X(\bar{\xi})]\{c\} \tag{20.5}$$

with the constants $\{c\}$.$[X(\bar{\xi})]$ satisfies Equation (20.3)

$$\bar{\xi}[X(\bar{\xi})]_{,\bar{\xi}} = -[Z][X(\bar{\xi})] - \bar{\xi}^2\begin{bmatrix} 0 & 0 \\ [M^0] & 0 \end{bmatrix}[X(\bar{\xi})] \tag{20.6}$$

To transform the coefficient matrix $[Z]$, the eigenvalue problem as in statics (Equation (16.7)) is formulated

$$[Z][\Phi] = -[\Phi]\lceil\Lambda\rfloor \tag{20.7}$$

with the diagonal eigenvalue matrix $-\lceil\Lambda\rfloor$ and the eigenvector matrix $[\Phi]$. For the Hamiltonian matrix $[Z]$ in Equation (20.4), the eigenvalues consist of two groups with opposite signs

$$\lceil\Lambda\rfloor = \begin{bmatrix} -\lceil\lambda_i\rfloor & \\ & \lceil\lambda_i\rfloor \end{bmatrix} \tag{20.8}$$

with $\text{Re}(\lambda_i) < 0$ [L1].

The eigenvalues $-\Lambda_i$ are arranged so that their real parts increase sequentially, i.e.

$$\text{Re}(-\Lambda_1) \leq \text{Re}(-\Lambda_2) \leq \quad \ldots \quad \leq \text{Re}(-\Lambda_N) \tag{20.9}$$

With the transformed variable $[W(\bar{\xi})]$ defined in

$$[X(\bar{\xi})] = [\Phi][W(\bar{\xi})] \tag{20.10}$$

substituting Equation (20.10) into (20.6), which is premultiplied by $[\Phi]^{-1}$, and using Equation (20.7), yields

$$\bar{\xi}[W(\bar{\xi})]_{,\bar{\xi}} = \lceil\Lambda\rfloor[W(\bar{\xi})] - \bar{\xi}^2[M][W(\bar{\xi})] \tag{20.11}$$

with

$$[M] = [\Phi]^{-1}\begin{bmatrix} 0 & 0 \\ [M^0] & 0 \end{bmatrix}[\Phi] \tag{20.12}$$

Proceeding as in Reference [G1, Section 10, Chapter 14], the solution of Equation (20.11) is written as the product of two square matrix functions in $\bar{\xi}$

$$[W(\bar{\xi})] = [A(\bar{\xi})][Y(\bar{\xi})] \tag{20.13}$$

The first matrix $[A(\bar{\xi})]$ is formulated as a power series in $\bar{\xi}$ as

$$[A(\bar{\xi})] = [A_0] + \bar{\xi}^2[A_1] + \bar{\xi}^4[A_2] + \quad \ldots \quad + \bar{\xi}^{2k}[A_k] + \quad \ldots \tag{20.14}$$

with

$$[A_0] = [I] \tag{20.15}$$

and the other coefficient matrices $[A_k]\, k = 1,\ 2, \ldots$ to be determined. The second matrix $[Y(\bar{\xi})]$ in Equation (20.13) satisfies the ordinary differential equations

$$\bar{\xi}[Y(\bar{\xi})]_{,\bar{\xi}} = [P^*(\bar{\xi})][Y(\bar{\xi})] \tag{20.16}$$

where $[P^*(\bar{\xi})]$ is again a power series in $\bar{\xi}$

$$[P^*(\bar{\xi})] = [P_0^*] + \bar{\xi}^2[P_1^*] + \bar{\xi}^4[P_2^*] + \quad \dots \quad + \bar{\xi}^{2k}[P_k^*] + \quad \dots \tag{20.17}$$

The coefficient matrices $[P_k^*] k = 0, 1, \dots$ will be chosen in the simplest form possible.
The derivative $[W(\bar{\xi})]_{,\bar{\xi}}$ equals (Equation (20.13))

$$\begin{aligned}[W(\bar{\xi})]_{,\bar{\xi}} &= [A(\bar{\xi})]_{,\bar{\xi}}[Y(\bar{\xi})] + [A(\bar{\xi})][Y(\bar{\xi})]_{,\bar{\xi}} \\ &= ([A(\bar{\xi})]_{,\bar{\xi}} + \frac{1}{\bar{\xi}}[A(\bar{\xi})][P^*(\bar{\xi})])[Y(\bar{\xi})]\end{aligned} \tag{20.18}$$

where Equation (20.16) is substituted. Substituting Equations (20.13) and (20.18) into (20.11) and post-multiplying by $[Y(\bar{\xi})]^{-1}$ results in

$$\bar{\xi}[A(\bar{\xi})]_{,\bar{\xi}} - \lceil\Lambda\rfloor[A(\bar{\xi})] + [A(\bar{\xi})][P^*(\bar{\xi})] + \bar{\xi}^2[M][A(\bar{\xi})] = 0 \tag{20.19}$$

The unknown coefficient matrices $[A_k]$, $[P_k^*]$ will be determined from this equation.
Substituting Equations (20.14) and (20.17) into (20.19) leads to a power series in $\bar{\xi}$

$$\begin{aligned}&[P_0^*] - \lceil\Lambda\rfloor + \bar{\xi}^2(-\lceil\Lambda\rfloor[A_1] + [A_1]\lceil\Lambda\rfloor + 2[A_1] + [P_1^*] + [M]) \\ &\quad + \quad \dots \quad + \bar{\xi}^{2k}(-\lceil\Lambda\rfloor[A_k] + [A_k]\lceil\Lambda\rfloor + 2k[A_k] + [P_k^*] + [C_k]) + \quad \dots \quad = 0\end{aligned} \tag{20.20}$$

where

$$[C_k] = [M][A_{k-1}] + \sum_{i=1}^{k-1}[A_i][P_{k-i}^*] \tag{20.21}$$

Note that $[C_k]$ is determined from $[A_i]$ and $[P_k^*](i = 0, 1, \dots, k-1)$.
To satisfy Equation (20.20), all the coefficients of the power series in $\bar{\xi}$ must vanish. This yields

$$[P_0^*] = \lceil\Lambda\rfloor \tag{20.22}$$

$$\lceil\Lambda\rfloor[A_1] - [A_1]\lceil\Lambda\rfloor - 2[A_1] = [P_1^*] + [M] \tag{20.23}$$

$$\lceil\Lambda\rfloor[A_k] - [A_k]\lceil\Lambda\rfloor - 2k[A_k] = [P_k^*] + [C_k] \quad k = 2, 3, \dots \tag{20.24}$$

Equation (20.22) determines $[P_0^*]$.
$[A_1]$ follows from Equation (20.23), with an appropriately chosen $[P_1^*]$. The matrix equation (Equation (20.23)) is replaced by the corresponding $(N \times N)$ scalar equations

$$(\Lambda_i - \Lambda_j - 2)[A_1]_{ij} = [P_1^*]_{ij} + [M]_{ij} \tag{20.25}$$

$[A_1]_{ij}$ denotes the element on the ith row and the jth column of the matrix $[A_1]$. Two cases must be distinguished. For $\Lambda_i - \Lambda_j - 2 \neq 0$, $[P_1^*]_{ij}$ is chosen as zero, yielding

$$[A_1]_{ij} = \frac{[M]_{ij}}{\Lambda_i - \Lambda_j - 2} \tag{20.26}$$

For $\Lambda_i - \Lambda_j - 2 = 0$,

$$[P_1^*]_{ij} = -[M]_{ij} \tag{20.27}$$

must be selected to be able to satisfy Equation (20.25). In this case, $[A_1]_{ij}$ is arbitrary.

After determining $[A_i]$ and $[P_i^*]$ up to $i = k - 1$, $[C_k]$ is calculated from Equation (20.21). To determine $[A_k]$ and $[P_k^*]$, Equation (20.24) is again formulated as scalar equations

$$(\Lambda_i - \Lambda_j - 2k)[A_k]_{ij} = [P_k^*]_{ij} + [C_k]_{ij} \tag{20.28}$$

For $\Lambda_i - \Lambda_j - 2k \neq 0$

$$[P_k^*]_{ij} = 0 \tag{20.29}$$

$$[A_k]_{ij} = \frac{[C_k]_{ij}}{\Lambda_i - \Lambda_j - 2k} \tag{20.30}$$

and for $\Lambda_i - \Lambda_j - 2k = 0$

$$[P_k^*]_{ij} = -[C_k]_{ij} \tag{20.31}$$

$$[A_k]_{ij} = \text{arbitrary} \tag{20.32}$$

are obtained.

After determining $[A_k]\,(k = 1, 2, \ldots)$, $[A(\bar{\xi})]$ follows from Equation (20.14).

As the eigenvalues $\Lambda_i\,(i = 1, 2, \ldots, N)$ are finite, a maximum k which satisfies $\Lambda_i - \Lambda_j - 2k = 0$ exists. Thus, the number of $[P_k^*]$ different from zero is finite. As $\Lambda_i - \Lambda_j$ is a fixed value for a specified pair i, j, $\Lambda_i - \Lambda_j - 2k = 0$ can only be satisfied for one k, if at all. It follows from Equation (20.17) that in the corresponding term $\bar{\xi}^{2k}[P_k^*]$ the power of $\bar{\xi}$ equals $2k$. $[P^*(\bar{\xi})]$ in Equation (20.17) can be written as a matrix

$$[P^*(\bar{\xi})] = [P_0^*] + \bar{\xi}^2[P_1^*] + \bar{\xi}^4[P_2^*] + \quad \ldots \quad + \bar{\xi}^{2k}[P_k^*] + \quad \ldots$$

$$= \begin{bmatrix} \Lambda_1 & [P^*_{(\Lambda_1-\Lambda_2)/2}]_{12}\bar{\xi}^{\Lambda_1-\Lambda_2} & [P^*_{(\Lambda_1-\Lambda_3)/2}]_{13}\bar{\xi}^{\Lambda_1-\Lambda_3} & \cdots & [P^*_{(\Lambda_1-\Lambda_N)/2}]_{1N}\bar{\xi}^{\Lambda_1-\Lambda_N} \\ & \Lambda_2 & [P^*_{(\Lambda_2-\Lambda_3)/2}]_{23}\bar{\xi}^{\Lambda_2-\Lambda_3} & \cdots & [P^*_{(\Lambda_2-\Lambda_N)/2}]_{2N}\bar{\xi}^{\Lambda_2-\Lambda_N} \\ & & \Lambda_3 & \cdots & [P^*_{(\Lambda_3-\Lambda_N)/2}]_{3N}\bar{\xi}^{\Lambda_3-\Lambda_N} \\ & & & \ddots & \vdots \\ & & & & \Lambda_N \end{bmatrix} \tag{20.33}$$

As the eigenvalues are arranged in the sequence determined by Equation (20.9), $[P^*(\bar{\xi})]$ is an upper-triangular matrix. The element $[P^*_{(\Lambda_i-\Lambda_j)/2}]_{ij}\bar{\xi}^{\Lambda_i-\Lambda_j}$ only differs from zero when $(\Lambda_i - \Lambda_j)/2$ is a positive integer. In passing, the computational effort required to calculate the sum in Equation (20.21) is limited corresponding to one multiplication of two matrices.

Defining the upper-triangular matrix with zeros on the diagonal

$$[U] = \begin{bmatrix} 0 & [P^*_{(\Lambda_1-\Lambda_2)/2}]_{12} & [P^*_{(\Lambda_1-\Lambda_3)/2}]_{13} & \cdots & [P^*_{(\Lambda_1-\Lambda_N)/2}]_{1N} \\ & 0 & [P^*_{(\Lambda_2-\Lambda_3)/2}]_{23} & \cdots & [P^*_{(\Lambda_2-\Lambda_N)/2}]_{2N} \\ & & 0 & \cdots & [P^*_{(\Lambda_3-\Lambda_N)/2}]_{3N} \\ & & & \ddots & \vdots \\ & & & & 0 \end{bmatrix} \tag{20.34}$$

and the matrix function

$$\bar{\xi}^{\lceil\Lambda\rfloor} = \begin{bmatrix} \bar{\xi}^{\Lambda_1} & & & \\ & \bar{\xi}^{\Lambda_2} & & \\ & & \ddots & \\ & & & \bar{\xi}^{\Lambda_N} \end{bmatrix} \tag{20.35}$$

Equation (20.33) is reformulated as

$$[P^*(\bar{\xi})] = \lceil\Lambda\rfloor + \bar{\xi}^{\lceil\Lambda\rfloor}[U]\bar{\xi}^{-\lceil\Lambda\rfloor} \tag{20.36}$$

Substituting Equation (20.36) into Equation (20.16) yields the ordinary differential equation for $[Y(\bar{\xi})]$

$$\bar{\xi}[Y(\bar{\xi})]_{,\bar{\xi}} = \left(\lceil\Lambda\rfloor + \bar{\xi}^{\lceil\Lambda\rfloor}[U]\bar{\xi}^{-\lceil\Lambda\rfloor}\right)[Y(\bar{\xi})] \tag{20.37}$$

Its analytical solution is obtained by substituting the transformation

$$[Y(\bar{\xi})] = \bar{\xi}^{\lceil\Lambda\rfloor}[\bar{Y}(\bar{\xi})] \tag{20.38}$$

yielding

$$\bar{\xi}[\bar{Y}(\bar{\xi})]_{,\bar{\xi}} = [U][\bar{Y}(\bar{\xi})] \tag{20.39}$$

The solution of this Euler–Cauchy equation equals the matrix function

$$[\bar{Y}(\bar{\xi})] = \bar{\xi}^{[U]} \tag{20.40}$$

The evaluation of matrix functions is discussed, for example, in Reference [G2, Chapter 11]. For the special structure of $[U]$ in Equation (20.34), the matrix function $\bar{\xi}^{[U]}$ is calculated with a Taylor expansion with a finite number of terms:

$$[\bar{Y}(\bar{\xi})] = \bar{\xi}^{[U]} = e^{[U]\ln\bar{\xi}} = [I] + [U]\ln\bar{\xi} + \frac{1}{2!}[U]^2(\ln\bar{\xi})^2 + \ \ldots \ + \frac{1}{m!}[U]^m(\ln\bar{\xi})^m \quad m < N \tag{20.41}$$

as $[U]^i = 0$ for $i \geq m + 1$.

Substituting $[\bar{Y}(\bar{\xi})]$ 'backwards' in Equations (20.38), (20.13) and (20.10) results in the analytical solution of Equation (20.6):

$$[X(\bar{\xi})] = [\Phi][A(\bar{\xi})]\bar{\xi}^{\lceil\Lambda\rfloor}\bar{\xi}^{[U]} \tag{20.42}$$

$\bar{\xi}^{[U]}$ in Equation (20.42) is calculated exactly with Equation (20.41) (without any truncation of terms). $[X(\bar{\xi})]$ is a series solution as $[A(\bar{\xi})]$ is an infinite series (Equation (20.14)).

After partitioning the special solution $[X(\bar{\xi})]$ in square submatrices of order $n \times n$

$$[X(\bar{\xi})] = \begin{bmatrix} [X_{11}(\bar{\xi})] & [X_{12}(\bar{\xi})] \\ [X_{21}(\bar{\xi})] & [X_{22}(\bar{\xi})] \end{bmatrix} \tag{20.43}$$

the amplitudes of the displacements and the internal nodal forces are formulated using Equations (20.43), (20.5), (20.2) as

$$\begin{Bmatrix} \{u(\xi)\} \\ \{Q(\xi)\} \end{Bmatrix} = \begin{bmatrix} \xi^{-(s-2)/2}[X_{11}(\bar{\xi})] & \xi^{-(s-2)/2}[X_{12}(\bar{\xi})] \\ \xi^{(s-2)/2}[X_{21}(\bar{\xi})] & \xi^{(s-2)/2}[X_{22}(\bar{\xi})] \end{bmatrix} \begin{Bmatrix} \{c_1\} \\ \{c_2\} \end{Bmatrix} \tag{20.44}$$

with integration constants $\{c_1\}$, $\{c_2\}$ partitioned conformably.

Thus, similar to the series solution of the Bessel function, the analytical solution of Equation (20.1) is formulated as a series in $\bar{\xi}$ (Equation (20.2)) or for a specified ω in the radial coordinate ξ. After determining the coefficient matrices of the series solution, two independent sets of solutions are obtained (Equation (20.44)). The two sets of integration constants of the general solution follow from the boundary conditions at two values of ξ denoted as ξ_i (i for interior) and ξ_e (e for exterior), as shown in Figure 20.1. For the bounded medium shown in Figure 3.1 the boundary conditions are

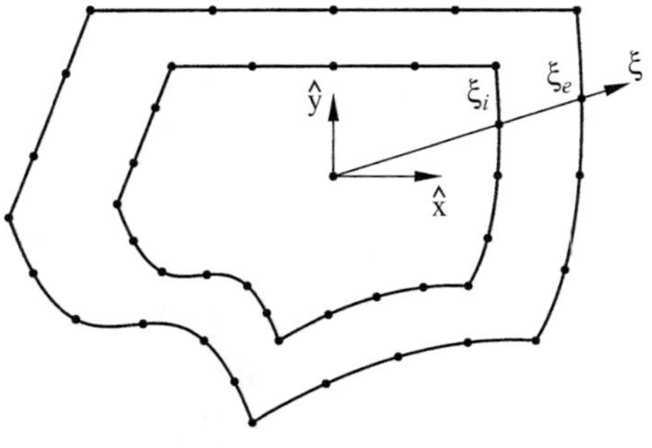

Figure 20.1 Domain between interior and exterior boundaries corresponding to $\xi_i \leq \xi \leq \xi_e$ (Reprinted from *Computer Methods in Applied Mechanics & Engineering*, Vol. 164, Song and Wolf 'The scaled boundary finite element method: analytical solution in frequency domain', pp. 249–264, © (1998), with permission from Elsevier Science)

formulated at $\xi_i = 0$ and $\xi_e = 1$ and for the unbounded medium in Figure 3.4 at $\xi_i = 1$ and $\xi_e \to \infty$.

The series solution (Equation (20.44)) is applied to calculate the dynamic-stiffness matrix of a bounded medium in section 20.2. A domain between two similar boundaries is also addressed, which is necessary to determine the dynamic-stiffness matrix of an unbounded medium in section 20.3.

20.2 DYNAMIC STIFFNESS OF BOUNDED MEDIUM

First, the domain between the two similar boundaries with ξ_i and ξ_e is addressed (Figure 20.1). Formulating Equation (20.44) for the two boundaries yields

$$\left\{ \begin{matrix} \{u_i\} \\ \{Q_i\} \end{matrix} \right\} = \left\{ \begin{matrix} \{u(\xi_i)\} \\ \{Q(\xi_i)\} \end{matrix} \right\} = \begin{bmatrix} \xi_i^{-0.5(s-2)}[X_{11}(\bar{\xi}_i)] & \xi_i^{-0.5(s-2)}[X_{12}(\bar{\xi}_i)] \\ \xi_i^{0.5(s-2)}[X_{21}(\bar{\xi}_i)] & \xi_i^{0.5(s-2)}[X_{22}(\bar{\xi}_i)] \end{bmatrix} \left\{ \begin{matrix} \{c_1\} \\ \{c_2\} \end{matrix} \right\} \tag{20.45}$$

$$\left\{ \begin{matrix} \{u_e\} \\ \{Q_e\} \end{matrix} \right\} = \left\{ \begin{matrix} \{u(\xi_e)\} \\ \{Q(\xi_e)\} \end{matrix} \right\} = \begin{bmatrix} \xi_e^{-0.5(s-2)}[X_{11}(\bar{\xi}_e)] & \xi_e^{-0.5(s-2)}[X_{12}(\bar{\xi}_e)] \\ \xi_e^{0.5(s-2)}[X_{21}(\bar{\xi}_e)] & \xi_e^{0.5(s-2)}[X_{22}(\bar{\xi}_e)] \end{bmatrix} \left\{ \begin{matrix} \{c_1\} \\ \{c_2\} \end{matrix} \right\} \tag{20.46}$$

which can also be written as

$$\left\{ \begin{matrix} \{u_i\} \\ \{u_e\} \end{matrix} \right\} = \begin{bmatrix} \xi_i^{-0.5(s-2)}[X_{11}(\bar{\xi}_i)] & \xi_i^{-0.5(s-2)}[X_{12}(\bar{\xi}_i)] \\ \xi_e^{-0.5(s-2)}[X_{11}(\bar{\xi}_e)] & \xi_e^{-0.5(s-2)}[X_{12}(\bar{\xi}_e)] \end{bmatrix} \left\{ \begin{matrix} \{c_1\} \\ \{c_2\} \end{matrix} \right\} \tag{20.47}$$

$$\left\{ \begin{matrix} \{Q_i\} \\ \{Q_e\} \end{matrix} \right\} = \begin{bmatrix} \xi_i^{0.5(s-2)}[X_{21}(\bar{\xi}_i)] & \xi_i^{0.5(s-2)}[X_{22}(\bar{\xi}_i)] \\ \xi_e^{0.5(s-2)}[X_{21}(\bar{\xi}_e)] & \xi_e^{0.5(s-2)}[X_{22}(\bar{\xi}_e)] \end{bmatrix} \left\{ \begin{matrix} \{c_1\} \\ \{c_2\} \end{matrix} \right\} \tag{20.48}$$

Introducing the nodal force amplitudes $\{P\}$ for the bounded medium in the global coordinate system on the interior boundary (negative face) and on the exterior boundary (positive face),

$$\left\{ \begin{matrix} \{P_i\} \\ \{P_e\} \end{matrix} \right\} = \left\{ \begin{matrix} -\{Q_i\} \\ \{Q_e\} \end{matrix} \right\} \tag{20.49}$$

applies (Equation (15.68)). Eliminating $\{c_1\}$, $\{c_2\}$ from Equations (20.47) and (20.48), and substituting Equation (20.49) leads to

$$\left\{ \begin{matrix} \{P_i\} \\ \{P_e\} \end{matrix} \right\} = \begin{bmatrix} [S_{ii}(\omega)] & [S_{ie}(\omega)] \\ [S_{ei}(\omega)] & [S_{ee}(\omega)] \end{bmatrix} \left\{ \begin{matrix} \{u_i\} \\ \{u_e\} \end{matrix} \right\} \tag{20.50}$$

with the dynamic-stiffness matrix with respect to the degrees of freedom on the interior and exterior boundaries

$$\begin{bmatrix}[S_{ii}(\omega)] & [S_{ie}(\omega)]\\ [S_{ei}(\omega)] & [S_{ee}(\omega)]\end{bmatrix} = \begin{bmatrix}-\xi_i^{0.5(s-2)}[X_{21}(\bar{\xi}_i)] & -\xi_i^{0.5(s-2)}[X_{22}(\bar{\xi}_i)]\\ \xi_e^{0.5(s-2)}[X_{21}(\bar{\xi}_e)] & \xi_e^{0.5(s-2)}[X_{22}(\bar{\xi}_e)]\end{bmatrix}$$
$$\times \begin{bmatrix}\xi_i^{-0.5(s-2)}[X_{11}(\bar{\xi}_i)] & \xi_i^{-0.5(s-2)}[X_{12}(\bar{\xi}_i)]\\ \xi_e^{-0.5(s-2)}[X_{11}(\bar{\xi}_e)] & \xi_e^{-0.5(s-2)}[X_{12}(\bar{\xi}_e)]\end{bmatrix}^{-1} \tag{20.51}$$

Secondly, the domain on the interior of the boundary $0 \leq \xi \leq 1$, i.e. a bounded medium, is examined (Figure 3.1), for which a more compact result is obtained.

The behaviour of Equations (20.42) and (20.5) at $\xi = 0$ is addressed. Partitioning and substituting Equation (20.8) yield

$$\{X(\bar{\xi})\} = \begin{bmatrix}[\Phi_{11}] & [\Phi_{12}]\\ [\Phi_{21}] & [\Phi_{22}]\end{bmatrix}$$
$$\times \begin{bmatrix}[A_{11}(\bar{\xi})] & [A_{12}(\bar{\xi})]\\ [A_{21}(\bar{\xi})] & [A_{22}(\bar{\xi})]\end{bmatrix} \begin{bmatrix}\bar{\xi}^{-\lceil\lambda\rfloor} & \\ & \bar{\xi}^{\lceil\lambda\rfloor}\end{bmatrix} \begin{bmatrix}[\bar{Y}_{11}(\bar{\xi})] & [\bar{Y}_{12}(\bar{\xi}))\\ 0 & [\bar{Y}_{22}(\bar{\xi})]\end{bmatrix} \begin{Bmatrix}\{c_1\}\\ \{c_2\}\end{Bmatrix} \tag{20.52}$$

For $\xi = 0$, $\bar{\xi} = 0$ follows, which leads to $[A] = [I]$ (Equation (20.14)). As the real parts of the eigenvalues $\lceil\lambda\rfloor$ are negative, for $\bar{\xi} \to 0$, $\bar{\xi}^{-\lceil\lambda\rfloor} \to 0$ and $\bar{\xi}^{\lceil\lambda\rfloor} \to \infty$ result. Examining the limit of the term $\bar{\xi}^{\lceil\lambda\rfloor}[\bar{Y}_{22}(\bar{\xi})]\{c_2\}$ in Equation (20.52) for $\bar{\xi} \to 0$, $\{c_2\}$ must be equal to zero for a finite solution as $[\bar{Y}_{22}(\bar{\xi})] \to \infty$ owing to the logarithmic functions in Equation (20.41).

To calculate the dynamic-stiffness matrix, Equation (20.52) with $\{c_2\} = 0$ leads to

$$\begin{Bmatrix}\{u(\xi)\}\\ \{Q(\xi)\}\end{Bmatrix} = \begin{Bmatrix}\xi^{-0.5(s-2)}([\Phi_{11}][A_{11}(\bar{\xi})] + [\Phi_{12}][A_{21}(\bar{\xi})])\,\bar{\xi}^{-\lceil\lambda\rfloor}[\bar{Y}_{11}(\bar{\xi})]\{c_1\}\\ \xi^{0.5(s-2)}([\Phi_{21}][A_{11}(\bar{\xi})] + [\Phi_{22}][A_{21}(\bar{\xi})])\,\bar{\xi}^{-\lceil\lambda\rfloor}[\bar{Y}_{11}(\bar{\xi})]\{c_1\}\end{Bmatrix} \tag{20.53}$$

Equation (20.53) is valid for arbitrary $\{c_1\}$. With $\{P(\xi)\} = \{Q(\xi)\}$ on the boundary (positive face), the force-displacement relationship is written as

$$\{P(\xi)\} = [S^{\mathrm{b}}(\xi)]\{u(\xi)\} \tag{20.54}$$

with the dynamic-stiffness matrix obtained by eliminating $\{c_1\}$ in Equation (20.53) as

$$[S^{\mathrm{b}}(\xi)] = \xi^{s-2}([\Phi_{21}][A_{11}(\bar{\xi})] + [\Phi_{22}][A_{21}(\bar{\xi})])([\Phi_{11}][A_{11}(\bar{\xi})] + [\Phi_{12}][A_{21}(\bar{\xi})])^{-1} \tag{20.55}$$

For the boundary $\xi = 1$ applies.

Note that, to calculate the dynamic-stiffness matrix of the bounded medium $0 \leq \xi \leq 1$, the matrix function $\xi^{[U]}$ in Equation (20.41) is not evaluated.

20.3 DYNAMIC STIFFNESS OF UNBOUNDED MEDIUM

The domain on the exterior of the boundary corresponding to $1 \leq \xi < \infty$, an unbounded medium, is examined (Figure 3.4). The radiation condition formulated at $\xi \to \infty$ must be enforced in Equation (20.44) when determining the integration constants $\{c_1\}$, $\{c_2\}$. No obvious choice of $\{c_1\}$ and $\{c_2\}$ in Equation (20.44) is identified, which satisfies the radiation condition. This is in contrast to the Hankel functions, which are readily defined from the first and second kind of Bessel functions for the scalar case. To calculate $[X(\xi)]$ for $\xi \to \infty$ would require an infinite number of terms if the series solution in Equation (20.42) were used. As an alternative, the high-frequency asymptotic expansion for the dynamic-stiffness matrix $[S^{\infty}(\omega)]$ of the unbounded medium (Chapter 18) is applied, which satisfies the radiation condition at infinity. The unbounded medium of Figure 3.4 is replaced by the bounded domain of Figure 20.1. On its exterior boundary ξ_e, the nodal force-displacement relationship determined by the high-frequency asymptotic expansion $[S_e^{\infty}(\omega)]$ (calculated for a large but finite $\bar{\xi}_e = \omega\xi_e$) serves as the boundary condition.

The asymptotic expansion for high frequency of the dynamic-stiffness matrix of the unbounded medium equals (Equation (18.12))

$$[S^{\infty}(\omega, \xi)] = \xi^{s-2}\left(\mathrm{i}\bar{\xi}[C_{\infty}] + [K_{\infty}] + \sum_{j=1}^{m} \frac{1}{(\mathrm{i}\bar{\xi})^j}[A_j]\right) \tag{20.56}$$

The coefficient matrices $[C_{\infty}]$, $[K_{\infty}]$, $[A_j]$ are specified in Equations (18.25), (18.28) and (18.31). The dashpot matrix $[C_{\infty}]$ can also be constructed directly based on impedances (Equation (18.40)).

To formulate the boundary condition of the bounded medium of Figure 20.1, the nodal force amplitudes of the unbounded medium exterior to the boundary ξ_e (negative face) are written as

$$\{R_e\} = [S_e^{\infty}(\omega)]\{u_e\} \tag{20.57}$$

with $[S_e^{\infty}(\omega)] = [S^{\infty}(\omega, \xi_e)]$ calculated from Equation (20.56). Formulating equilibrium, the nodal force amplitudes $\{R_e\}$ are related to the nodal force amplitudes $\{P_e\}$ of Equation (20.50) as

$$\{R_e\} + \{P_e\} = 0 \tag{20.58}$$

On the boundary $\xi = \xi_i = 1$ the nodal force-displacement relationship of the unbounded domain is defined as

$$\{R_i\} = [S^{\infty}(\omega)]\{u_i\} \tag{20.59}$$

with $[S_i^{\infty}(\omega)] = [S^{\infty}(\omega)]$ and

$$\{R_i\} = \{P_i\} \tag{20.60}$$

Eliminating $\{P\}$ and $\{R\}$ by substituting Equations (20.50), (20.59) and (20.57) into Equations (20.60) and (20.58) leads to

$$\begin{bmatrix} [S_{ii}(\omega)] & [S_{ie}(\omega)] \\ [S_{ei}(\omega)] & [S_{ee}(\omega)] \end{bmatrix} \begin{Bmatrix} \{u_i\} \\ \{u_e\} \end{Bmatrix} = \begin{bmatrix} [S^{\infty}(\omega)] & \\ & -[S_e^{\infty}(\omega)] \end{bmatrix} \begin{Bmatrix} \{u_i\} \\ \{u_e\} \end{Bmatrix} \tag{20.61}$$

The relation linking $[S^{\infty}(\omega)]$ and $[S_e^{\infty}(\omega)]$ can be derived by eliminating $\{u_e\}$ from Equation (20.61) for an arbitrary $\{u_i\}$, resulting in

$$[S^{\infty}(\omega)] = [S_{ii}(\omega)] - [S_{ie}(\omega)]([S_e^{\infty}(\omega)] + [S_{ee}(\omega)])^{-1}[S_{ei}(\omega)] \tag{20.62}$$

To calculate the dynamic-stiffness matrix $[S^{\infty}(\omega)]$ of the unbounded medium at the boundary $\xi = \xi_i = 1$ for a specific ω, a sufficiently large $\bar{\xi}_e$ is selected for which the high-frequency asymptotic expansion of Equation (20.56) yields an accurate $[S_e^{\infty}(\omega)]$. The ξ_e then follows as

$$\xi_e = \frac{\bar{\xi}_e}{\omega} \tag{20.63}$$

and $\bar{\xi}_i = \omega$. The submatrices $[S_{ii}(\omega)]$, $[S_{ie}(\omega)]$, $[S_{ei}(\omega)]$, $[S_{ee}(\omega)]$ are determined from Equation (20.51). $[S^{\infty}(\omega)]$ is then calculated from Equation (20.62).

In passing, it is worth mentioning that Equation (20.62) represents the key equation in the dynamic condensation method mentioned towards the end of section 19.1. In this case, the dynamic-stiffness matrix of the domain between the two similar boundaries, the finite-element cell, is calculated numerically and not analytically.

20.4 FAR-FIELD DISPLACEMENT

The displacements in the far-field ($\xi > \xi_e$), where $\bar{\xi} > \bar{\xi}_e$, can also be calculated analytically. In this range, the high-frequency asymptotic expansion of the dynamic-stiffness matrix (Equation (20.56)) applies. Selecting the first two terms on the right-hand side and substituting in Equation (15.115) yields

$$[E^0]\bar{\xi}\{u(\bar{\xi})\}_{,\bar{\xi}} + ([E^1]^T + \mathrm{i}\bar{\xi}[C_{\infty}] + [K_{\infty}])\{u(\bar{\xi})\} = 0 \tag{20.64}$$

These first-order ordinary differential equations can be solved with a series expansion analogous to the procedure of section 20.1 addressing Equation (20.3). The boundary condition $\{u(\bar{\xi}_e)\}$ follows from the series expansion in Equation (20.44).

Summarising, the analytical solution of this scaled boundary finite element equation in displacement is obtained as a power series for any frequency. This solution applies directly to a bounded domain. For an unbounded domain a high-frequency asymptotic expansion of dynamic stiffness, which represents an approximation, satisfying the radiation condition at infinity is used. Thus, it is possible to calculate analytically the response throughout the unbounded medium avoiding numerical discretisation in the radial direction.

21 Extensions

Important extensions are discussed in this chapter, which allow the scaled boundary finite element method to be applied to solve additional problems efficiently. The scope of addressing especially unbounded media is thus enlarged. Section 21.1 examines the formulation for certain variations of the material properties in the radial direction, which permits, for instance, unbounded soil with material properties varying with depth to be analysed. Section 21.2 addresses incompressible elasticity, which can be used, for instance, to model undrained saturated soil as a one-phase material. Section 21.3 develops a reduced set of base functions to be used to represent the degrees of freedom on the boundary of an unbounded medium, reducing the order of the fully coupled dynamic-stiffness matrix. Realistic dynamic unbounded medium-structure interaction problems can thus be calculated efficiently. Sections 21.4 and 21.5 streamline the formulation for the case where the scaling centre is located at infinity. A two-dimensional stratified unbounded medium and a prismatic unbounded medium with an inhomogeneous section can be examined concisely, as the governing equations become simpler.

21.1 VARIATION OF MATERIAL PROPERTIES IN RADIAL DIRECTION

This extension is discussed for the scaled boundary finite element equation in dynamic stiffness based on the mechanically-based derivation in Reference [W9].

In all derivations up to now, the unbounded medium is assumed to be homogeneous in the radial direction towards infinity, i.e. the elasticity matrix $[D]$ and the mass density ρ are constant in the radial direction r. As an extension, $[D]$ and ρ are assumed to be power functions of r. The derivation is discussed for the case where the Young's modulus and the shear modulus vary with the same power g

$$E(\xi) = E_0\left(\frac{r}{r_0}\right)^g = E_0\xi^g \tag{21.1a}$$

$$G(\xi) = G_0\left(\frac{r}{r_0}\right)^g = G_0\xi^g \tag{21.1b}$$

with a constant Poisson's ratio ν, yielding

$$[D(\xi)] = [D_0]\xi^g \tag{21.2}$$

The mass density varies as

$$\rho(\xi) = \rho_0 \left(\frac{r}{r_0}\right)^m = \rho_0 \xi^m \tag{21.3}$$

The subscript 0 denotes the material constants on the boundary (($\xi = 1,\ r = r_0$). The powers g and m are real numbers which can be selected as positive or negative.

The material properties can of course vary in the circumferential directions on the boundary, from element to element, but also within an element. As the integrations are performed numerically, the values can be specified in each Gauss point ($[D_0] = [D_0(\eta, \zeta)],\ \rho_0 = \rho_0(\eta, \zeta)$). The powers g and m are constants and cannot vary in the circumferential directions.

The scaled boundary finite element equation in displacement is derived based on the virtual work formulation, as demonstrated in section 15.7. A three-dimensional unbounded medium is addressed ($s = 3,\ 1 \le \xi < \infty$).

The virtual work statement of Equation (15.163), repeated for convenience, equals

$$\int_V \{\delta\varepsilon(\xi, \eta, \zeta)\}^T \{\sigma(\xi, \eta, \zeta)\} \mathrm{d}V - \int_V \{\delta u(\xi, \eta, \zeta)\}^T \rho(\xi)\omega^2 \{u(\xi, \eta, \zeta)\} \mathrm{d}V$$
$$- \int_V \{\delta u(\xi, \eta, \zeta)\}^T \{p(\xi, \eta, \zeta)\} \mathrm{d}V - \int_S \{\delta u(\xi = 1, \eta, \zeta)\}^T \{t(\eta, \zeta)\} \mathrm{d}S = 0 \tag{21.4}$$

Substituting Equations (15.162), (15.161), $\mathrm{d}V$ from Equation (15.29) and with the stress amplitudes (Equation (15.47) with Equation (21.2))

$$\{\sigma(\xi, \eta, \zeta)\} = [D_0](\xi^g [B^1]\{u(\xi)\}_{,\xi} + \xi^{g-1}[B^2]\{u(\xi)\}) \tag{21.5}$$

in the first term of Equation (21.4) representing the internal work yields

$$\int_V \left([B^1]\{\delta u(\xi)\}_{,\xi} + \frac{1}{\xi}[B^2]\{\delta u(\xi)\}\right)^T$$
$$\times [D_0](\xi^g [B^1]\{u(\xi)\}_{,\xi} + \xi^{g-1}[B^2]\{u(\xi)\})\xi^2 |J| \mathrm{d}\xi \mathrm{d}\eta \mathrm{d}\zeta =$$
$$\int_1^\infty \{\delta u(\xi)\}_{,\xi}^T \int_S [B^1]^T [D_0][B^1]|J| \mathrm{d}\eta \mathrm{d}\zeta\, \xi^{g+2} \{u(\xi)\}_{,\xi} \mathrm{d}\xi$$
$$+ \int_1^\infty \{\delta u(\xi)\}_{,\xi}^T \int_S [B^1]^T [D_0][B^2]|J| \mathrm{d}\eta \mathrm{d}\zeta\, \xi^{g+1} \{u(\xi)\} \mathrm{d}\xi \tag{21.6}$$
$$+ \int_1^\infty \{\delta u(\xi)\}^T \int_S [B^2]^T [D_0][B^1]|J| \mathrm{d}\eta \mathrm{d}\zeta\, \xi^{g+1} \{u(\xi)\}_{,\xi} \mathrm{d}\xi$$
$$+ \int_1^\infty \{\delta u(\xi)\}^T \int_S [B^2]^T [D_0][B^2]|J| \mathrm{d}\eta \mathrm{d}\zeta\, \xi^{g} \{u(\xi)\} \mathrm{d}\xi$$

Integration by parts in the radial coordinate ξ is performed for all terms containing derivatives of the virtual displacements with respect to ξ, $\{\delta u(\xi)\}_{,\xi}$. This Green's theorem leads to integrals over the boundary S. Equation (21.6) is reformulated as

$$\begin{aligned}
&\{\delta u(\xi)\}^T \int_S [B^1]^T [D_0] [B^1] |J| \mathrm{d}\eta \mathrm{d}\zeta\, \xi^{g+2} \{u(\xi)\}_{,\xi}\Big|_{\xi=1} \\
&- \int_1^\infty \{\delta u(\xi)\}^T \int_S [B^1]^T [D_0] [B^1] |J| \mathrm{d}\eta \mathrm{d}\zeta\, ((g+2)\xi^{g+1}\{u(\xi)\}_{,\xi} + \xi^{g+2}\{u(\xi)\}_{,\xi\xi}) \mathrm{d}\xi \\
&+ \{\delta u(\xi)\}^T \int_S [B^1]^T [D_0] [B^2] |J| \mathrm{d}\eta \mathrm{d}\zeta\, \xi^{g+1} \{u(\xi)\}|_{\xi=1} \\
&- \int_1^\infty \{\delta u(\xi)\}^T \int_S [B^1]^T [D_0] [B^2] |J| \mathrm{d}\eta \mathrm{d}\zeta\, ((g+1)\xi^{g}\{u(\xi)\} + \xi^{g+1}\{u(\xi)\}_{,\xi}) \mathrm{d}\xi \\
&+ \int_1^\infty \{\delta u(\xi)\}^T \int_S [B^2]^T [D_0] [B^1] |J| \mathrm{d}\eta \mathrm{d}\zeta\, \xi^{g+1} \{u(\xi)\}_{,\xi} \mathrm{d}\xi \\
&+ \int_1^\infty \{\delta u(\xi)\}^T \int_S [B^2]^T [D_0] [B^2] |J| \mathrm{d}\eta \mathrm{d}\zeta\, \xi^{g} \{u(\xi)\} \mathrm{d}\xi
\end{aligned} \tag{21.7}$$

The coefficient matrices $[E^0]$, $[E^1]$, $E^2]$ are defined as in Equation (15.166), but with $[D_0]$ replacing $[D]$.

Using $\{u\}$ to represent $\{u(\xi = 1)\}$ and so forth, Equation (21.7) is expressed as

$$\begin{aligned}
\int_V \{\delta\varepsilon\}^T \{\sigma\} \mathrm{d}V &= \{\delta u\}^T (-[E^0]\{u\}_{,\xi} - [E^1]^T \{u\}) \\
&- \int_1^\infty \{\delta u(\xi)\}^T ([E^0]\xi^{g+2}\{u(\xi)\}_{,\xi\xi} \\
&+ ((g+2)[E^0] + [E^1]^T - [E^1])\xi^{g+1}\{u(\xi)\}_{,\xi} \\
&+ ((g+1)[E^1]^T - [E^2])\xi^{g}\{u(\xi)\}) \mathrm{d}\xi
\end{aligned} \tag{21.8}$$

Substituting Equations (15.161), (21.3), (15.159) and (15.29) in the second term of Equation (21.4) representing the work of the inertial loads yields

$$-\omega^2 \int_1^\infty \{\delta u(\xi)\}^T \int_S [N]^T \rho_0 [N] |J| \mathrm{d}\eta \mathrm{d}\zeta\, \xi^{m+2} \{u(\xi)\} \mathrm{d}\xi \tag{21.9}$$

With the coefficient matrix $[M^0]$ defined in Equation (15.168), but with ρ_0 replacing ρ, Equation (21.9) equals

$$-\omega^2 \int_1^\infty \{\delta u(\xi)\}^T [M^0] \xi^{m+2} \{u(\xi)\} d\xi \tag{21.10}$$

The third and fourth terms representing the external work of the body loads and surface tractions are specified in Equation (15.171) with Equation (15.172) and in Equation (15.173).

Substituting Equations (21.8), (21.10), (15.171) and (15.173) into Equation (21.4) results with the equivalent nodal force amplitudes $\{R\}$ due to the boundary tractions in

$$\begin{aligned}
&\{\delta u\}^T(-[E^0]\{u\}_{,\xi} - [E^1]^T\{u\} - \{R\}) \\
&- \int_1^\infty \{\delta u(\xi)\}^T([E^0]\xi^{g+2}\{u(\xi)\}_{,\xi\xi} \\
&+ ((g+2)[E^0] + [E^1]^T - [E^1])\xi^{g+1}\{u(\xi)\}_{,\xi} \\
&+ ((g+1)[E^1]^T - [E^2])\xi^g\{u(\xi)\} \\
&+\omega^2[M^0]\xi^{m+2}\{u(\xi)\} + \xi^2\{F^b(\xi)\})d\xi = 0
\end{aligned} \tag{21.11}$$

In order to satisfy Equation (21.11) for all $\{\delta u(\xi)\}$ with $1 \leq \xi < \infty$, the coefficient matrix of $\{\delta u\}$ at $\xi = 1$ and the coefficient matrix of $\{\delta u(\xi)\}$ must vanish

$$\{R\} = -[E^0]\{u\}_{,\xi} - [E^1]^T\{u\} \tag{21.12}$$

$$\begin{aligned}
&[E^0]\xi^{g+2}\{u(\xi)\}_{,\xi\xi} + ((g+2)[E^0] - [E^1] + [E^1]^T)\xi^{g+1}\{u(\xi)\}_{,\xi} \\
&+ ((g+1)[E^1]^T - [E^2])\xi^g\{u(\xi)\} + \omega^2[M^0]\xi^{m+2}\{u(\xi)\} \\
&+ \xi^2\{F^b(\xi)\} = 0
\end{aligned} \tag{21.13}$$

Equation (21.12) represents the nodal force displacement relationship formulated at the boundary $\xi = 1$. It coincides for $g = 0$ with Equation (15.67) with (15.68) formulated with the negative sign for an unbounded medium. The derivation of Equation (21.7) also leads to the expression for any ξ

$$\{R(\xi)\} = -[E^0]\xi^{g+2}\{u(\xi)\}_{,\xi} - [E^1]^T\xi^{g+1}\{u(\xi)\} \tag{21.14}$$

Equation (21.13) is the scaled boundary finite element equation in displacement, coinciding for $g = 0$ and $m = 0$ with Equation (15.57) using Equation (15.56).

For a two-dimensional unbounded medium ($s = 2$, $1 \leq \xi < \infty$) the derivation is analogous. The infinitesimal area dV is specified in Equation (15.92). All integrations over the boundary S are one dimensional with the coefficient matrices $[E^0]$, $[E^1]$, $[E^2]$ (using $[D_0]$ instead of $[D]$) and $[M^0]$ (using ρ_0 instead of ρ) specified in Equations (15.98) and (15.99), respectively. Equation (21.12) still applies and the nodal force-displacement relationship for any ξ equals

$$\{R(\xi)\} = -[E^0]\xi^{g+1}\{u(\xi)\}_{,\xi} - [E^1]^T\xi^g\{u(\xi)\} \tag{21.15}$$

The *scaled boundary finite element equation in displacement* for two dimensions is equal to

$$[E^0]\xi^{g+2}\{u(\xi)\}_{,\xi\xi} + ((g+1)[E^0] - [E^1] + [E^1]^T)\xi^{g+1}\{u(\xi)\}_{,\xi} + (g[E^1]^T - [E^2])\xi^g\{u(\xi)\} + \omega^2[M^0]\xi^{m+2}\{u(\xi)\} + \xi^2\{F^b(\xi)\} = 0 \tag{21.16}$$

Unifying the three- and two-dimensional formulations with the spatial dimension s yields, for an unbounded medium *for varying material properties in the radial direction*

$$\{R(\xi)\} = -[E^0]\xi^{g+s-1}\{u(\xi)\}_{,\xi} - [E^1]^T\xi^{g+s-2}\{u(\xi)\} \tag{21.17}$$

and

$$[E^0]\xi^{g+2}\{u(\xi)\}_{,\xi\xi} + ((g+s-1)[E^0] - [E^1] + [E^1]^T)\xi^{g+1}\{u(\xi)\}_{,\xi} + ((g+s-2)[E^1]^T - [E^2])\xi^g\{u(\xi)\} + \omega^2[M^0]\xi^{m+2}\{u(\xi)\} + \xi^2\{F^b(\xi)\} = 0 \tag{21.18}$$

For a bounded medium, the signs on the right-hand side of Equation (21.17) are reversed, while Equation (21.18) remains unchanged.

Note that all coefficient matrices $[E^0]$, $[E^1]$, $[E^2]$, $[M^0]$ are evaluated with the material properties at the boundary $\xi = 1$. The general form of Equations (21.17) and (21.18) is the same as for the homogeneous case, with differences arising in the powers of ξ and in certain coefficients.

The derivation of the scaled boundary finite element equation in dynamic stiffness for varying material properties is analogous to that for the homogeneous case performed in section 15.4. Equation (21.17) corresponds to Equations (15.67) and (15.68), which is combined with Equation (21.18). The dimensionless dynamic-stiffness matrix corresponding to Equation (15.75) (and for the two-dimensional case to Equation (15.107)) is introduced as

$$[S^\infty(\omega, \xi)] = G_0 r_0^{s-2}\xi^{g+s-2}[\bar{S}^\infty(a)] \tag{21.19}$$

where the dimensionless frequency for any ξ (Equation 15.77)

$$a = \frac{\omega r_0}{c_{s0}}\xi^{1 - \frac{g}{2} + \frac{m}{2}} \tag{21.20}$$

with the shear-wave velocity at the boundary $c_{s0} = \sqrt{G_0/\rho_0}$.

For vanishing nodal loads due to the body load the *scaled boundary finite element equation in dynamic stiffness for an unbounded medium* at the boundary $\xi = 1$ equals, *for varying material properties* in the radial direction,

$$([S^\infty(\omega)] + [E^1])[E^0]^{-1}([S^\infty(\omega)] + [E^1]^T) - (g+s-2)[S^\infty(\omega)] - \left(1 - \frac{g}{2} + \frac{m}{2}\right)\omega[S^\infty(\omega)]_{,\omega} - [E^2] + \omega^2[M^0] = 0 \tag{21.21}$$

For a bounded medium, the corresponding relationship follows substituting $[S^{\infty}(\omega)] = -[S^{b}(\omega)]$.

Equation (21.21) is in the same form as for a medium with constant material properties in the radial direction (Equation (15.112) with $\xi = 1$, $\bar{\xi} = \omega$). As the difference is limited to two scalar coefficients, the two cases can be treated analogously. The scaled boundary finite element equation in the time domain for a unit-impulse (velocity) response can straightforwardly be derived as in section 19.2.1. Most analytical and numerical solution procedures discussed in previous chapters also apply directly. This also applies when solving the scaled boundary finite element equation in displacement (Equation (21.18)).

For statics ($\omega = 0$), after dividing by ξ^g, the scaled boundary finite element equation in displacement for vanishing body load equals (Equation (21.18))

$$\begin{aligned}&[E^0]\xi^2\{u(\xi)\}_{,\xi\xi} + ((g+s-1)[E^0] - [E^1] + [E^1]^T)\xi\{u(\xi)\}_{,\xi}\\&((g+s-2)[E^1]^T - [E^2])\{u(\xi)\} = 0\end{aligned} \tag{21.22}$$

This Euler–Cauchy equation, which differs from that of the homogeneous case in two scalar coefficients, can be solved analytically (section 16.1).

The extensions of the variation of the material properties to the diffusion equation is straightforward (section 15.6).

The variation of material properties described in Equations (21.2) and (21.3) refers to the radial direction ξ only with constant powers g and m. However, as the material properties along the circumference of the boundary ($[D_0(\eta,\zeta)]$, $\rho_0(\eta,\zeta)$) can also vary, with different values in all Gauss points of integration of the curved finite elements on the boundary, more general variations of material properties can be calculated. For instance, the important case of varying properties as a function of a Cartesian coordinate can be analysed. This permits certain variations of the soil properties of a site with depth to be examined, as pioneered in Reference [D7].

As an example, a foundation embedded in a site, an unbounded medium, is addressed. Figure 21.1 presents a section with the boundary (structure–soil interface), but

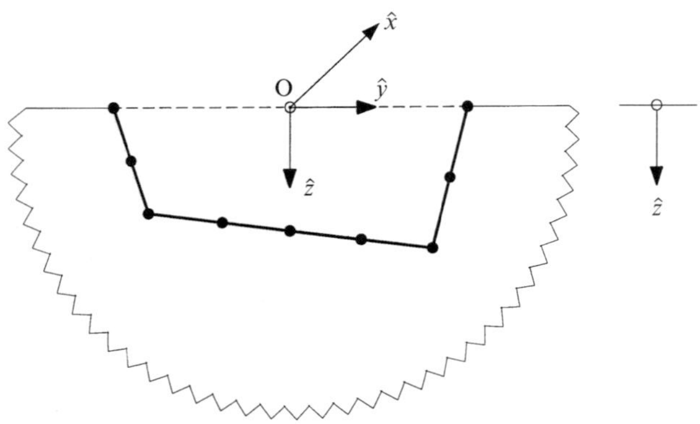

Figure 21.1 Foundation embedded in site with material properties varying with depth

not the free surface, discretised. The scaling centre O is selected on the extension of the free surface of the half space which represents the side-faces. The material properties are polynomials of the coordinate $\hat{z}$ denoting the depth

$$[D(\hat{z})] = [\bar{D}]\hat{z}^g \tag{21.23}$$

$$\rho(\hat{z}) = \bar{\rho}\,\hat{z}^m \tag{21.24}$$

where $[\bar{D}]$ and $\bar{\rho}$ are constants corresponding to the elasticity matrix and the mass density. Typically, $g < 1$ and $m << 1$.

To demonstrate that this variation of the material properties corresponds to power functions in the radial direction and a specified variation in the circumferential directions, the transformation to the scaled boundary coordinates is performed (Figure 15.1). The scaling equation (Equation (15.2)) is formulated

$$\hat{z}(\xi, \eta, \zeta) = \xi\, z(\eta, \zeta) \tag{21.25}$$

where the point on the boundary is mapped as (Equation (15.1))

$$z(\eta, \zeta) = [N(\eta, \zeta)]\{z\} \tag{21.26}$$

with the $\hat{z}$-coordinates of the nodes in $\{z\}$. Substituting Equation (21.25) into (21.23) yields the elasticity matrix

$$[D(\hat{z})] = [\bar{D}]\xi^g z^g(\eta, \zeta) = [D_0(\eta, \zeta)]\xi^g \tag{21.27}$$

with

$$[D_0(\eta, \zeta)] = [\bar{D}]z^g(\eta, \zeta) \tag{21.28}$$

Equation (21.27) is analogous to Equation (21.2), with the variation of the elasticity matrix in the circumferential directions described in Equation (21.28). Analogously, the mass density (Equation (21.24)) is transformed into

$$\rho(\hat{z}) = \bar{\rho}\,\xi^m z^m(\eta, \zeta) = \rho_0(\eta, \zeta)\xi^m \tag{21.29}$$

with its variation in the circumferential direction

$$\rho_0(\eta, \zeta) = \bar{\rho}\,z^m(\eta, \zeta) \tag{21.30}$$

Equation (21.29) has the same form as Equation (21.3).

21.2 INCOMPRESSIBLE ELASTODYNAMICS

Many important physical problems involve incompressible materials, i.e. their volumes are preserved locally during deformations. Rubber is often modelled as an incompressible elastic material. In civil engineering, the saturated unbounded soil when undrained behaves as a nearly incompressible material in a one-phase formulation of a soil–structure interaction analysis.

It is well known that the case of incompressible elasticity requires a special approach in the finite element method. The same also applies to the scaled boundary finite element method.

An in-depth derivation of the fundamental equations will not be provided in this section. This is justified, as a compact detailed discussion covering all aspects is readily available in Reference [S6] and in Chapter 6 of Reference [W9]. The interested reader can follow this work easily, as only the changes compared to the compressible case are developed, and the basis has been developed in previous chapters. No recent research has been conducted, which thus does not call for a revisit of the subject. In particular, only the scaled boundary finite element equation in dynamic stiffness, but not in displacement, has been addressed.

Certain highlights are, however, mentioned. As in the finite element method, the elasticity matrix is decomposed into the shear and volumetric parts. However, in the derivation for the scaled boundary finite element method, no selective reduced integration with a Poisson's ratio very close to 0.5 is introduced, as the limit of 0.5 is enforced analytically. A certain arbitrariness in choosing Poisson's ratio is thus eliminated. Only the coefficient matrices $[E^0]$, $[E^1]$, $[E^2]$ are affected. In addition, the final scaled boundary finite element equation in dynamic stiffness (and the corresponding relationship in the time domain for the unit-impulse response function) is slightly modified in this streamlined formulation. Concentrated masses appear in the high-frequency limit, which are not present in the compressible case, in addition to modified dashpots. This leads in the time domain to an instantaneous response over the entire domain, which is caused by the infinite dilatational-wave velocity.

21.3 REDUCED SET OF BASE FUNCTIONS

For an unbounded medium with vanishing body load calculated in the frequency domain, the dynamic-stiffness matrix $[S^\infty]$ fully describes its dynamic behaviour. This coupled symmetric matrix is of the order of the number of physical degrees of freedom on the boundary, which represents the structure–soil interface in an unbounded medium-structure interaction analysis. (No discretisation and thus degrees of freedom are present on the free surface (side-faces), Figure 3.4b.) The order of the system of the ordinary differential equations in the scaled boundary finite element equation in displacement (Equation 15.111) is also equal to the same number of degrees of freedom $\{u(\xi)\}$. For vanishing body load, this equation equals

$$[E^0]\xi^2\{u(\xi)\}_{,\xi\xi} + ((s-1)[E^0] - [E^1] + [E^1]^T)\xi\{u(\xi)\}_{,\xi} + ((s-2)[E^1]^T - [E^2])\{u(\xi)\} + \omega^2[M^0]\xi^2\{u(\xi)\} = 0 \quad (21.31)$$

To improve the computational efficiency, the displacement amplitudes $\{u(\xi)\}$ are represented by a reduced set of base functions $[\Phi]$ and corresponding amplitudes $\{v(\xi)\}$ [W15]:

$$\{u(\xi)\} = [\Phi]\{v(\xi)\} \tag{21.32}$$

The order of $\{v(\xi)\}$ will be chosen significantly smaller than that of $\{u(\xi)\}$. Substituting Equation (21.32) into Equation (21.31) and premultiplying by $[\Phi]^T$ yields

$$\begin{aligned}&[e^0]\xi^2\{v(\xi)\}_{,\xi\xi} + ((s-1)[e^0] - [e^1] + [e^1]^T)\xi\{v(\xi)\}_{,\xi}\\&+((s-2)[e^1]^T - [e^2])\{v(\xi)\} + \omega^2[m^0]\xi^2\{v(\xi)\} = 0\end{aligned} \tag{21.33}$$

with the reduced coefficient matrices

$$\begin{aligned}[e^0] &= [\Phi]^T[E^0][\Phi]\\ [e^1] &= [\Phi]^T[E^1][\Phi]\\ [e^2] &= [\Phi]^T[E^2][\Phi]\end{aligned} \tag{21.34}$$

and

$$[m^0] = [\Phi]^T[M^0][\Phi] \tag{21.35}$$

The reduced scaled boundary finite element equation in $v(\xi)$ (Equation (21.33)) is solved analytically or numerically analogously as the unreduced original equation in $\{u(\xi)\}$ (Equation (21.31)). The same applies to all operations concerning the dynamic-stiffness matrix.

Various possibilities exist to determine the reduced set of base functions $[\Phi]$. For instance, for certain unbounded medium-structure interaction analyses, such as those addressing dynamic soil–structure interaction, the low-frequency behaviour can dominate. It is thus appropriate to select the reduced set of base functions $[\Phi]$ from the solution of Equation (21.31) for statics (Equation (16.2)). This yields the eigenvalue problem (Equation (16.7) with (16.6))

$$\begin{bmatrix}[E^0]^{-1}[E^1]^T - 0.5(s-2)[I] & -[E^0]^{-1}\\ -[E^2] + [E^1][E^0]^{-1}[E^1]^T & -([E^1][E^0]^{-1} - 0.5(s-2)[I])\end{bmatrix}[\Phi] = -[\Phi]\lceil\Lambda\rfloor \tag{21.36}$$

The first few eigenvalues λ_i (in ascending order) of the eigenvalues $\lceil\Lambda\rfloor$ (Equation (16.8)) are calculated. The corresponding eigenvectors determine the reduced set $[\Phi]$. When the λ_i are complex conjugates, both are included. The imaginary parts caused by the complex conjugate eigenvectors in the pre- and post-multiplications of Equations (21.34) and (21.35) will cancel, yielding real reduced coefficient matrices. As discussed in connection with Equation (16.17), the eigenvectors represent the modal displacements at the boundary nodes and λ_i can be identified as a modal scaling factor in the radial direction.

This reduction in size from $\{u(\xi)\}$ to $\{v(\xi)\}$ enables realistic dynamic unbounded medium-structure interaction problems with many degrees of freedom to be analysed.

21.4 TWO-DIMENSIONAL LAYERED UNBOUNDED MEDIUM

A layered unbounded medium is often encountered in practice (Figure 21.2). The semi-infinite medium is enclosed by two parallel boundaries extending to infinity (e.g. free and fixed) and by the boundary which can be curved. Often, the material properties vary in the direction perpendicular to the boundaries extending to infinity, leading to stratification.

A typical example is a horizontally stratified unbounded soil layer resting on rigid rock. The boundary represents the structure–soil interface.

For this two-dimensional case, the scaling centre O is located at infinity ($x \to -\infty$). As an approximation, the scaling centre can be chosen at a large distance, which permits the scaled boundary finite element method to be applied without any modification. Typically, the distance to the scaling centre equals 100 d with the depth d of the layer. Note that only the boundary is discretised, but not the free and fixed surfaces (side-faces) as well as the parallel interfaces between the different materials, which have extensions passing through the scaling centre at infinity.

A rigorous streamlined formulation with the scaling centre located at infinity can be derived. In this special case, the so-called consistent boundary method, also called the thin-layer method (see References [W1], [L3], [K1], and the references cited in these publications), leads to the same relations. The dimensionless frequency is defined as

$$a_0 = \frac{\omega d}{c_s} \tag{21.37}$$

with the (constant) depth d of the soil layers. a_0 is independent of the location of the boundary which simplifies the fundamental equations significantly.

The derivation of the fundamental equations is analogous to the case where the scaling centre is at a finite distance but much simpler. It is thus sufficient just to state the final equations (for the mechanically-based derivation, Reference [W9] can be consulted).

The scaled boundary finite element method in displacement for vanishing body load equals

$$\begin{aligned}[E^0]\{u(\xi)\}_{,\xi\xi} &+ ([E^0] - [E^1] + [E^1]^T)\{u(\xi)\}_{,\xi} \\ &- [E^2]\{u(\xi)\} + \omega^2[M^0]\{u(\xi)\} = 0\end{aligned} \tag{21.38}$$

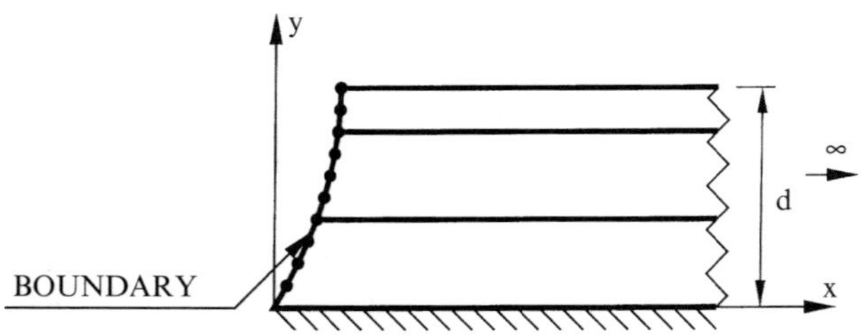

Figure 21.2 Two-dimensional layered semi-infinite medium with boundary discretised with line finite elemzents (Reproduced by permission of John Wiley & Sons Ltd)

which is a system of linear second-order ordinary differential equations with constant coefficients (compared to Equation (15.103) with variable coefficients). This represents the simplest system of ordinary differential equations, which is very easy to solve analytically.

The scaled boundary finite element method in dynamic stiffness is formulated as

$$([S^{\infty}(\omega)] + [E^1])[E^0]^{-1}([S^{\infty}(\omega)] + [E^1]^T) - [E^2] + \omega^2[M^0] = 0 \qquad (21.39)$$

Compared to Equation (15.110), the term with the derivative $\omega[S^{\infty}(\omega)]_{,\omega}$ is missing. This permits $[S^{\infty}(\omega)]$ for distinct frequencies to be determined directly, solving a quadratic eigenvalue problem.

An inverse Fourier transformation of Equation (21.39) into the time domain is possible, leading to simpler expressions. For a derivation of the unit-impulse response function, Reference [W9] can be consulted.

21.5 THREE-DIMENSIONAL PRISMATIC UNBOUNDED MEDIUM WITH INHOMOGENEOUS SECTION

As another case where the scaling centre is located at infinity, a three-dimensional prismatic unbounded medium with an inhomogeneous section can be examined. The free and fixed surfaces as well as the material interfaces, when present, are parallel extending to infinity in this three-dimensional semi-infinite case (Figure 21.3). The spatial discretisation with two-dimensional finite elements is limited to the cross-section S, which can be inhomogeneous. In this figure, only the nodes but not the actual finite elements are shown. The free and fixed surfaces and the material interfaces are side-faces with extensions passing through the scaling centre O located at infinity.

The dimensionless frequency is defined as

$$a_0 = \frac{\omega d}{c_s} \qquad (21.40)$$

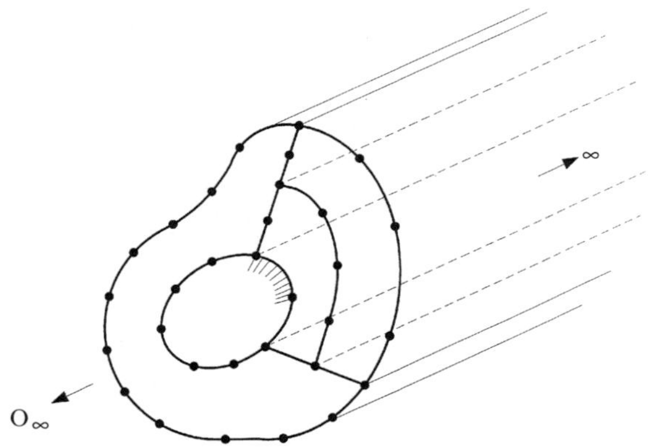

Figure 21.3 Hollow prismatic semi-infinite medium with inhomogeneous section, inner boundary free and fixed, outer boundary free

with d denoting a characteristic length of the cross-section, e.g. the largest diameter.

As a_0 remains constant along the prismatic unbounded medium, the formulation of the scaled boundary finite element method can be streamlined. This derivation for the three-dimensional case is analogous to that of the two-dimensional layered unbounded medium mentioned in section 21.4. The scaled boundary finite element equation in displacement for vanishing body load equals

$$\begin{aligned}[E^0]\{u(\xi)\}_{,\xi\xi} &+ (2[E^0] - [E^1] + [E^1]^T)\{u(\xi)\}_{,\xi} \\ &+ ([E^1]^T - [E^2])\{u(\xi)\} + \omega^2[M^0]\{u(\xi)\} = 0\end{aligned} \tag{21.41}$$

which is a system of linear second-order differential equations with constant coefficients (compared to Equation (15.60) with variable coefficients). The scaled boundary finite element equation in dynamic stiffness is formulated as

$$\begin{aligned}([S^\infty(\omega)] &+ [E^1])[E^0]^{-1}([S^\infty(\omega)] + [E^1]^T) \\ &- [S^\infty(\omega)] - [E^2] + \omega^2[M^0] = 0\end{aligned} \tag{21.42}$$

Compared to Equation (15.85), the term with the derivative $\omega[S^\infty(\omega)]_{,\omega}$ is missing. This permits $[S^\infty(\omega)]$ for distinct frequencies to be determined directly, solving a quadratic eigenvalue problem. Performing the inverse Fourier transformation into the time domain of Equation (21.42) yields an expression for the unit-impulse response function which can be solved numerically.

22 Substructuring

At first sight, the scaled boundary finite element method seems to place severe restrictions on the geometry and material variation. For instance, the entire boundary must be visible from the scaling centre. For a bounded domain, however, this limitation is lifted completely by using substructuring (Figure 3.2), i.e. by subdividing the total domain into subdomains, also called substructures or superelements. Each subdomain has its own scaling centre. This introduces additional boundaries between adjacent subdomains which, in principle, have to be discretised, leading to an increase in the number of degrees of freedom. (In some cases such as statics and dynamics performed in the frequency domain, the additional degrees of freedom can be eliminated after assemblage of the subdomains.) For an unbounded domain, the use of substructuring is limited, as discussed below.

The disadvantage of substructuring, the increase in the number of degrees of freedom of the total system which is modest, is more than compensated for by the many advantages which exist. In many cases, it is even computationally efficient to introduce substructuring to make full use of these advantages in cases where, strictly speaking, it is not necessary.

Five reasons for substructuring can be identified, which are illustrated in Figure 22.1. The first reason that the entire boundary must be visible from the scaling centre has already been mentioned. In Figure 22.1a, a domain with re-entrant corners is examined. On the left, the parts of the boundary which are not visible from the potential scaling centre are shown. On the right, the two subdomains, which do not have to be convex, are shown with their scaling centres O and the common boundary.

The second reason stems from an increase of computational efficiency for certain problems. When the number of operations increases more than linearly with the number of unknowns, substructuring can be advantageous. For instance, in statics (section 16.1) a quadratic eigenvalue problem is solved, with the computational expense rising rapidly as the number of degrees of freedom increases. It can thus be advantageous to break the domain into a few subdomains. Each subdomain interacts only on its boundary, which will be discretised with a small number of degrees of freedom. Thus, a few quadratic eigenvalue problems of modest size are calculated instead of one quadratic eigenvalue problem of large size. Figure 22.1b illustrates the substructuring of a domain into two subdomains.

The third reason is linked to adequate modelling of dynamics in a bounded domain, when the static-stiffness and mass matrices are used (Chapter 17). Since degrees of freedom on the boundary only are present, it can be necessary to create more subdomains which introduces additional degrees of freedom on the boundaries between

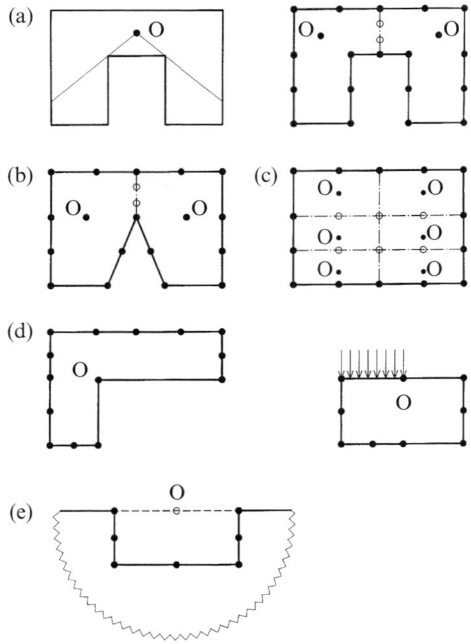

Figure 22.1 Reasons for substructuring and optimal location of scaling centres. (a) Parts of boundary not visible from scaling centre, (b) computational efficiency, (c) insufficient number of degrees of freedom, (d) discontinuities of boundary geometry or boundary conditions with stress singularities or discontinuities, (e) side-faces without discretisation

adjacent subdomains. At least four to ten nodal values per wave length are required. In Figure 22.1c, six subdomains are introduced which results in a significant number of degrees of freedom in the interior of the original domain. High efficiency results when subdomains are selected with similar boundaries (see the discussion in connection with Figure 17.1).

The fourth reason is based on the fact that at the scaling centre, the stress can be singular or discontinuous which is a very special property of the scaled boundary finite element method. As many problems contain more than one singular or discontinuous point, substructuring into subdomains, each with its own scaling centre, is advantageous. On the left of Figure 22.1d the scaling centre is placed at the re-entrant corner to capture the stress singularity. To be able to consistently model the two re-entrant corners of the domain shown in Figure 22.1a, two subdomains, each with the scaling centre at the re-entrant corner should be introduced. On the right of Figure 22.1d, the scaling centre is located at the end of the loaded zone to represent the stress discontinuity.

The fifth reason, which is in many cases linked with the fourth, stems from the fact that on the side-faces (straight segments which intersect (Figure 3.1b), or with extensions which intersect at the scaling centre (Figure 3.4b)) without any discretisation, surface tractions or displacements are exactly modelled. This is a very powerful feature of the scaled boundary finite element method. Again, many problems contain multiple boundaries where it is desirable to enforce the boundary conditions exactly, which will require substructuring with the corresponding location of the scaling centres to generate side-faces. In Figure 22.1e, the free surface of the unbounded domain representing side-faces does not have to be discretised, as the scaling centre lies on the extension on the side-faces. Note that in Figures 22.1d, no discretisation is required on the two straight segments of the boundary intersecting at the scaling centre. Summarising reasons four and five, to maintain the outstanding features of the scaled boundary finite element method, scaling centres should be located at discontinuities of the geometry of the boundary or of the boundary conditions, and the boundary should be modelled as far as possible with side-faces. To be able to do this, substructuring is necessary.

Of course, if possible, the reasons outlined above should be combined. As already mentioned, the domain with the two re-entrant corners of Figure 22.1a has to be modelled with two substructures. The scaling centres should be located at the reentrant corners which permits the stress singularities in these points to be represented. No discretisation is required on the adjacent straight segments of the boundary (side-faces) where the free surface boundary conditions are satisfied exactly.

A practical example [D6] of soil mechanics in two-dimensional statics is shown in Figure 22.2. A strip footing is excavated into the side of a long slope, which also changes the inclination. The vertical face of the excavation is supported rigidly in the horizontal direction, but is free to move vertically. A part of the horizontal face is loaded.

The simplest but not the most accurate model of the scaled boundary finite element method would represent the unbounded medium with the scaling centre O_1 at the intersection of the extensions of the free surfaces of the slope and the horizontal and vertical faces of the excavation forming the boundary. The conditions on this boundary can then be enforced once the static-stiffness matrix is calculated. This approach

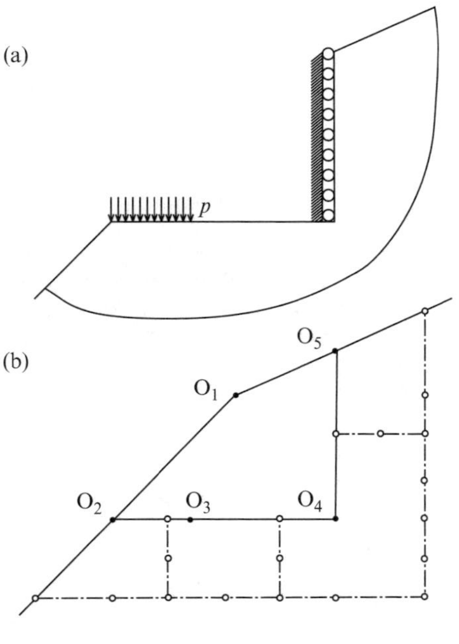

Figure 22.2 Excavated foundation with prescribed surface tractions and displacements in side of slope: (a) geometry and loading, (b) scaled boundary finite element model (Reproduced by permission of John Wiley & Sons Ltd)

does not allow the potential stress singularities and discontinuities in O_2, O_3 (end of loaded zone), O_4, O_5 to be modelled in an optimum manner. This is, however, possible by selecting four bounded subdomains in the near field with scaling centres O_2, O_3, O_4, O_5 and one unbounded subdomain in the far field with scaling centre O_1, but with the discretised boundary moved further outwards than in the other model. The advantages discussed in connection with Figures 22.1d and 22.1e apply. Note that, not only is the slope with the two inclinations not discretised, but neither are the horizontal and vertical faces of the excavation (with the exception of nodes between adjacent subdomains).

As another example, multiple interior cracks in fracture mechanics are addressed (Figure 22.3). For each crack, a subdomain is introduced, with the scaling centre located at the crack tip. Note that the straight crack faces are not discretised. The stress intensity factors will be very accurate, as the singularities are calculated analytically and no discretisation is present in the vicinity of the crack tips. Only the exterior boundary of each subdomain is discretised!

As already mentioned, the use of substructuring for unbounded domains is limited. A two-dimensional example consisting of a unbounded medium fixed at its base, with parallel and inclined free surfaces is shown in Figure 22.4. This could represent a soil site. Nodes 1, 2, 3 define the boundary, which in a soil-structure interaction problem would be equal to the structure-soil interface, where the dynamic-stiffness matrix is to

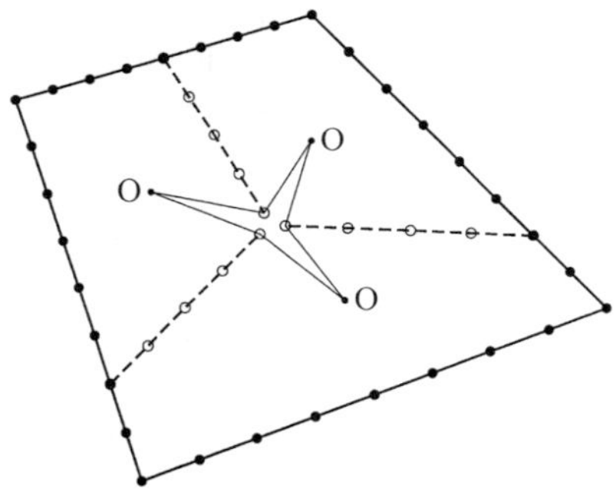

Figure 22.3 Subdivision into subdomains for multiple interior crack (Reprinted from *Computers and Structures*, Vol. 80, Song and Wolf, 'Semi-analytical representation of stress singularities as occurring in cracks in anisotropic multi-materials with the scaled boundary finite element method', pp. 183–197, © (2002), with permission from Elsevier Science)

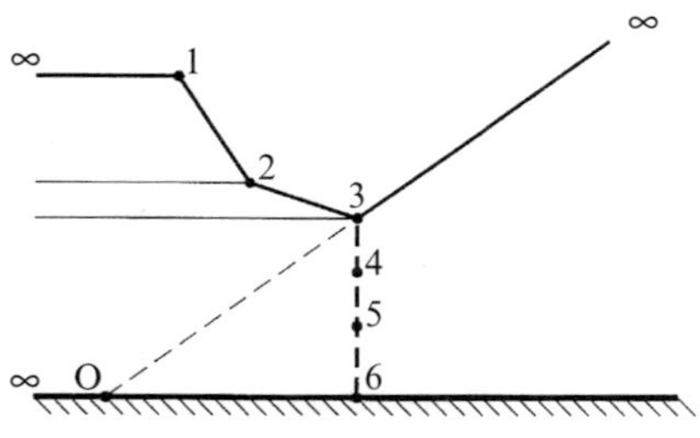

Figure 22.4 Substructuring in an unbounded medium

be calculated. Two subdomains are selected, the first domain with parallel layers on the left of line 1, 2, 3, 4, 5, 6 with the scaling centre located on the right at infinity, and the second subdomain on the right of the line 3, 4, 5, 61 with the inclined free surface and fixed base determining its scaling centre *O*. Additional degrees of freedom are introduced on the common boundary in nodes 4, 5, 6. An analysis of the scaled boundary finite element equation in dynamic stiffness for the first subdomaion yields the dynamic-stiffness with degrees of freedom in nodes 1 to 6, an analogous computation for the second subdomain leads to the dynamic-stiffness matrix with degrees of freedom in nodes 3 to 6. Assembling the two dynamic-stiffness matrices and eliminating the degrees of freedom in nodes 4, 5, 6 yields the final result, the dynamic-stiffness matrix of the site at the structure-soil interface with nodes 1, 2, 3.

23 Examples for Bounded Media

Besides analysing unbounded media, the scaled boundary finite element method is very well suited to calculating bounded media with stress singularities (or discontinuities). Using substructuring (Chapter 22), no restrictions of geometry or material variation exist. Each subdomain will have its own scaling centre where a singular or discontinuous point can occur. In addition, the boundary condition on the straight segments of the boundary adjacent to the scaling centre, called side-faces, are represented exactly without discretisation. Thus, bounded media with multiple stress singularities or discontinuities and multiple boundary segments where the boundary conditions are to be enforced exactly, can be processed. It will be demonstrated in Chapter 25 that, for this type of problem, for a given discretisation on the boundary, the scaled boundary finite element method yields much more accurate results than the finite element method, or that for a specified target error, the number of degrees of freedom and the computational effort are significantly smaller for the scaled boundary finite element method than for the finite element method.

To demonstrate the versatility, problems with stress singularities, as occur in fracture mechanics, will be mostly addressed. To be able to demonstrate the superb performance, it is sufficient to investigate bounded media with one stress singularity, which corresponds to one subdomain if several singularities are present. For an example with two stress singularities occurring at the two re-entrant corners, the plate with square holes in section 25.7.1 can be studied.

For fracture mechanics problems, the scaling centre is chosen at the crack tip, where a stress singularity with an *a priori* unknown power occurs. The two adjacent straight crack faces and the interfaces between different materials passing through the scaling centre are not discretised. The analytical results yielding the stress intensity factors are determined at the crack tip. As no discretisation is present in the adjacent region (and boundary), high accuracy will result.

Sections 23.1 to 23.5 calculate static stress-intensity factors in two-dimensional elastostatics for increasing complexity [S13]. Section 23.6 examines a dynamic stress-intensity factor in elastodynamics [S12]. Section 23.7 addresses a singularity in flux for the diffusion equation [S9]. Finally, section 23.8 evaluates a stress discontinuity in elastostatics in a bounded subdomain with an unbounded subdomain also present [D6].

23.1 EDGE-CRACKED ORTHOTROPIC PLATE WITH ROTATED MATERIAL AXES UNDER SHEAR

A homogeneous orthotropic plate in plane stress with a crack perpendicular to the edge is examined (Figure 23.1). The material properties [T1] in the principal directions which form an angle Ψ with the x-axis equal $E_1 = 144.8\text{GPa}$, $E_2 = 11.7\text{GPa}$, $G_{12} = 9.66\text{GPa}$, $\nu_{12} = 0.21$. A uniformly distributed shear stress τ_0 is applied to the plate fixed at its base.

The fundamental equations of two-dimensional statics for a bounded medium apply (Chapter 16). The displacements are specified in Equation (16.12a) and the stresses in Equation (15.94).

The scaling centre O is selected at the crack tip. The crack faces are not discretised. The discretisation of the boundary with 18 cubic elements using a non-uniform mesh is shown in Figure 23.1. The stress intensity factors of mode I and mode II are defined using stresses as

$$K_{\text{I}} = \lim_{\xi \to 0}\left(\sqrt{\pi b \xi}\varsigma_y\,(\xi, \theta = 0^\circ)\right) \tag{23.1}$$

$$K_{\text{II}} = \lim_{\xi \to 0}\left(\sqrt{\pi b \xi}\tau_{xy}\,(\xi, \theta = 0^\circ)\right) \tag{23.2}$$

respectively. As the stress singularities proportional to $1/\sqrt{\xi}$ are represented analytically in the scaled boundary finite element method, K_{I} and K_{II} follow straightforwardly, without addressing singular functions.

The convergence of the scaled boundary finite element is investigated for $\Psi = 60^\circ$ with uniform meshes of quadratic and cubic elements on the boundary. The stress intensity factors of mode I and mode II are shown in Figure 23.2 for different number of nodes on the boundary. The results with both quadratic and cubic elements converge. Cubic elements lead to a better rate of convergence than quadratic elements do.

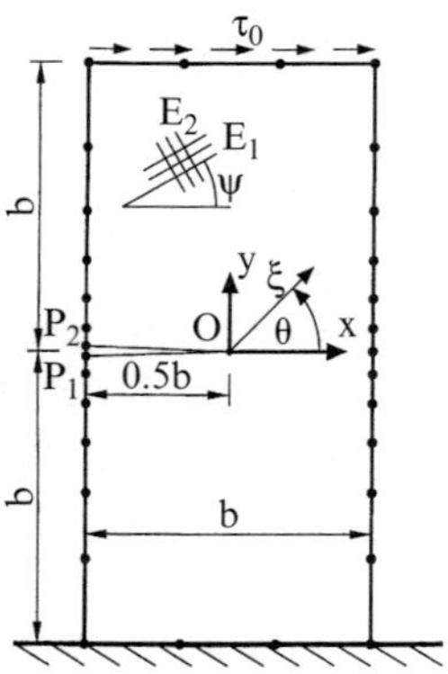

Figure 23.1 Edge-cracked orthotropic plate with rotated material axes (Reprinted from *Computers and Structures*, Vol. 80, Song and Wolf, 'Semi-analytical representation of stress singularities as occurring in cracks in anisotropic multi-materials with the scaled boundary finite element method', pp. 183–197, © (2002), with permission from Elsevier Science)

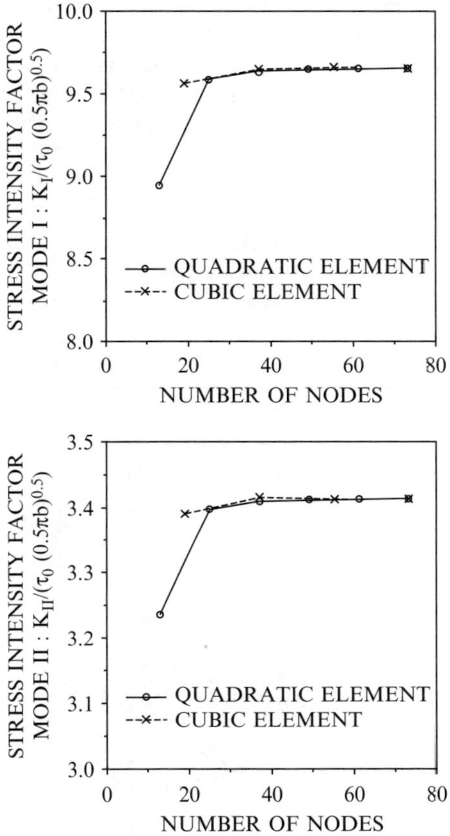

Figure 23.2 Stress-intensity factor of edge-cracked orthotropic plate for principal material direction $\Psi = 60°$ (Reprinted from *Computers and Structures*, Vol. 80, Song and Wolf, 'Semi-analytical representation of stress singularities as occurring in cracks in anisotropic multi-materials with the scaled boundary finite element method', pp. 183–197, © (2002), with permission from Elsevier Science)

The difference between the stress intensity factors obtained with 37 nodes from that with 73 nodes is less than 0.15% for quadratic elements and 0.03% for cubic elements, respectively.

Good agreement with the results of a boundary-element solution using so-called traction-singular elements [T1] involving the discretisation of the crack faces is achieved for different directions of the principal material axis (Table 23.1).

23.2 EDGE-CRACKED BIMATERIAL PLATE UNDER TENSION

A rectangular plate composed of two isotropic materials characterised by Young's modulus E and Poisson's ratio v with an interface crack of length a for plane stress perpendicular to the boundary is addressed (Figure 23.3).

Table 23.1 Stress intensity factor of edge-cracked orthotropic plate with rotated material axes (Reprinted from *Computers and Structures*, Vol. 80, Song and Wolf, 'Semi-analytical representation of stress singularities as occurring in cracks in anisotropic multi-materials with the scaled boundary finite element method', pp. 183–197, © (2002), with permission from Elsevier Science)

			Boundary elements	
Angle Ψ	$\frac{K_I}{\tau_0\sqrt{0.5\pi b}}$	$\frac{K_{II}}{\tau_0\sqrt{0.5\pi b}}$	$\frac{K_I}{\tau_0\sqrt{0.5\pi b}}$	$\frac{K_{II}}{\tau_0\sqrt{0.5\pi b}}$
0°	8.821	1.341	8.789	1.458
30°	9.852	5.066	9.924	5.122
60°	9.645	3.407	9.697	3.405
90°	8.871	1.029	8.890	1.044

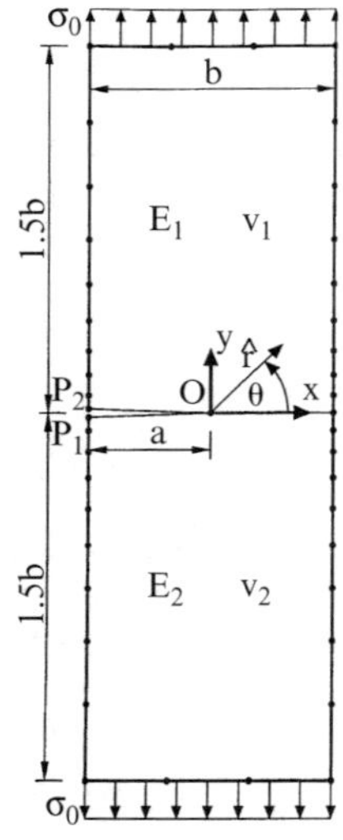

Figure 23.3 Edge-cracked bimaterial plate (Reprinted from *Computers and Structures*, Vol. 80, Song and Wolf, 'Semi-analytical representation of stress singularities as occurring in cracks in anisotropic multi-materials with the scaled boundary finite element method', pp. 183–197, © (2002), with permission from Elsevier Science)

As already mentioned, the scaling centre O is chosen at the crack tip. The straight traction free crack faces are not discretised. The same also applies to the material interface. 4-node (cubic) line elements are used. On each half of the cracked edge, three line elements are selected. The distances between two adjacent nodes are geometrically proportional with a ratio of 1.2. On the other edge, one line element is chosen. Fourteen line elements are used to discretise the boundary.

The stress intensity factor as a function of the relative crack face displacements is formulated as

$$K_{\mathrm{I}} + \mathrm{i}K_{\mathrm{II}} = \frac{E_1 E_2(1+2\mathrm{i}\varepsilon)\cosh(\pi\varepsilon)}{4(E_1+E_2)} \lim_{\hat{r}\to 0}\left(\sqrt{\frac{2\pi}{\hat{r}}}\left(\frac{2a}{\hat{r}}\right)^{\mathrm{i}\varepsilon}(\delta_y(\hat{r}) + \mathrm{i}\delta_x(\hat{r}))\right) \tag{23.3}$$

where the relative crack displacements $\delta_x(\hat{r})$, $\delta_y(\hat{r})$ are defined as

$$\delta_x(\hat{r}) = u_x(\hat{r}, \theta = \pi) - u_x(\hat{r}, \theta = -\pi) \tag{23.4a}$$

$$\delta_y(\hat{r}) = u_y(\hat{r}, \theta = \pi) - u_y(\hat{r}, \theta = -\pi) \tag{23.4b}$$

The bimaterial constant equals

$$\varepsilon = \frac{1}{2\pi} \ln \frac{\frac{3-\nu_1}{1+\nu_1}\frac{G_2}{G_1} + 1}{\frac{3-\nu_2}{1+\nu_2} + \frac{G_2}{G_1}} \tag{23.5}$$

with the shear modulus $G = E/(2(1 + \nu))$.

The displacements u_x, u_y follow from Equation (16.12a).

The stress intensity factors $K_I + iK_{II}$ non-dimensionalised with $\sigma_0\sqrt{\pi a}$ are calculated for $\nu_1 = \nu_2 = 0.3$ varying the crack length a/b (characteristic length of plate b) and the ratio of Young's moduli E_1/E_2. They are listed in Table 23.2. Good agreement is achieved with the results of the boundary-element method [Z2], [M1], [Y4].

23.3 ORTHOTROPIC BIMATERIAL PLATE WITH CRACK NORMAL TO AND TERMINATING AT INTERFACE

An orthotropic bimaterial plate in plane stress with a crack normal to and terminating at the interface (Figure 23.4) is addressed. In the scaled boundary finite element method,the scaling centre O again coincides with the crack tip. The discretisation is shown in Figure 23.4. Note that no discretisation is required not only on the crack faces, but also on

Table 23.2 Stress intensity factor of edge-cracked bimaterial plate (Reprinted from *Computers and Structures*, Vol. 80, Song and Wolf, 'Semi-analytical representation of stress singularities as occurring in cracks in anisotropic multi-materials with the scaled boundary finite element method', pp. 183–197, © (2002), with permission from Elsevier Science)

Crack length a/b	Boundary elements	Ratio of Young's Moduli (E_1/E_2) 1	2	4	10
0.4		2.111	2.108-i0.198	2.100-i0.359	2.089-i0.495
	[Z2]	2.112	2.106-i0.211	2.099-i0.361	2.083-i0.496
	[M1]	2.113	2.109-i0.197	2.100-i0.358	2.089-i0.493
	[Y4]	2.110	2.108-i0.198	2.100-i0.359	2.088-i0.495
0.5		2.824	2.818-i0.267	2.805-i0.484	2.787-i0.664
	[Z2]	2.820	2.820-i0.268	2.825-i0.489	2.795-i0.671
	[M1]	2.836	2.819-i0.267	2.806-i0.483	2.787-i0.661
	[Y4]	2.810	2.820-i0.268	2.807-i0.483	2.788-i0.664
0.6		4.032	4.022-i0.397	3.998-i0.716	3.967-i0.979
	[Z2]	4.036	4.020-i0.398	3.998-i0.716	3.947-i0.979
	[M1]	4.037	4.027-i0.396	4.001-i0.714	3.971-i0.974
	[Y4]	4.033	4.024-i0.398	4.000-i0.716	3.966-i0.980

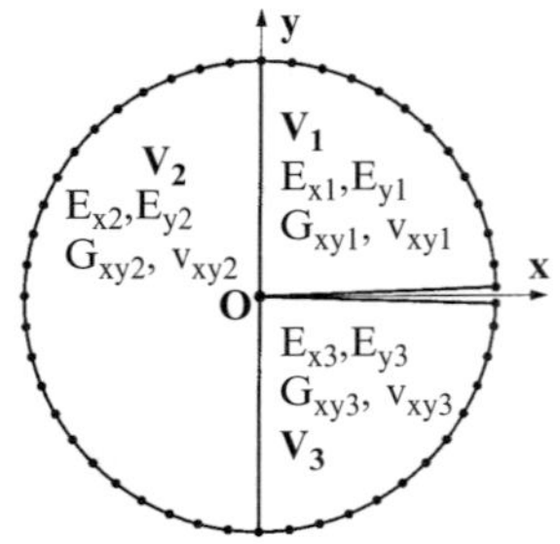

Figure 23.4 Orthotropic bimaterial plate with crack normal to and terminating at interface (Reprinted from *Computers and Structures*, Vol. 80, Song and Wolf, 'Semi-analytical representation of stress singularities as occurring in cracks in anisotropic multi-materials with the scaled boundary finite element method', pp. 183–197, © (2002), with permission from Elsevier Science)

Table 23.3 Powers of singularity at crack tip of orthotropic bimaterial plate with crack normal to and terminating at interface (Reprinted from *Computers and Structures*, Vol. 80, Song and Wolf, 'Semi-analytical representation of stress singularities as occurring in cracks in anisotropic multi-materials with the scaled boundary finite element method', pp. 183–197, © (2002), with permission from Elsevier Science)

No. of materials				Analytical	
V_1, V_3	V_2	λ_I	λ_{II}	λ_I	λ_{II}
1	2	0.44942	0.62362	0.44948	0.62368
3	2	0.34301	0.46762	0.34301	0.46760
7	8	0.25474	0.46589	0.25477	0.46589
2	1	0.57752	0.35329	0.57744	0.35327
2	3	0.63088	0.54417	0.63088	0.54417
9	10	0.38447	0.48739	0.38447	0.48739
10	9	0.56860	0.50487	0.56860	0.50487

the material interface. The analytical solution of the displacement around the crack tip is expressed as a series of power functions [C1], as for the scaled boundary finite element method. For various combinations of materials specified by four constants $E_x, E_y, G_{xy}, \nu_{xy}$ [C1] with the same properties for V_1 and V_3, the two powers of singularity are compared in Table 23.3 with the analytical solution. Excellent agreement is achieved.

23.4 ROTATING CIRCULAR ORTHOTROPIC PLATE WITH CENTRAL CRACK

As an example with body loads, a circular plate of radius R with a central crack of length $2a$ rotating around its centre C is examined (Figure 23.5). The material properties of the orthotropic two-dimensional plate in plane stress are defined in the Cartesian coordinate system x', y' as $E_x = 131\text{GPa}$, $E_y = 13\text{GPa}$, $G = 6.4\text{GPa}$, $\nu_{xy} = 0.038$. The stress intensity factors for various crack lengths a are calculated.

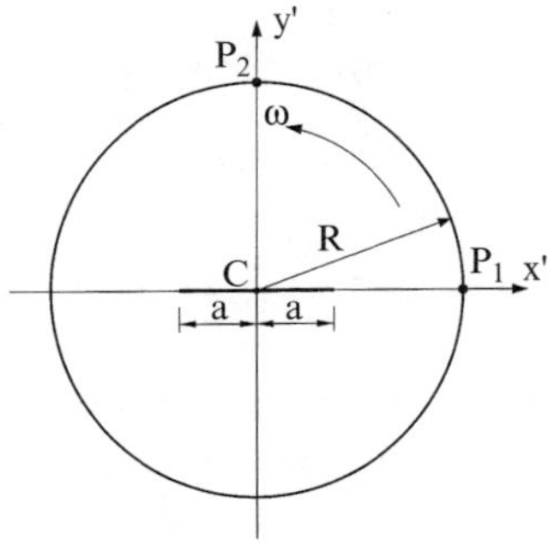

Figure 23.5 Rotating circular orthotropic plate with central crack (Reprinted from *Computers and Structures*, Vol. 80, Song and Wolf, 'Semi-analytical representation of stress singularities as occurring in cracks in anisotropic multi-materials with the scaled boundary finite element method', pp. 183–197, © (2002), with permission from Elsevier Science)

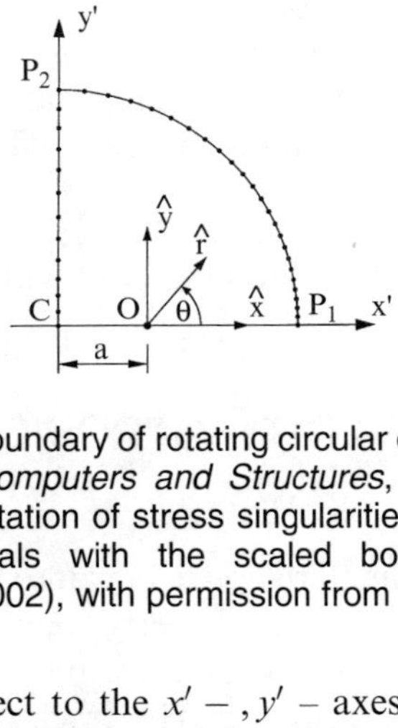

Figure 23.6 Discretisation on boundary of rotating circular orthotropic plate with central crack (Reprinted from *Computers and Structures*, Vol. 80, Song and Wolf, 'Semi-analytical representation of stress singularities as occurring in cracks in anisotropic multi-materials with the scaled boundary finite element method', pp. 183–197, © (2002), with permission from Elsevier Science)

Due to symmetry with respect to the $x' -, y' -$ axes, only a quarter of the plate is addressed (Figure 23.6). The scaling centre O is chosen at the crack tip. The arc $P_1 - P_2$ and the line $P_2 - C$ are discretised with 12 and 6 three–node line elements, respectively. As the scaling centre O lies on the straight line $C - P_1$, no discretisation is required on this part of the boundary, i.e. on the crack face C-O and on $O - P_1$, where the boundary conditions are enforced rigorously.

The centrifugal load is formulated as

$$\{p(x',y')\} = \rho\omega^2 \begin{Bmatrix} x' \\ y' \end{Bmatrix} \tag{23.6}$$

The scaled boundary coordinates ξ, η are introduced as

$$x' = a + \hat{x}(\xi, \eta) = a + \xi x(\eta) \tag{23.7a}$$

$$y' = \hat{y}(\xi, \eta) = \xi y(\eta) \tag{23.7b}$$

$\hat{x}(\xi, \eta)$, $\hat{y}(\xi, \eta)$ are the coordinates in the system with its origin coinciding with the scaling centre. $x(\eta)$, $y(\eta)$ are the coordinates on the boundary ($\xi = 1$) defined in the same coordinate system. Substituting Equation (23.7) into Equation (23.6) yields the load formulated in ξ, η:

$$\{p(\xi, \eta)\} = \rho\omega^2 \left(\begin{Bmatrix} a \\ 0 \end{Bmatrix} + \xi \begin{Bmatrix} x(\eta) \\ y(\eta) \end{Bmatrix} \right) \tag{23.8}$$

Note that $\{p(\xi, \eta)\}$ varies as a power function of ξ, which permits the integration in the ξ-direction to be performed analytically (section 16.4.2).

The stress intensity factor for mode I is defined as

$$K_{\mathrm{I}} = \sqrt{2\pi} \lim_{\hat{r} \to 0} \left(\sqrt{\hat{r}} \sigma_y(\hat{r}, \theta = 0^\circ) \right) \tag{23.9}$$

with the origin of the polar coordinates $\hat{r}, \theta$ coinciding with the crack tip. The stress σ_y follows from Equation (15.94). In the scaled boundary finite element method, an analytical expression of $\{u(\xi)\}$ exists for the body load in Equation (23.8). $\{u(\xi)\}$ consists of power functions and logarithmic functions (Equations (16.127) and (16.128)). An analytical expression thus also applies for the stress σ_y. With $\hat{r} = \xi r$, where for $\theta = 0^\circ$ $r = R - a$, the limit $\xi \to 0$ in Equation (23.9) can be determined analytically.

The stress intensity factor K_{I} is non-dimensionalised with $K_a = \sigma_0\sqrt{\pi a}$, where σ_0 is equal to the circumferential normal stress σ_θ at the centre C in the absence of the crack ($= 0.3974\rho\omega^2 R^2$, see Reference [L2]). K_{I}/K_a is listed for various dimensionless crack lengths a/R in Table 23.4. These values agree well with the boundary element results presented in Figure 12 of Reference [Z1].

23.5 CENTRAL-CRACKED ORTHOTROPIC BIMATERIAL PLATE UNDER TENSION

A central-cracked orthotropic bimaterial plate in plane stress with uniformly distributed normal stress σ_0 is examined. Due to symmetry, only half of the plate is addressed (Figure 23.7). The materials of each region are defined by E_1, E_2, G_{12} and ν_{12}. The scaling centre O is chosen at the crack tip. The crack faces and the material interface are not discretised. The boundary is discretised with 18 four-node (cubic) line elements. On each half of the sides perpendicular to the crack face, four line elements are selected. The distances between two adjacent nodes are geometrically proportional with a ratio of 1.15. On the other sides, one line element is chosen. This example, together with a definition of the stress-intensity factors for interface cracks in dissimilar anisotropic

Table 23.4 Stress intensity factor of rotating circular orthotropic plate with central crack (Reprinted from *Computers and Structures*, Vol. 80, Song and Wolf, 'Semi-analytical representation of stress singularities as occurring in cracks in anisotropic multi-materials with the scaled boundary finite element method', pp. 183–197, © (2002), with permission from Elsevier Science)

a/R	0.125	0.250	0.375	0.500	0.625	0.750
K_{I}/K_a	1.007	1.038	1.086	1.160	1.279	1.495

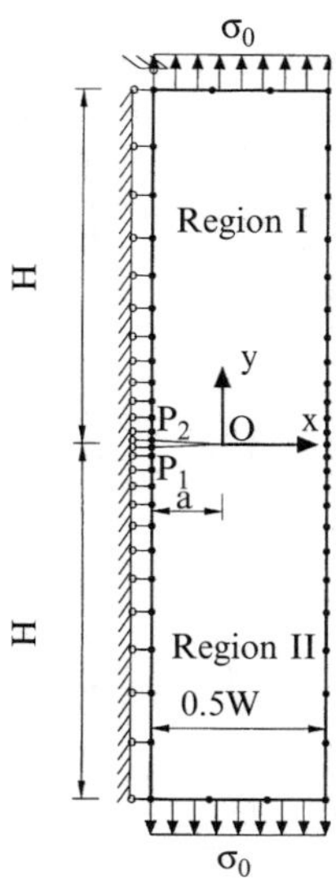

Figure 23.7 Central-cracked orthotropic bimaterial plate (Reprinted from *Computers and Structures*, Vol. 80, Song and Wolf, 'Semi-analytical representation of stress singularities as occurring in cracks in anisotropic multi-materials with the scaled boundary finite element method', pp. 183–197, © (2002), with permission from Elsevier Science)

materials, is discussed in Reference [C3]. The material properties $(E_1)_{\text{I}} = (E_1)_{\text{II}} = 100\text{N/mm}^2; (\nu_{12})_{\text{I}} = (\nu_{12})_{\text{II}} = 0.3$ (subscripts I and II for regions I and II of the plate) are chosen as constants. The stress-intensity factors normalised with $\sigma_0\sqrt{\pi a}$ are calculated for $(E_2/E_1)_{\text{II}} = 0.5, 0.3$ and 0.1, with $(G_{12})_{\text{I}}$ and $(G_{12})_{\text{II}}$ determined by a constant ρ relating material properties as

$$\rho = \frac{-\dfrac{2\nu_{12}}{E_1} + \dfrac{1}{G_{12}}}{2\sqrt{\dfrac{1}{E_1 E_2}}} \tag{23.10}$$

For the definitions of the various terms K_1, K_2, F_i describing the stress-intensity factors, Reference [C3] should be consulted. The results agree well with those of the boundary element method (Table 3, Reference [C3]), except for K_2/K_1 at $(E_2/E_1)_{\text{II}} = 0.1$ and $\rho = 3.32$ (Table 23.5).

23.6 DYNAMIC STRESS-INTENSITY FACTOR

Transient dynamic analysis of fracture mechanics problems can efficiently be performed with the scaled boundary finite element method. This is demonstrated by calculating the dynamic stress-intensity factor of a central-cracked plate under tension applied as a Heaviside step function. The advantages of the novel computational procedure discussed for statics in the previous sections remain valid for dynamics. The stress-intensity factor is calculated also without any increase in computational effort for anisotropic

Table 23.5 Stress intensity factor of central-cracked orthotropic bimaterial plate (Reprinted from *Computers and Structures*, Vol. 80, Song and Wolf, 'Semi-analytical representation of stress singularities as occurring in cracks in anisotropic multi-materials with the scaled boundary finite element method', pp. 183–197, © (2002), with permission from Elsevier Science)

		$\rho = 3.32$		$\rho = 1$		$\rho = 0.49$	
$(E_2/E_1)_{II}$	Boundary elements	F_i	K_2/K_1	F_i	K_2/K_1	F_i	K_2/K_1
0.5		1.303		1.319		1.334	
	[C3]	1.300		1.316		1.331	
0.3		1.399	−0.02815	1.419	−0.04154	1.4345	−0.04862
	[C3]	1.392	−0.02813	1.415	−0.04142		−0.04579
0.1		1.718	−0.08313	1.751	−0.124	1.775	−0.146
	[C3]	1.697	−0.01407	1.746	−0.124	1.772	−0.143

material, directly from its definition using stresses or using the relative displacements describing the crack opening at the crack tip from the analytical solution without any *a priori* assumption. No discretisation of straight crack faces and material interfaces (passing through the crack tip) is required, for which the boundary conditions are satisfied exactly. The static-stiffness and mass matrices of the bounded medium are obtained with the scaled boundary finite element method, and the dynamic response is calculated directly in the time domain using a standard time-integration scheme, as in the finite element method. To improve the accuracy at high frequency, the medium is divided in subdomains introducing additional interior degrees of freedom (a more efficient procedure would result if substructures with similar boundaries as discussed in connection with Figure 17.1 were applied).

A rectangular plate in plane strain with a central crack and the dimensions specified in Figure 23.8 is examined. The crack length is $2a = 4.8$ mm. The material is characterised by shear modulus $G = 76.923$ GPa, Poisson's ratio $\nu = 0.3$ and mass density $\rho = 5000\,\text{kg/m}^3$. The plate is loaded along opposite boundaries at time $t = 0$ by uniform tension with Heaviside function time dependence $p(t) = P\text{H}(t)$. Due to symmetry, only a quarter of the plate is discretised, as shown in Figure 23.8. Twelve subdomains are introduced. The boundary of each subdomain is discretised with three-node line finite elements. Note that the scaling centre O_1 of the subdomain containing the crack is chosen at the crack tip. The free crack face is not discretised. The scaling centre O_2 is placed on the boundary representing the axis of symmetry, which is again not discretised.

The time integration scheme of the assembled discretised equations of motion with the static-stiffness and mass matrices of each subdomain determined as described in Chapter 17 yields the displacements as a function of time in the nodes of the subdomain containing the crack tip. These determine the integration constants $\{c_1\}$ in the displacements $\{u(\xi)\}$ in Equation (16.12a), with ξ measured from the crack tip. The spatial variation of the displacements in the radial direction is represented by $\xi^{-\lambda_i}$ with the eigenvalue (λ_i determined in Equation (16.7)). The eigenvalue leading to the stress singularity is calculated as $\lambda = -0.499998$, which is very close to the exact value

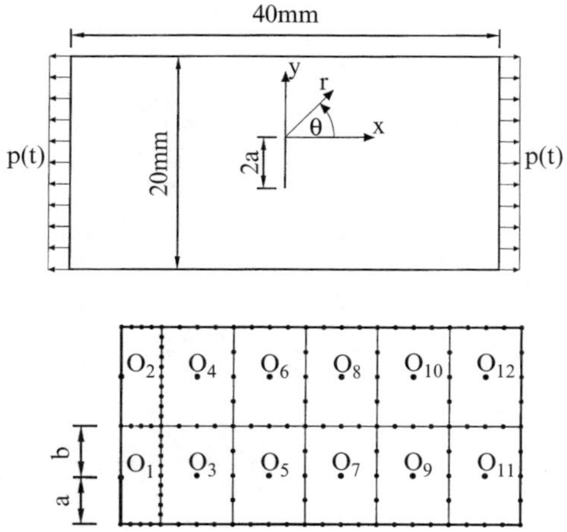

Figure 23.8 Central-cracked plate and scaled boundary finite element mesh (Reprinted from Proceedings First Asian-Pacific Congress on Computational Mechanics, Vol. 2, Song and Wolf, 'Semi-analytical evaluation of dynamic stress-intensity factors', pp. 1041–1046, © (2001), with permission from Elsevier Science)

$$K_{\mathrm{I}} = \sqrt{2\pi} \lim_{\xi \to 0} (r^{1+\lambda} \sigma_x(\xi, \theta = 90°)) \tag{23.11}$$

$\lambda = -0.5$. The mode I stress-intensity factor is defined with the stress at the front of the crack as with $r = \xi b$ (Figure 23.8). As the stress singularity, proportional to $\xi^{-1-\lambda}$ (Equation (15.94)), is represented analytically in the scaled boundary finite element method, K_{I} follows straightforwardly.

The Newmark method with $\alpha = 0.5$ and $\beta = 0.25$ is used for the time integration of the equation of motion. A time step $\Delta t = 0.32 \cdot 10^{-3}$s is selected. The mode I dynamic stress-intensity factor calculated with Equation (23.11) is shown in Figure 23.9, normalised with the static value. Good agreement with the finite-difference results ([C2] shown as a dashed line) is achieved.

23.7 BIMATERIAL PLATE WITH SINGULARITY OF FLUX IN DIFFUSION

An interface problem where a singularity may occur governed by the diffusion equation is discussed. The steady-state problem of a bimaterial circular plate with radius r is addressed (Figure 23.10) [Y3]. The two sectors $V_1(0 \leq \theta < \pi/2)$ and $V_2(\pi/2 \leq \theta < 2\pi)$ are composed of isotropic materials characterised by their material constants κ_1 and κ_2, respectively. $\kappa_1 = 10\kappa_2$ is selected in the following analysis.

An exact solution of this interface problem [Y3] is available in polar coordinates $\hat{r}, \theta$. This solution is formulated in the dimensionless radial coordinate $\xi = \hat{r}/r$ ($\xi = 1$ on boundary) and angle θ as

$$u(\xi, \theta) = u_0((2\xi)^{-\lambda_1} h_1(\theta) + (2\xi)^{-\lambda_2} h_2(\theta)) \tag{23.12}$$

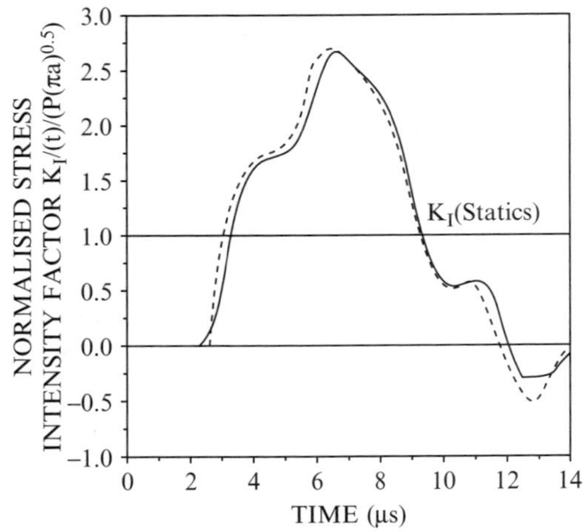

Figure 23.9 Normalised dynamic stress-intensity factor for central-cracked plate (Reprinted from Proceedings First Asian-Pacific Congress on Computational Mechanics, Vol. 2, Song and Wolf, 'Semi-analytical evaluation of dynamic stress-intensity factors', pp. 1041–1046, © (2001), with permission from Elsevier Science)

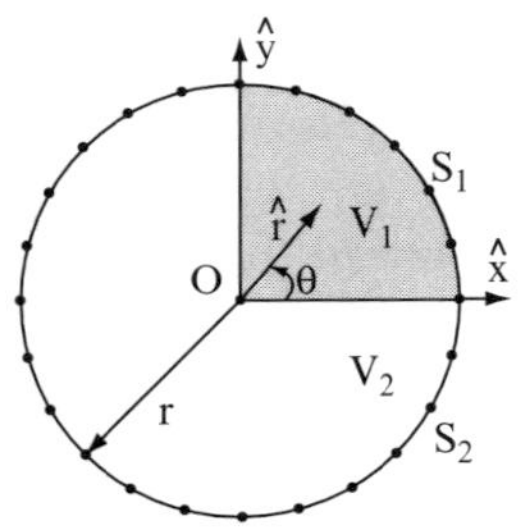

Figure 23.10 Bimaterial plate and scaled boundary finite element mesh (Reproduced by permission of John Wiley & Sons Ltd)

where u_0 is a constant with the dimension of the function $u(\xi, \theta)$, and $\lambda_1 = -0.731691779$ and $\lambda_2 = -1.268308221$ are constants. $h_1(\theta)$ and $h_2(\theta)$ are functions of θ

$$h_1(\theta) = \begin{cases} \cos((1-a)\theta) + c_3 \sin((1-a)\theta) & \text{for } 0 \le \theta < \pi/2 \\ c_1 \cos((1-a)\theta) + c_2 c_3 \sin((1-a)\theta) & \text{for } \pi/2 \le \theta < 2\pi \end{cases} \tag{23.13}$$

$$h_2(\theta) = \begin{cases} \cos((1+a)\theta) - c_3 \sin((1+a)\theta) & \text{for } 0 \le \theta < \pi/2 \\ c_1 \cos((1+a)\theta) - c_2 c_3 \sin((1+a)\theta) & \text{for } \pi/2 \le \theta < 2\pi \end{cases} \tag{23.14}$$

with the constants $c_1 = 6.31818181818182$, $c_2 = -2.6818181818181818$, $c_3 = -0.64757612580273$ and $a = 0.26830822130025$. Equation (23.12) satisfies the diffusion equation for the two materials and the boundary conditions at the interfaces of the materials. The normal flux on a boundary with a constant dimensionless radial coordinate ξ is obtained from Equation (23.12) as

$$q_n = -\frac{u_0}{r}\xi^{-1}(\lambda_1(2\xi)^{-\lambda_1}h_1(\theta) + \lambda_2(2\xi)^{-\lambda_2}h_2(\theta)) \times \begin{Bmatrix} \kappa_1 \text{ for } 0 \leq \theta < \pi/2 \\ \kappa_2 \text{ for } \pi/2 \leq \theta < 2\pi \end{Bmatrix} \tag{23.15}$$

To verify the accuracy of the scaled boundary finite element method, the normal flux q_n in Equation (23.15) is prescribed on the boundaries S_1 and $S_2(\xi = 1)$. The unique solution (up to an additive constant) for this interface problem is represented by Equation (23.12).

The scaling centre O is chosen at the centre of the circular plate (Figure 23.10). No discretisation is necessary on the material interfaces, as they pass through the scaling centre. The boundaries S_1 and S_2 are discretised with four-node line finite elements of equal length. Meshes of 4, 8, 12 and 16 elements on the boundary are used in the calculations.

Analogous to the static case for elastodynamics (Equation (16.12a)), the steady state solution for diffusion of the function for the two-dimensional case for a bounded medium in the scaled boundary finite element method equals

$$\{u(\xi)\} = \sum_{i=0} c_i \xi^{-\lambda_i}\{\Phi_i\} \tag{23.16}$$

with the integration constants c_i, eigenvalues λ_i arranged in descending order of their real parts and with the corresponding subvectors $\{\Phi_i\}$ of the matrix $[\Phi_{11}]$. Equation (23.16) is a series of power functions of the dimensionless radial coordiante ξ, analogous to the analytical solution (Equation (23.12)).

The first term ($i = 0$) in the series solution (Equation (23.16)) corresponds to the rigid body motion (the additive constant) with the eigenvalue $\lambda_0 = 0$. The eigenvalues λ_1 and λ_2 in the second ($i = 1$) and third ($i = 2$) terms of the solution are compared with the exact values in Table 23.6. Excellent accuracy is observed. Thus, the result in the radial direction for any ξ of the scaled boundary finite element method (Equation (23.16)) agrees well with the exact solution (Equation (23.12)).

The functions $h_1(\theta)$ and $h_2(\theta)$ (Equations (23.13) and (23.14)) in the exact solution represent the variation of the function in the circumferential direction. The same task is accomplished in the discretised form of the series solution (Equation (23.16)). Comparing Equation (23.12) with (23.16) leads to the numerical results of the functions $h_1(\theta)$ and $h_2(\theta)$, which are listed in Table 23.7 for different values of the angle θ. Highly accurate results are obtained. The deviation is smaller than 0.5% even with only 4 elements.

The sum of the remaining terms ($i \geq 3$) in Equation (23.16) is 10^{-5} times smaller than that of the second ($i = 1$) and the third ($i = 2$) terms.

Table 23.6 Eigenvalues of bimaterial plate in steady state of diffusion (Reproduced by permission of John Wiley & Sons Ltd)

Elements	4	8	12	16	Exact
λ_1	−0.731658	−0.731691	−0.731692	−0.731692	−0.731692
λ_2	−1.268532	−1.268313	−1.268308	−1.268308	−1.268308

Table 23.7 Function variation in circumferential direction of bimaterial plate in steady state of diffusion (Reproduced by permission of John Wiley & Sons Ltd)

	Elements	Angle θ (degree)					
		60	90	120	150	180	210
$h_1(\theta)$	4	1.16943	0.99929	−1.49432	−3.76811	−5.49611	−6.43186
	8	1.16956	0.99996	−1.49314	−3.76990	−5.49976	−6.43260
	12	1.16957	0.99999	−1.49318	−3.76986	−5.49995	−6.43261
	16	1.16958	1.00000	−1.49320	−3.76989	−5.49999	−6.43266
	Exact	1.16958	1.00000	−1.49320	−3.76989	−5.50000	−6.43267
$h_2(\theta)$	4	−0.38787	−0.99891	−4.76152	−6.50692	−5.49402	−2.13326
	8	−0.38827	−0.99993	−4.77794	−6.52514	−5.49959	−2.13547
	12	−0.38835	−0.99999	−4.77867	−6.52624	−5.49992	−2.13592
	16	−0.38835	−1.00000	−4.77869	−6.52627	−5.49997	−2.13593
	Exact	−0.38836	−1.00000	−4.77874	−6.52634	−5.50000	−2.13595

The flux is specified in Equation (15.131). Substituting Equation (23.16) leads to the variation of the flux in the radial direction, which for each mode is proportional to $\xi^{-\lambda_i-1}$. As $\lambda_1 > -1$ (Table 23.6), $\lambda_1 - 1 < 0$, yielding a flux at $\xi = 0$ which is singular. The scaled boundary finite element method represents this singularity very accurately with its analytical solution (Equation 23.16).

23.8 SURFACE FOUNDATION ON HALF-PLANE WITH DISCONTINUOUS LOADING

To demonstrate the ability of the scaled boundary finite element method to model accurately a stress discontinuity, a surface foundation on a half-plane with Poisson's ratio $\nu = 0.3$ with a constant loading p acting on the length $2a$ is investigated. Half of the symmetric foundation is shown in Figure 23.11. A discontinuity of the vertical stress at a distance a from the axis of symmetry on the free surface occurs.

Substructuring with two subdomains is applied. The scaling centre O_1 of the bounded subdomain is located at the discontinuity of the loading, allowing the stress discontinuity present at this point to be modelled accurately. Three-node quadratic elements are used with the end nodes (but not the middle nodes), shown in Figure 23.11. Note that no discretisation of the free surface, which is a side-face, is present. The scaling centre O_2 of the unbounded subdomain is located on the axis of symmetry at the free surface. Again, the free surface and axis of symmetry, being side-faces, are not discretised, the discretisation of the unbounded subdomain being limited to the interface with the other subdomain. (The side-face coinciding with the axis of symmetry of the unbounded subdomain is restrained in the other direction, which leads to one logarithmic mode only, section 16.2, corresponds to Figure 16.3b.) There are 41 degrees of freedom present.

The results are very accurate. This is due to the location of the scaling centre of the bounded subdomain at the point of the stress discontinuity. The accurate modelling of

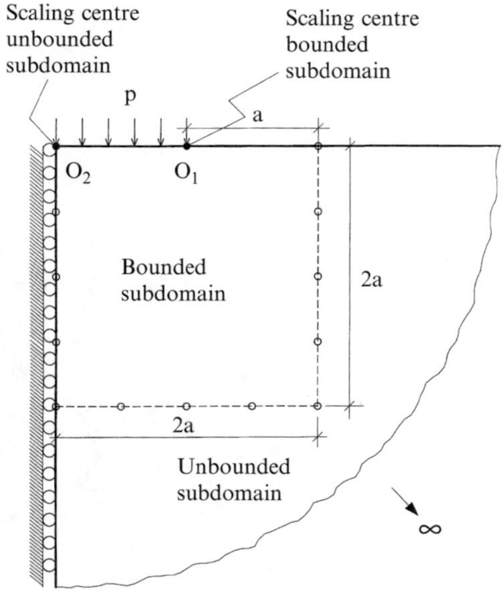

Figure 23.11 Surface foundation on half-plane with model using symmetry and two subdomains (Reproduced by permission of John Wiley & Sons Ltd)

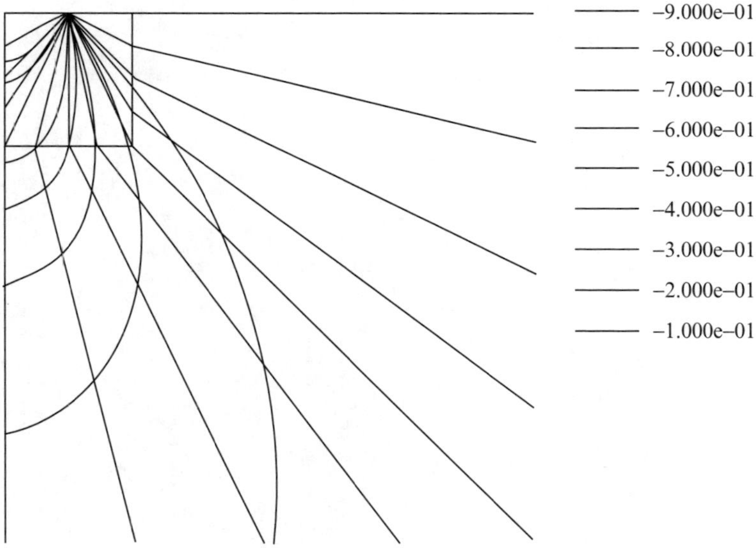

Figure 23.12 Vertical stresses for surface foundation on half-plane (Reproduced by permission of John Wiley & Sons Ltd)

this feature is dramatically illustrated by the passing of all the vertical stress contours through a single point, as plotted in Figure 23.12 (stress non-dimensionalised by p). The stresses are indistinguishable from the well-known analytical expressions.

24 Examples for Unbounded Media

Besides analysing bounded media with stress singularities, the classical problem addressed by the scaled boundary finite element method (and other preliminary versions based on similarity (scaling)) is to calculate unbounded media (infinite or semi-infinite). The novel computational procedure permits a rigorous enforcement of the boundary conditions at infinity, in elastodynamics called the radiation condition. This satisfaction, based on the high-frequency asymptotic expansion of the dynamic stiffness in analytical form in the radial direction, is consistent with the spatial discretisation in the circumferential direction (Chapter 18). In particular, the boundary conditions on free and fixed surfaces, as well as material interfaces which are side-faces extending to infinity, are enforced exactly.

For an unbounded medium, certain restrictions on geometry and variation of material properties exist. These cannot be eliminated completely by applying substructuring, as discussed at the end of Chapter 22. To be able to enforce scaling, the boundary of the unbounded medium can be moved outwards towards infinity (Chapter 26, Figure 26.1). This introduces a slight modification of the mechanical model, which, however, is analysed consistently.

As the title of the book [W9] indicates, many unbounded media are analysed in this text. Two- and three-dimensional problems in elastodynamics and diffusion are calculated in the frequency and time domains, as well as in statics. Certain extensions (Chapter 21), such as the variation of the material properties in the radial direction, incompressible material and a two-dimensional layered unbounded medium, are also addressed. These results, which are easily accessible [W9], will not be reproduced in the following. Only new aspects are illustrated.

Section 24.1 examines a prism foundation embedded in a half-space, calculating numerically the dynamic stiffness and the displacements in the unbounded medium [W3]. Section 24.2 discusses a spherical cavity embedded in a full space, applying the analytical solution in the frequency domain [S8]. Section 24.3 demonstrates the computational efficiency achieved applying a reduced set of base functions, addressing the in-plane motion of a semi-infinite wedge [W15].

24.1 PRISM FOUNDATION EMBEDDED IN HALF-SPACE

To demonstrate the numerical solution in the frequency domain (Chapter 19), a truly three-dimensional problem, a square prism of length $2b$ embedded with depth e in a homogeneous half-space governed by the vector-wave equation of elastodynamics, is addressed. This represents, for instance, a foundation embedded in homogeneous soil. Due to symmetry, only a quarter of the embedded prism shown in Figure 24.1 is modelled. The boundary identified as the structure-medium interface (basemat and side-walls) is rigid. A vertical harmonic load $P(\omega)$ of frequency ω acts in the centre of the basemat. The vertical dynamic-stiffness coefficient is to be determined. The vertical displacements of the basemat (Point P_1), below the foundation (Points P_2 and P_3) and of the free surface (Point P_4) are to be calculated.

The dimensions of the foundation follow from $b = 0.5\,\text{m}$, $e = 0.5\,\text{m}$. The material properties of the half-space equal shear-wave velocity $c_s = 250\,\text{m/s}$, Poisson's ratio $\nu = 0.4$ and mass density $\rho = 2000\,\text{kg/m}^3$, resulting in a dilatational-wave velocity $c_p = c_s\sqrt{2(1-\nu)/(1-2\nu)} = 2.45c_s = 612\text{m/s}$ and a Rayleigh-wave velocity $c_R = 0.94\ c_s = 235\,\text{m/s}$. The frequency f of the harmonic load varies from 0 to 200 Hz in increments of 10 Hz ($\omega = 2\pi f$). The result points are specified by the coordinates $P_1(0, 0, 0.5)$, $P_2(0, 0, 1.5)$, $P_3(0, 0, 3)$, $P_4(3, 0, 0)$.

For comparison, the results of a boundary element method analysis in the frequency domain [F1] are used. Constant boundary elements with the fundamental solution of the full space are applied. Besides the structure–medium interface, the free surface must also be discretised up to a distance of 6 m, with boundary elements of lengths varying from 0.25–0.4 m. For a quarter of the embedded foundation, almost 300 boundary elements are introduced.

In the scaled boundary finite element method, the scaling centre O is chosen at the origin of the coordinate system (Figure 24.1a). This permits the free surface of the half-space to be identified as a side-face without any discretisation. The spatial discretisation is thus limited to the structure-medium interface of the foundation. Due to symmetry, only a quarter is analysed. The finite element mesh of this structure–medium interface using 12 8-node elements is shown in Figure 24.1b. This leads to 49 nodes with 129 degrees of freedom after enforcing the symmetry boundary conditions. The shortest wave length calculated for the highest frequency $f = 200\,\text{Hz}$ and with c_s equals $c_s/f = 1.25\,\text{m}$. With the distance between two adjacent nodes equal to 0.125 m, 10

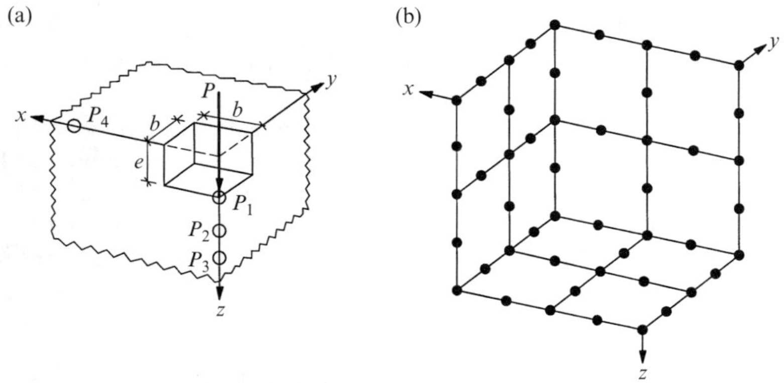

Figure 24.1 One quarter of square prism embedded in half-space: (a) geometry with scaling centre, (b) finite element discretisation of structure–medium interface (Reproduced by permission of John Wiley & Sons Ltd)

nodes per wave length are thus present. (An analysis performed using a finer mesh with 27 8-node elements, yielding 274 degrees of freedom essentially confirms the results.)

The numerical solution procedure of sections 19.1 and 19.3 is applied.

The analysis for this three-dimensional case ($s = 3$) proceeds as follows. The order of the vector $\{u(\bar{\xi})\}$ is equal to the number of degrees of freedom 129, and the coefficient matrices $[E^0], [E^1], [E^2], [M^0]$, and dynamic-stiffness matrix $[S^\infty]$ are of order 129×129. Solving the eigenvalue problem (Equation (18.13)) permits $[C_\infty]$ (Equation (18.25)), $[K_\infty], [A_1](m = 1)$ to be calculated. The high-frequency asymptotic expansion of the dynamic-stiffness matrix evaluated at $\bar{\xi}_h = 30c_s/b$ yields the boundary condition $(1/\bar{\xi})[S^\infty]$ (Equation (19.2)). This allows Equation (19.1) to be solved numerically for decreasing $\bar{\xi}$ down to $\bar{\xi} = 0$, leading to $(1/\bar{\xi})[S^\infty]$ as a function of $\bar{\xi}$ for all $\bar{\xi}$. In particular, at the structure-soil interface $\xi = 1$ $[S^\infty]$ of order 129×129 is a function of ω corresponding to the dynamic-stiffness matrix of a flexible interface. Enforcing the rigid interface constrains the degrees of freedom. This condition of all vertical displacement amplitudes being equal to the scalar $u_0(\omega)$ and all horizontal displacement amplitudes vanishing defines the vector $\{\Phi\}$, leading to

$$\{u(\omega)\} = \{\Phi\}u_0(\omega) \tag{24.1}$$

The corresponding dynamic-stiffness coefficient of the rigid interface follows as

$$S^\infty(\omega) = \{\Phi\}^T[S^\infty]\{\Phi\} \tag{24.2}$$

It is customary to introduce the dimensionless frequency at the structure-soil interface $\xi = 1$, $a_0 = \omega b/c_s$, to non-dimensionalise S^∞ with the static-stiffness coefficient $K^\infty(= S^\infty(\omega = 0))$ and to apply the following decomposition

$$S^\infty(a_0) = K^\infty(k(a_0) + \mathrm{i}a_0c(a_0)) \tag{24.3}$$

with the dimensionless spring coefficient $k(a_0)$ and damping coefficient $c(a_0)$.

With $K^\infty \approx 6.86 \times 10^8 \mathrm{N/m}, k(a_0)$ and $c(a_0)$ are plotted as a function of a_0 in Figure 24.2. In addition, the results avoiding solving the eigenvalue problem (Equation (18.13)) and with $[K_\infty] = [A_1] = 0$ in the asymptotic high-frequency expansion (Equation (19.2)) are presented. In this case, $[C_\infty]$ is directly constructed from the impedances (Chapter 18). Distributed dashpots with the coefficients ρc_p (Equation (18.39)) for the degrees of freedom perpendicular to the finite element, and ρc_s for those in the plane of the finite element act at the structure–medium interface. The $[C_\infty]$-matrix of each finite element involves integrations of the products of shape functions. It is thus equal to the standard two-dimensional mass matrix, but multiplied by the corresponding wave velocity. Equation (19.1) is again solved numerically. Good agreement of the dynamic-stiffness coefficient as presented in Figure 24.2 is achieved.

It is worth noting that Equation (19.1) with the boundary condition (Equation (19.2)) involves as independent variable $\bar{\xi}$, which is the product of ξ with ω. For the various ω, this operation has to be performed only once. Each coefficient of $(1/\bar{\xi})[S^\infty]$ could be decomposed and plotted as in Figure 24.2 as a function of $\bar{\xi}$ or of the dimensionless frequency at ξ, $\omega b\xi/c_s = b\bar{\xi}/c_s$.

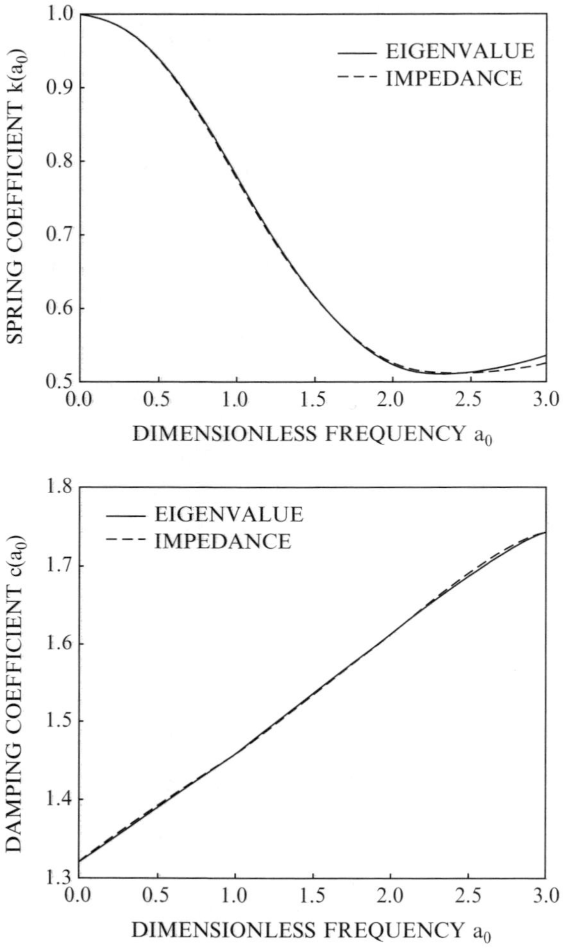

Figure 24.2 Vertical dynamic-stiffness coefficient (Reproduced by permission of John Wiley & Sons Ltd)

Applying the load with the amplitude $P(\omega)$ to the prism foundation with a rigid interface results in the vertical displacement amplitude $u_0(\omega)$ of the foundation

$$u_0(\omega) = \frac{P(\omega)}{S^{\infty}(\omega)} \tag{24.4}$$

Equation (24.1) then yields the boundary condition at the structure-soil interface $\bar{\xi} = \omega$, enabling the integration of Equation (19.27) for increasing $\bar{\xi}$ up to the value corresponding to the result point. $(1/\xi)[S^{\infty}]$ for each $\bar{\xi}$ has been determined before by solving Equation (19.1). The corresponding displacment amplitudes $u(\bar{\xi})$ are complex.

The magnitudes of the vertical displacement $|u_z(\omega)| = \sqrt{(\mathrm{Re}u_z(\omega))^2 + (\mathrm{Im}u_z(\omega))^2}$, non-dimensionalised with $K^\infty/P(\omega)$, are plotted as a function of the dimensionless frequency a_0 in the four result points in Figure 24.3. The result in Point P_1 (basemat) corresponds to $u_0(\omega)$, and thus to the inverse of $S^\infty(\omega)$ (Equation (24.4)). Good agreement with the boundary element solution in all points and for all frequencies is achieved.

The variation of the vertical displacement for a specific frequency along the vertical z-axis and the horizontal x-axis is represented in Figures 24.4 and 24.5. The real and imaginary parts are non-dimensionalised by multiplying with the factor $K^\infty/P(\omega)$. The vertical and horizontal distances are also non-dimensional. The frequencies $f = 100$ Hz and $f = 200$ Hz are processed. Note that the variation both in the horizontal and vertical directions is governed by the Rayleigh-wave velocity c_R. The corresponding wave length equals c_R/f which, for $f = 200$ Hz, for example, results in 1.17 m, or in non-dimensional form by dividing by e or b in 2.35. This is clearly visible in Figures 24.4 and 24.5. Thus, for this intermediate to high-frequency range, the motion is caused by surface waves.

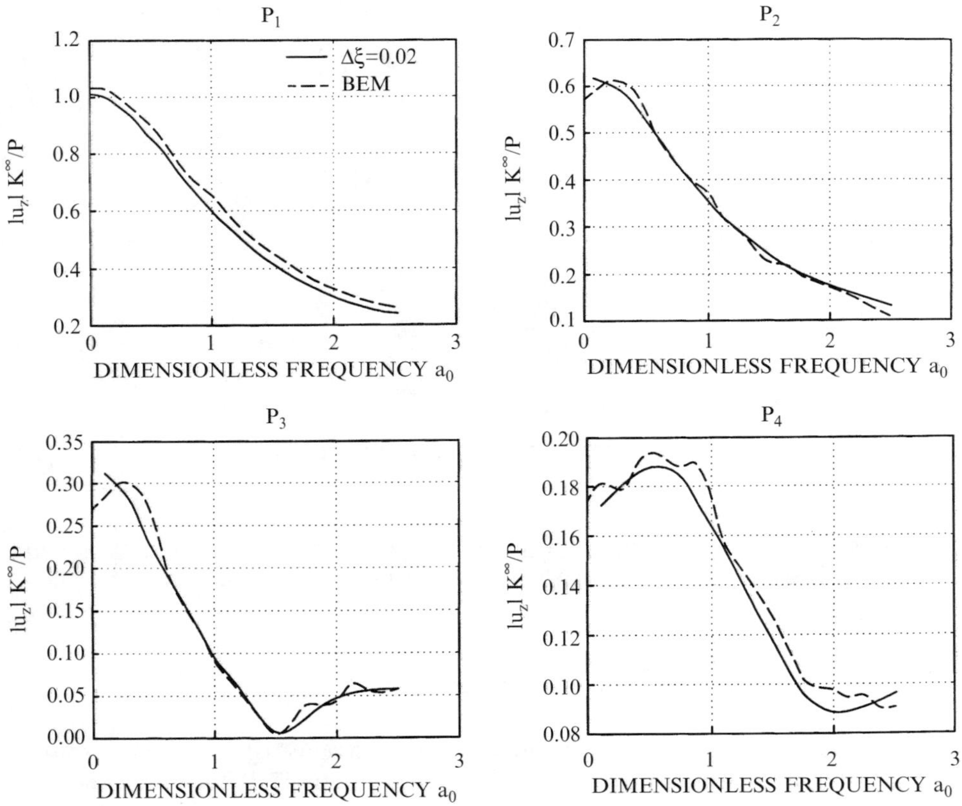

Figure 24.3 Magnitude of vertical displacement in half-space (Reproduced by permission of John Wiley & Sons Ltd)

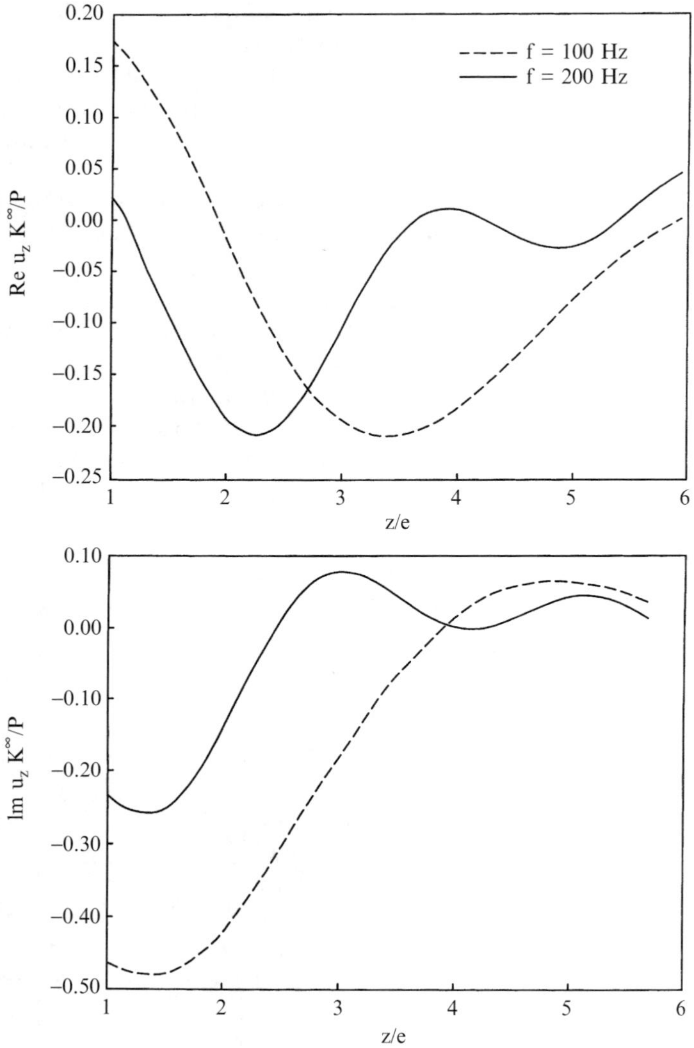

Figure 24.4 Real and imaginary parts of amplitude of vertical displacement along vertical z-axis (below basemat) (Reproduced by permission of John Wiley & Sons Ltd)

24.2 SPHERICAL CAVITY EMBEDDED IN FULL SPACE

To discuss the analytical solution in the frequency domain (Chapter 20), a spherical cavity of radius r embedded in a full space with shear moduls G, Poisson's ratio $\nu = 0.25$ and mass density ρ is examined (Figure 24.6). A uniform normal displacement u_0 on the wall of the cavity is enforced which yields a one-dimensional problem. The wall representing the boundary of the unbounded medium corresponds to the structure–medium interface in a typical unbounded medium–structure interaction analysis. The dynamic-stiffness coefficient at the boundary is to be calculated.

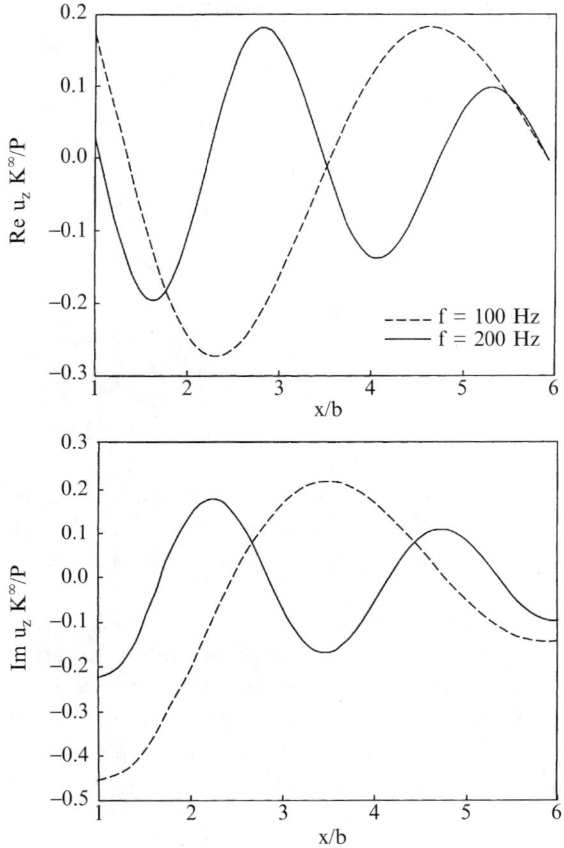

Figure 24.5 Real and imaginary parts of amplitude of vertical displacement along horizontal *x*-axis (free surface) (Reproduced by permission of John Wiley & Sons Ltd)

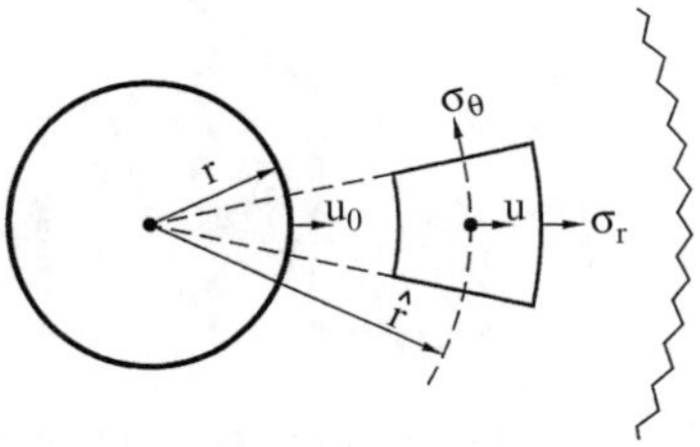

Figure 24.6 Spherical cavity embedded in full space (section) with symmetric waves (Reprinted from *Computer Methods in Applied Mechanics & Engineering*, Vol. 164, Song and Wolf, 'The scaled boundary finite element method: analytical solution in frequency domain', pp. 249–264, © (1998), with permission from Elsevier Science)

The analytical solution of the dynamic-stiffness coefficient equals

$$S^\infty(a_0) = 16\pi Gr(k(a_0) + \mathrm{i}a_0 c(a_0)) \tag{24.5}$$

with the dimensionless spring and damping coefficients

$$k(a_0) = 1 - \frac{1-\nu}{2(1-2\nu)}\frac{a_0^2}{1+a_0^2} \tag{24.6a}$$

$$c(a_0) = \frac{1-\nu}{2(1-2\nu)}\frac{a_0^2}{1+a_0^2} \tag{24.6b}$$

and the dimensionless frequency

$$a_0 = \frac{\omega r}{c_p} \tag{24.7}$$

where c_p is the dilatational-wave velocity ($= \sqrt{G/\rho}\sqrt{2(1-\nu)/(1-2\nu)}$).

The spherical cavity is solved with the scaled boundary finite element method as a three-dimensional problem with nine-node surface finite elements. The mesh of one octant consists of three finite elements as shown in Figure 24.7. On the boundary, 294 degrees of freedom are introduced. The analytical solution procedure of section 20.3 is applied.

Next, the bounded domain between the interior boundary with radius $r(\xi_i = 1)$ and the exterior boundary with $2r(\xi_e = 2)$, a hollow sphere, is addressed. In the series solution of Equation (20.14), 18 terms ($k = 17$) are selected. The submatrices $[S_{ii}(\omega)], [S_{ie}(\omega)], [S_{ei}(\omega)], [S_{ee}(\omega)]$ are of order 294×294. After enforcing the uniform radial displacement, the four dynamic-stiffness coefficients (Equation (20.51)) follow. An analytical solution exists for comparison.

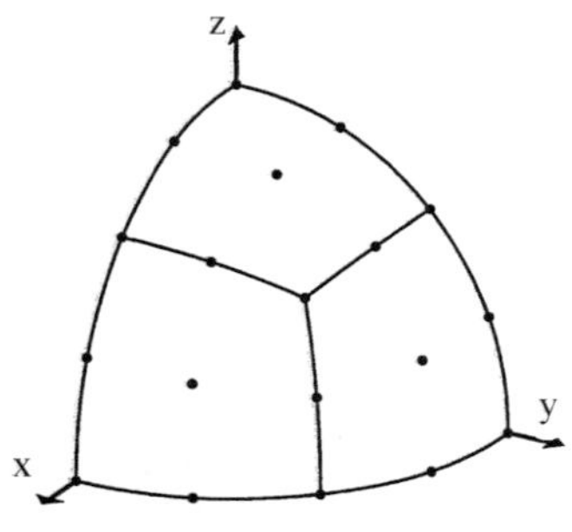

Figure 24.7 Finite element mesh of one octant of structure-medium interface (wall of cavity) (Reprinted from *Computer Methods in Applied Mechanics & Engineering*, Vol. 164, Song and Wolf, 'The scaled boundary finite element method: analytical solution in frequency domain', pp. 249–264, © (1998), with permission from Elsevier Science)

$$\begin{bmatrix} S_{ii}(a_0) & S_{ie}(a_0) \\ S_{ei}(a_0) & S_{ee}(a_0) \end{bmatrix} = \frac{8\pi}{1-2\nu} Gr$$
$$\times \begin{bmatrix} -((1-\nu)\xi_i a_0 \mathrm{j}_1(\xi_i a_0),_{a_0} + 2\nu\xi_i \mathrm{j}_1(\xi_i a_0)) & -((1-\nu)\xi_i a_0 \mathrm{y}_1(\xi_i a_0),_{a_0} + 2\nu\xi_i \mathrm{y}_1(\xi_i a_0)) \\ (1-\nu)\xi_e a_0 \mathrm{j}_1(\xi_e a_0),_{a_0} + 2\nu\xi_e \mathrm{j}_1(\xi_e a_0) & (1-\nu)\xi_e a_0 \mathrm{y}_1(\xi_e a_0),_{a_0} + 2\nu\xi_e \mathrm{y}_1(\xi_e a_0) \end{bmatrix}$$
$$\times \begin{bmatrix} \mathrm{j}_1(\xi_i a_0) & \mathrm{y}_1(\xi_i a_0) \\ \mathrm{j}_1(\xi_e a_0) & \mathrm{y}_1(\xi_e a_0) \end{bmatrix}^{-1} \tag{24.8}$$

with the first-order spherical Bessel functions of the first kind j_1 and the second kind y_1

$$\mathrm{j}_1\left(\frac{\omega\hat{r}}{c_p}\right) = \left(\frac{c_p}{\omega\hat{r}}\right)^2 \left(\sin\left(\frac{\omega\hat{r}}{c_p}\right) - \frac{\omega\hat{r}}{c_p}\cos\left(\frac{\omega\hat{r}}{c_p}\right)\right) \tag{24.9a}$$

$$\mathrm{y}_1\left(\frac{\omega\hat{r}}{c_p}\right) = -\left(\frac{c_p}{\omega\hat{r}}\right)^2 \left(\cos\left(\frac{\omega\hat{r}}{c_p}\right) + \frac{\omega\hat{r}}{c_p}\sin\left(\frac{\omega\hat{r}}{c_p}\right)\right) \tag{24.9b}$$

Excellent agreement, for example, for $S_{ii}(a_0)$ is achieved up to $a_0 = 6$, as shown in Figure 24.8, although a strong dependency on a_0 exists.

The high-frequency asymptotic expansion for $(1/\xi_e)[S_e^{\infty}(\omega, \xi_e)]$ of order 294 × 294 (Equation (20.56)) is evaluated at $\bar{\xi}_e = 3c_p/r$ with $m = 1$. The dynamic-stiffness matrix $[S^{\infty}(\omega)]$ of order 294 × 294 follows from Equation (20.62) with the dynamic-stiffness matrix $[S_{ii}(\omega)]$, $[S_{ie}(\omega)]$, $[S_{ei}(\omega)]$, $[S_{ee}(\omega)]$ of the corresponding hollow sphere (with $\xi_i = 1$) calculated with seven terms ($k = 6$) in the power series of Equation (20.14). The

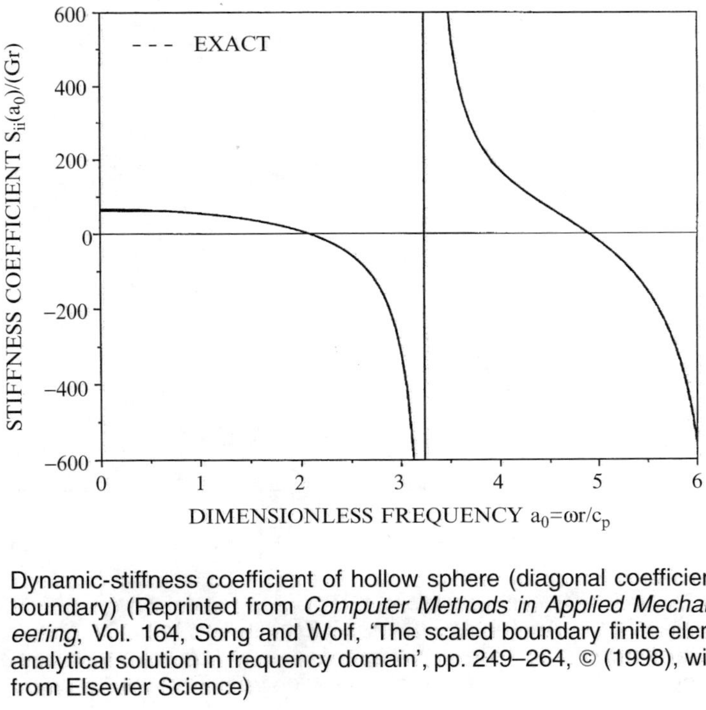

Figure 24.8 Dynamic-stiffness coefficient of hollow sphere (diagonal coefficient for interior-boundary) (Reprinted from *Computer Methods in Applied Mechanics & Engineering*, Vol. 164, Song and Wolf, 'The scaled boundary finite element method: analytical solution in frequency domain', pp. 249–264, © (1998), with permission from Elsevier Science)

corresponding dynamic-stiffness coefficient of the spherical cavity $S^{\infty}(a_0)$ with enforced uniform radial symmetry agrees well with the analytical solution (Equations (24.5) and (24.6)) as is visible in Figure 24.9. It can be concluded that $S^{\infty}(a_0)$ can be directly calculated accurately for any a_0 based on the high-frequency asymptotic expansion. Note that the error decreases for diminishing a_0.

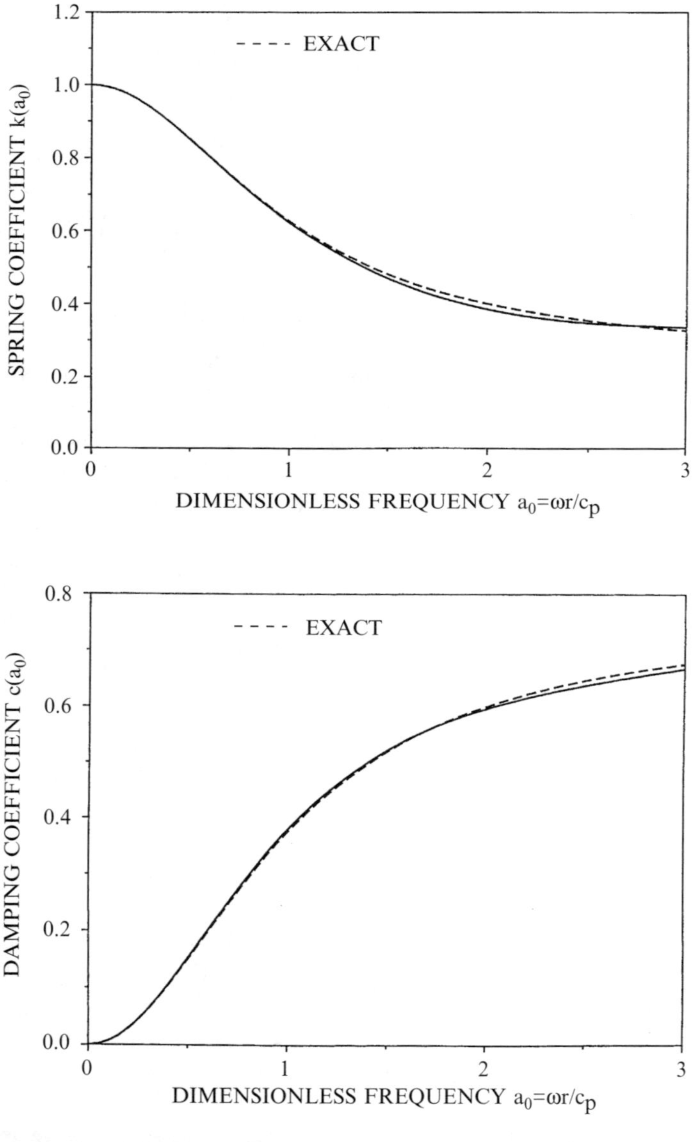

Figure 24.9 Dynamic-stiffness coefficient of spherical cavity (Reprinted from *Computer Methods in Applied Mechanics & Engineering*, Vol. 164, Song and Wolf, 'The scaled boundary finite element method: analytical solution in frequency domain', pp. 249–264, © (1998), with permission from Elsevier Science)

24.3 IN-PLANE MOTION OF SEMI-INFINITE WEDGE

To demonstrate the efficient use of a reduced set of base functions (section 21.3), the in-plane motion of a semi-infinite wedge (Figure 24.10) with shear modulus G, Poisson's ratio $\nu = 0.25$ and mass density ρ is addressed. One of the boundaries extending to infinity is fixed, and the other is a free surface. The boundary (structure–medium interface) is an arc of radius r_0 with an opening angle $\alpha = 30°$. On the structure–medium interface a linear function in the circumferential direction of the horizontal motion $u_0(\theta)$ and zero vertical motion are prescribed. The boundary is discretised with eight three-node line elements, which leads to 32 degrees of freedom. The equivalent spring and damping coefficients defined as in Equation (24.3), with $K^{\infty} = 0.752G$ and $a_0 = \omega r_0/c_s$ are plotted in Figure 24.11 as solid lines. The coefficients obtained with reduced numbers of base functions (Equation (21.33)) are also shown. The result with only four base functions is close to that with the full set of base functions for $a_0 < 2$.

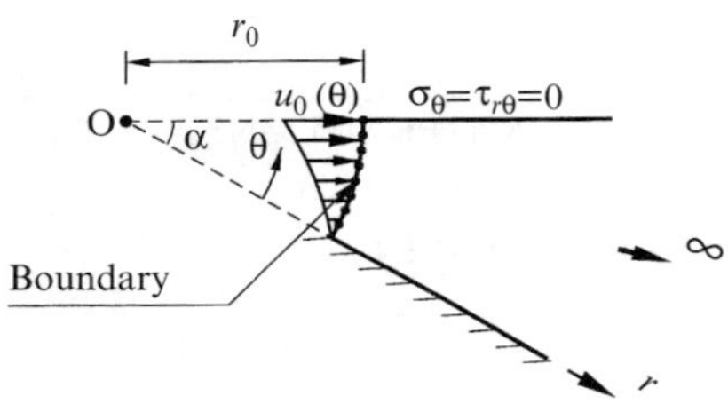

Figure 24.10 In-plane motion of semi-infinite wedge with prescribed horizontal displacement (Reprinted from *Engineering Structures*, Vol. 34, Wolf and Song, 'Some corner stones of dynamic soil-structure interaction', pp. 13–28, © (2002), with permission from Elsevier Science)

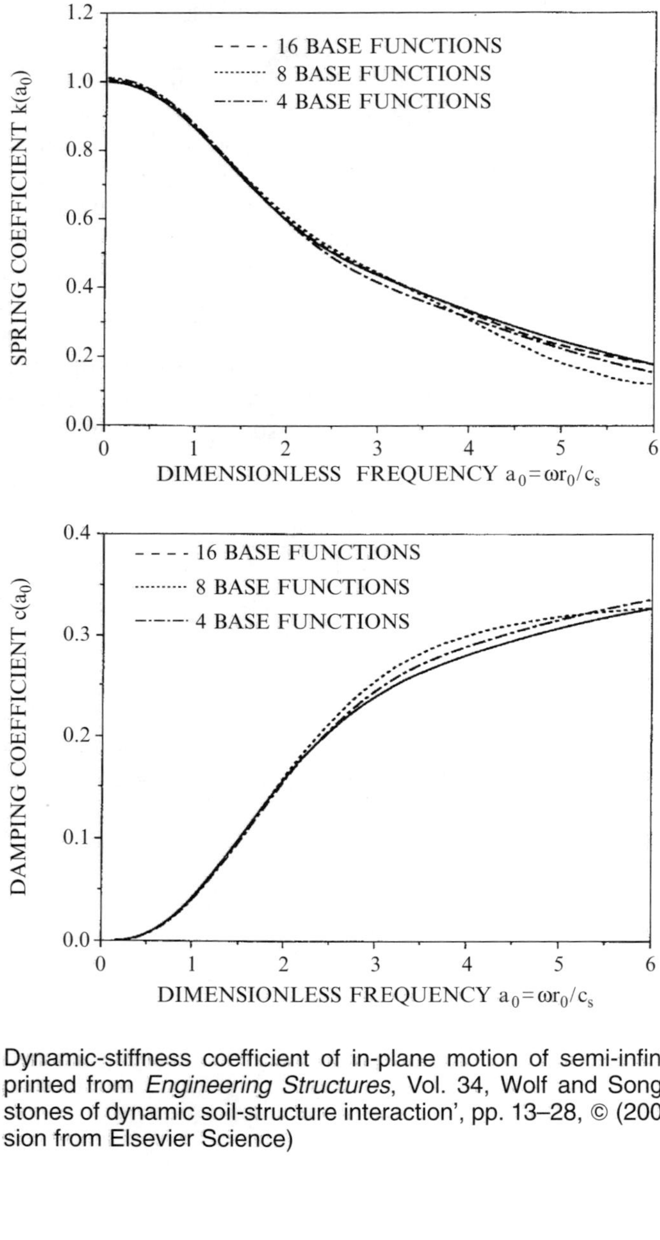

Figure 24.11 Dynamic-stiffness coefficient of in-plane motion of semi-infinite wedge (Reprinted from *Engineering Structures*, Vol. 34, Wolf and Song, 'Some corner stones of dynamic soil-structure interaction', pp. 13–28, © (2002), with permission from Elsevier Science)

25 Error Estimation and Adaptivity

The examples presented in Chapters 23 and 24 demonstrate, based on comparisons with exact solutions and the results of other numerical procedures, that the scaled boundary finite element method is accurate. This statement is based on selected results in certain points of the domain. No information on the overall accuracy and, in particular, no guidance on how to proceed to increase the accuracy to a specified level without excessive computational effort, is available. This also excludes the possibility of evaluating the computational cost of competing numerical methods of similar overall accuracy.

As an analytical solution is obtained in the radial direction, the scaled boundary finite element method is particularly useful in situations involving stress singularities and for unbounded domains. For these two key applications, an eigenvalue problem has to be solved, the computational expense of which rises rapidly as the number of degrees of freedom increases. The most systematic and satisfactory method of achieving the accuracy desired with the minimum number of degrees of freedom is to use adaptive techniques. This can, for instance, be achieved by so-called h-hierarchical methods, in which the size of the elements is reduced in regions of low accuracy (high error). However, adaptive techniques require reliable and accurate error estimation of the overall results of an analysis. One way to determine such an *a posteriori* error estimator is based on calculating an improved solution by some smoothing process using derivative quantities of the function. In elastostatics, this procedure is called stress recovery. The differences between the 'recovered' stresses and the 'raw' stresses indicate the level of error.

The analytical solution mentioned above leads to no spatial discretisation in the radial direction. As will be demonstrated, the stress recovery and error estimation technique preserve this advantage of the scaled boundary finite element method, as all computations in the radial coordinate, also for unbounded domains, are performed analytically. The h-hierarchical adaptive procedure can be viewed as substructuring (Chapter 22), which is well suited for the scaled boundary finite element method. In particular, the variation of the stresses in the radial direction is determined from the eigenvalues, and is thus part of the solution. No *a priori* knowledge of the power of the stress singularity is required. This is a significant advantage when solving problems where the power is unknown, such as for anisotropic material. Substructuring also permits discretisation of certain straight segments adjacent to the scaling centre (side-faces) to be avoided, but enforcing the boundary conditions rigorously.

The stress-recovery error estimator technique developed in this chapter for the scaled boundary finite element method is fully compatible with that for the finite element method. The same also applies to the h-hierarchical adaptive procedures. Thus, fair and direct comparisons of the computational efficiencies of the two numerical procedures with similar overall accuracy are possible. The developed stress-recovery error estimator is generally applicable and accurate, applies without any modifications to bounded and unbounded domains with equal ease and preserves all the advantages of the scaled boundary finite element method.

Error estimation and adaptivity are vast areas of research in numerical analysis. A detailed review of the significant achievements in finite elements and their application to the scaled boundary finite element method lie outside the scope of this chapter. Only stress-recovery error estimators and h-hierarchical adaptivity are addressed. To demonstrate their essential features, two-dimensional elastostatics is addressed for the sake of simplicity. An extension to three dimensions of the procedures is straightforward.

This chapter is organised as follows. An introduction to error estimation for numerical methods, with particular attention paid to stress-recovery error estimation is provided in section 25.1. Stress recovery in the finite element method is examined in section 25.2. This review is necessary as the procedure is extended to the scaled boundary finite element method in section 25.3. The stress recovery method removes the discontinuity of the raw stresses in the circumferential direction, while maintaining the analytical variation in the radial direction. The so-called superconvergent patch recovery technique pioneered in References [Z3], [Z4] is applied, but smoothing is only required on the boundary. The resulting recovered stress field is continuous throughout the domain. An error estimator is determined based on the recovered stress field in section 25.4. The error estimator corresponds to the energy norm estimator of References [Z3], [Z4] with the integrations determined semi-analytically, that is numerically in the circumferential direction and analytically in the radial direction. The error estimation is expressed in a unified form for both bounded and unbounded domains. It is important to realise that the smoothing of the stresses and the calculation of the error estimator are performed for each mode in the scaled boundary finite element method separately, which permits the numerical integrations to be restricted to the boundary. To illustrate the stress recovery procedure, and to demonstrate the accuracy and convergence of the error estimator, a bounded plate of infinite extent with a circular hole is examined as an example in section 25.5. An h-hierarchical adaptivity procedure is developed in section 25.6. Based on the error estimator, the model with a specific discretisation is adapted automatically to meet some specified level of accuracy. This is achieved by recursive substructuring, i.e. by element subdivision. In section 25.7 examples clearly demonstrate that the computational performance of the scaled boundary finite element method is superior to that of the finite element method for the same level of accuracy, especially when stress singularities occur or unbounded problems are analysed.

This chapter follows closely References [D3] and [D4]. The latter reference also discusses an object-oriented implementation of recursive h-hierarchical adaptivity.

25.1 ERROR ESTIMATION IN NUMERICAL ANALYSIS

The accuracy of the approximate solution of a numerical analysis is of interest to determine the reliability of the solution, and to allow adaptive schemes to be developed.

Since in elastostatics the exact solution of the displacements $\{u_{ex}(x, y)\}$ (subscript *ex* for exact) is, in general, not known, the error in the approximate solution must be estimated in some way. The procedure based on a *recovery-based estimator* is applied. An improved solution is obtained by some smoothing process using derivatives of the function, i.e. of the displacements. This corresponds to stresses. The differences between the raw stresses and the recovered (smoothed) stresses indicate the level of error.

Although the error in the displacements varies over the domain, a scalar norm of the error may be used to characterise the accuracy of the solution over the entire domain or part of the domain. A commonly used norm is the so-called energy norm, which can be interpreted as a weighted root-mean-square of the stresses:

$$\|\{\sigma\}\|_V = \left(\int_V \{\sigma(x,y)\}^T \{\varepsilon(x,y)\} \mathrm{d}V \right)^{0.5} = \left(\int_V \{\sigma(x,y)\}^T [D]^{-1} \{\sigma(x,y)\} \mathrm{d}V \right)^{0.5} \tag{25.1}$$

If the local stress error in a point is defined as

$$\{e_\sigma(x,y)\} = \{\sigma_{ex}(x,y)\} - \{\sigma(x,y)\} \tag{25.2}$$

the energy norm of the stress error is

$$\|e_\sigma\|_V = \left(\int_V \{e_\sigma(x,y)\}^T [D]^{-1} \{e_\sigma(x,y)\} \mathrm{d}V \right)^{0.5} \tag{25.3}$$

If this energy norm of the stress error is then normalised by dividing by the energy norm of the total stresses, a relative energy norm is obtained. Converting the fraction to a percentage, this *relative energy norm of the stress error* represents a weighted per cent root-mean-square of the stress error

$$\eta = \frac{\|\{e_\sigma\}\|_V}{\|\{\sigma\}\|_V} \times 100\% \tag{25.4}$$

Since, in general, the exact stresses are not known, η cannot be computed directly. Instead, a recovery procedure is used to obtain an approximation to the exact stresses, which is significantly better than the raw stresses computed from the displacements. This 'better' approximation (denoted by a superscript asterisk) can be used in place of the exact stresses to form an approximation of the error

$$\{\sigma(x,y)\} \approx \{\sigma^*(x,y)\} \tag{25.5}$$

The local stress error can be approximated by

$$\{e_\sigma(x,y)\} \approx \{e_\sigma^*(x,y)\} = \{\sigma^*(x,y)\} - \{\sigma(x,y)\} \tag{25.6}$$

and the relative energy norm of the stress error then equals

$$\eta^* = \frac{\|\{e_\sigma^*\}\|_V}{\|\{\sigma^*\}\|_V} \times 100\% \tag{25.7}$$

The volume integral can either be defined to be a small region of the domain to give a measure of local error, or to be the entire domain to provide a measure of overall (global) error.

This simple recovery error estimator, referred to the Zienkiewicz–Zhu error estimator, is often applied in the finite element method, but can be defined in any numerical analysis procedure, provided a recovered stress field can be determined which is reliably more accurate than the raw stress field.

25.2 STRESS RECOVERY IN FINITE ELEMENT METHOD

For a finite element, the approximate solution of the displacements $\{u(x, y)\}$ is specified as a linear combination of predetermined shape functions $[N(x, y)]$

$$\{u(x, y)\} = [N(x, y)]\{u\} \tag{25.8}$$

with the nodal displacements $\{u\}$. The approximate stresses are then with the elasticity matrix $[D]$ determined as

$$\{\sigma(x, y)\} = [D][B(x, y)]\{u\} \tag{25.9}$$

with the strain-nodal displacement matrix $[B(x, y)]$.

Since the shape functions used in the finite element are defined on a local element basis, the displacement field satisfies continuity across the element boundaries. Consequently, the raw stress field represented by Equation (25.9) will exhibit discontinuities at the element boundaries and at the nodes. To present a solution that appears more realistic, nodal averaging of these stresses is often used. This represents a very simple method of stress recovery. Once the recovered nodal stresses $\{\sigma^*\}$ have been obtained, the stress field can be interpolated between the nodes using the shape functions

$$\{\sigma^*(x, y)\} = [N(x, y)]\{\sigma^*\} \tag{25.10}$$

However, a much better method of recovering the nodal stresses is the superconvergent patch recovery technique which is simple to implement, and is very cost effective. Since the *superconvergent patch recovery* technique will be used in the stress recovery process developed for the scaled boundary finite element method, it will be reviewed in some detail here.

The superconvergent patch recovery procedure is based on the premise that within an element there are points at which the stresses are superconvergent, i.e. points at which the stresses converge at a similar rate as the displacement field does, or perform even better and at a higher rate than at any other points within the element. For many elements, the stresses are superconvergent at the Gauss points. The procedure aims to use the stresses at these points to recover nodal stresses that are also superconvergent. Equation (25.10) can then be used to interpolate the recovered stress field between the nodes. The stress field recovered in this way is both continuous between elements and superconvergent. It is thus ideally suited for use in a recovery error estimator.

The procedure uses fully connected corner nodes as patch assembly points. An element patch consists of all elements connecting to the assembly point. In the one-dimensional case, there will be only two elements. In Figure 25.1, two three-noded elements with quadratic shape functions are shown. The stresses varying linearly over the elements will be discontinuous at the boundary, which is equal to the patch assembly point. In each element there are two superconvergent points coinciding with the Gauss points used for numerical integration. A polynomial function of the same order as the shape functions (for the three-noded elements a quadratic function) is fitted to the raw stresses computed at the superconvergent points contained within the patch. The fit is done using a least squares approach, and must be performed for each stress component individually.

If the fitted stress function for a particular stress component is represented by $\sigma_p(x)$ (subscript p for polynomial), for the one-dimensional situation illustrated in Figure 25.1:

$$\sigma_p(x) = [P]\{a\} \tag{25.11}$$

where

$$[P] = [\,1 \quad x \quad x^2\,] \tag{25.12}$$

with the unknown coefficients

$$\{a\} = \begin{Bmatrix} a_1 \\ a_2 \\ a_3 \end{Bmatrix} \tag{25.13}$$

The raw values of the particular stress component at the four superconvergent points contained by the patch are $\{\sigma_r\}$ (subscript r for raw), where subscripts are used to indicate the four points:

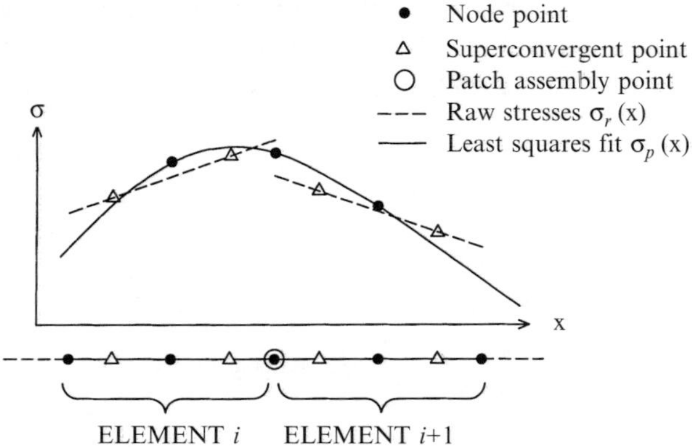

Figure 25.1 Patch of one-dimensional three-noded elements with quadratic shape functions (Reproduced by permission of John Wiley & Sons Ltd)

$$\{\sigma_r\} = \begin{Bmatrix} \sigma_{r1} \\ \sigma_{r2} \\ \sigma_{r3} \\ \sigma_{r4} \end{Bmatrix} \tag{25.14}$$

The coordinates of the four points can be used to determine

$$[\bar{P}] = \begin{bmatrix} 1 & x_1 & x_1^2 \\ 1 & x_2 & x_2^2 \\ 1 & x_3 & x_3^2 \\ 1 & x_4 & x_4^2 \end{bmatrix} \tag{25.15}$$

The stresses in the superconvergent points calculated with the polynomial function are equal to $[\bar{P}]\{a\}$

A least squares fit using $[\bar{P}]^T$ as the weighting function can be formed by requiring that

$$[\bar{P}]^T(\{\sigma_r\} - [\bar{P}]\{a\}) = 0 \tag{25.16}$$

Solving for $\{a\}$ yields

$$\{a\} = [[\bar{P}]^T[\bar{P}]]^{-1}[\bar{P}]^T\{\sigma_r\} = [A]\{\sigma_r\} \tag{25.17}$$

The construction of the $[A]$ matrix has to be done only once for each patch. Best fit coefficients for each stress component can then be obtained by substituting the appropriate raw values at the superconvergent points. Once the coefficients have been computed, Equation (25.11) can be used to obtain the fitted stress at any point in the patch. The recovered stresses at the patch assembly node, and any other nodes contained within the patch which are not covered by an adjacent patch, are determined in this way. Mid-side and mid-element nodes are normally covered by more than one patch. In these cases, the recovered stresses from all patches containing the nodes are averaged. Recovered nodal stresses $\{\sigma^*\}$ at all nodes are thus obtained.

For the case of two-dimensional elements, the element patches become two-dimensional, as illustrated in Figure 25.2. The fitted polynomial contains terms in both x and y, and possibly cross terms. However, the basic procedure is unchanged.

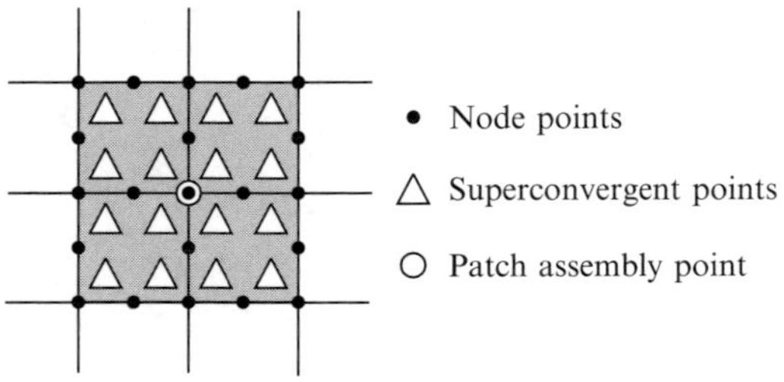

Figure 25.2 Patch of two-dimensional eight-noded elements with quadratic shape functions (Reproduced by permission of John Wiley & Sons Ltd)

The nodal stresses recovered by the superconvergent patch recovery technique can be substituted into Equation (25.10), leading to a complete superconvergent stress field. The resulting stress field is smooth and continuous over the domain. This stress field can be used in Equation (25.7) to obtain an error estimator for the finite element solution.

25.3 STRESS RECOVERY IN SCALED BOUNDARY FINITE ELEMENT METHOD

Displacements computed with the scaled boundary finite element method are continuous in the radial and circumferential directions. The displacements are also differentiable in the radial direction, but the first derivatives in the circumferential direction are, in general, discontinuous across the element boundaries. Consequently, the stresses are continuous along radial lines from the scaling centre, but are discontinuous across the radial lines corresponding to inter-element nodes at the boundary. The stresses in the circumferential direction must be smoothened. The variation of stress in the radial direction is much more accurate (equilibrium is satisfied in this direction), and so no smoothing is needed in this direction.

The procedure is discussed for a bounded medium. The displacements are specified for the two-dimensional case ($s = 2$) as (Equations (16.12a), (16.17) and (15.93))

$$\{u(\xi, \eta)\} = [N(\eta)]\{u(\xi)\} \tag{25.18}$$

where

$$\{u(\xi)\} = [\Phi_{11}] \lceil \xi^{-\lambda_i} \rfloor \{c\} = \sum_{i=1}^{n} c_i \xi^{-\lambda_i} \{\Phi_i\} \tag{25.19}$$

with $\{\Phi_i\}$ equal to the ith column of $[\Phi_{11}]$ and n denoting the number of modes. The stresses are equal to (Equation (15.94))

$$\{\sigma(\xi, \eta)\} = [D]\left([B^1(\eta)]\{u(\xi)\}_{,\xi} + \frac{1}{\xi}[B^2(\eta)]\{u(\xi)\} \right) \tag{25.20}$$

Substituting Equation (25.19) into (25.20) yields

$$\{\sigma(\xi, \eta)\} = \sum_{i=1}^{n} c_i \{\sigma_i(\xi, \eta)\} \tag{25.21}$$

where the stresses for the ith mode are

$$\{\sigma_i(\xi, \eta)\} = [D](-\lambda_i[B^1(\eta)] + [B^2(\eta)])\{\Phi_i\}\xi^{-\lambda_i - 1} \tag{25.22}$$

Each mode may vary at a different rate in the radial direction. Clearly, smoothing the total stresses (Equation (25.21)) at any particular value of ξ will not result in the stresses being continuous for all ξ, and so smoothing the total stresses on the boundary is not an adequate procedure. Instead, the stresses must be smoothed for each mode independently. This can be done at the boundary, since Equation (25.22) shows that, for each mode, the stress variations around the circumference at different values of ξ are just scaled versions of the stress variation around the boundary. Consequently, if a continuous circumferential stress variation is obtained at the boundary, the stress variation will be continuous at every value of ξ for that particular mode. Any linear combination of the smoothened modal stress contributions must also be continuous, as desired.

Since the circumferential discretisation is the same as in the finite element method, an improved stress distribution in this direction can be sought using the shape functions to interpolate between nodal stress values. As the radial stress variation is obtained analytically, it should not be altered by this interpolation. Consequently, recovered modal stresses are sought in the form

$$\{\sigma_i^*(\xi,\eta)\} = [N(\eta)]\{\sigma_i^*\}\xi^{-\lambda_i-1} \tag{25.23}$$

where $\{\sigma_i^*\}$ are the recovered modal stresses at the boundary nodes.

Along the boundary, for each mode

$$\{\sigma_i^*(\eta)\} = [N(\eta)]\{\sigma_i^*\} \tag{25.24}$$

applies. Nodal stress values may be recovered by simple averaging. The superconvergent patch recovery technique may also be used along the boundary. In this case, since the recovered stresses are superconvergent on the boundary, they are also superconvergent for every value of ξ. However, superconvergent patch recovery cannot be used at sharp boundary corners, as the true stress variation is not smooth in the circumferential direction around the corner. Such a corner point can be treated as an end point for the two adjoining boundary sections, and two independent estimates for stress at the corner point obtained from the boundary sections using the superconvergent patch recovery technique. The two estimates can then be averaged.

Stress recovery must be performed for each component of stress and for each mode independently, but the coefficient matrix used to perform the least-squares best-fit operation for a patch remains constant for all modes and all stress components. Thus, stress recovery can be done quite efficiently.

The total recovered stress field is formed by multiplying each recovered modal stress field by the corresponding modal participation factor, and summing

$$\{\sigma^*(\xi,\eta)\} = \sum_{i=1}^{n} c_i\{\sigma_i^*(\xi,\eta)\} = \sum_{i=1}^{n} c_i\xi^{-\lambda_i-1}[N(\eta)]\{\sigma_i^*\} \tag{25.25}$$

Since each of the modal stress fields are continuous, the summation is also continuous. Also, since the recovery procedure is performed at the same superconvergent points for each mode, the summation is also superconvergent.

The stress recovery procedure presented in this section need only be performed on the boundary, in contrast to the finite element method, where recovery must be performed over the entire domain. Scaling ensures that the stresses recovered are superconvergent throughout the domain.

Unlike the finite element method, the recovery procedure can be performed on the modes before application of the boundary conditions and computation of the final displacement field. The recovered stresses for multiple load cases can therefore be obtained at the same computational cost as for a single load case. In the finite element method, stress recovery can only be performed after computation of the displacement field, and so must be done separately for each load case.

For an unbounded medium, the stress recovery procedure is analogous, with only minor changes. The sign of λ_i is changed, $\{\Phi_i\}$ denotes the i-th column of $[\Phi_{12}]$. For instance, the stresses for the i-th mode equal

$$\{\sigma_i(\xi,\eta)\} = [D](\lambda_i[B^1(\eta)] + [B^2(\eta)])\{\Phi_i\}\xi^{\lambda_i-1} \tag{25.26}$$

and the recovered modal stresses

$$\{\sigma_i^*(\xi,\eta)\} = [N(\eta)]\{\sigma_i^*\}\xi^{\lambda_i-1} \tag{25.27}$$

25.4 ERROR ESTIMATION IN SCALED BOUNDARY FINITE ELEMENT METHOD

The simple recovery error estimator described by Equation (25.7) is applied to the scaled boundary finite element method. A bounded medium is addressed. Substituting Equations (25.25) and (25.22) into (25.6) yields

$$\{e_\sigma^*(\xi,\eta)\} = \{\sigma^*(\xi,\eta)\} - \{\sigma(\xi,\eta)\}$$

$$= \sum_{i=1}^{n} c_i\xi^{-\lambda_i-1}[N(\eta)]\{\sigma_i^*\} - [D]\sum_{i=1}^{n} c_i\xi^{-\lambda_i-1}[[B^2(\eta)] - \lambda_i[B^1(\eta)]]\{\Phi_i\} \tag{25.28}$$

Since the variation of the raw and recovered stresses is the same in the radial direction within each mode, this expression can be simplified as

$$\{e_\sigma^*(\xi,\eta)\} = \sum_{i=1}^{n} c_i\xi^{-\lambda_i-1}\{[N(\eta)]\{\sigma_i^*\} - [D]\,[[B^2(\eta)] - \lambda_i[B^1(\eta)]]\{\Phi_i\}\}$$

$$= \sum_{i=1}^{n} c_i\xi^{-\lambda_i-1}\{e_{\sigma i}^*(\eta)\} \tag{25.29}$$

where the modal stress error on the boundary equals

$$\{e_{\sigma i}^*(\eta)\} = [N(\eta)]\{\sigma_i^*\} - [D]\,[[B^2(\eta)] - \lambda_i[B^1(\eta)]]\{\Phi_i\} \tag{25.30}$$

The energy norm of the stress error (Equation (25.3)) evaluated with the recovered stresses becomes

$$\|\{e_\sigma^*\}\|_V = \left(\int_V \left\{\sum_{i=1}^{n} c_i\xi^{-\lambda_i-1}\{e_{\sigma i}^*(\eta)\}\right\}^T [D]^{-1}\left\{\sum_{i=1}^{n} c_i\xi^{-\lambda_i-1}\{e_{\sigma i}^*(\eta)\}\right\}\mathrm{d}V\right)^{0.5} \tag{25.31}$$

Multiplying the summations and substituting Equation (15.92) gives

$$\begin{aligned}\|\{e_\sigma^*\}\|_V &= \left(\int_{\xi_0}^{\xi_1}\int_S \sum_{i=1}^{n}\sum_{j=1}^{n} c_i c_j \xi^{-\lambda_i-1}\xi^{-\lambda_j-1}\{e_{\sigma i}^*(\eta)\}^T[D]^{-1}\{e_{\sigma j}^*(\eta)\}\xi|J|\mathrm{d}\xi\mathrm{d}\eta\right)^{0.5}\\ &= \left(\sum_{i=1}^{n}\sum_{j=1}^{n} c_i c_j \int_{\xi_0}^{\xi_1} \xi^{-\lambda_i-\lambda_j-1}\left(\int_S \{e_{\sigma i}^*(\eta)\}^T[D]^{-1}\{e_{\sigma j}^*(\eta)\}|J|\mathrm{d}\eta\right)\mathrm{d}\xi\right)^{0.5}\end{aligned} \tag{25.32}$$

Integrating in the radial direction yields

$$\|\{e_\sigma^*\}\|_V = \left(\sum_{i=1}^{n}\sum_{j=1}^{n} c_i c_j \left[\frac{1}{-\lambda_i-\lambda_j}\xi^{-\lambda_i-\lambda_j}\right]_{\xi_0}^{\xi_1}\int_S \{e_{\sigma i}^*(\eta)\}^T[D]^{-1}\{e_{\sigma j}^*(\eta)\}|J|\mathrm{d}\eta\right)^{0.5} \tag{25.33}$$

For the bounded case the integration limits are $\xi_0 = 0$ and $\xi_1 = 1$. The real parts of λ_i are all non-positive. Consequently,

$$\left[\frac{1}{-\lambda_i-\lambda_j}\xi^{-\lambda_i-\lambda_j}\right]_{\xi_0}^{\xi_1} = \frac{1}{-\lambda_i-\lambda_j} \tag{25.34}$$

The integrations in the radial direction are thus performed analytically, and are very simple, as the variation of the stresses is polynomial.

Addressing the unbounded case, using Equations (25.26) and (25.27) yields, for Equation (25.32),

$$\|\{e_\sigma^*\}\|_V = \left(\sum_{i=1}^{n}\sum_{j=1}^{n} c_i c_j \int_{\xi_0}^{\xi_1} \xi^{\lambda_i+\lambda_j-1}\left(\int_S \{e_{\sigma i}^*(\eta)\}^T[D]^{-1}\{e_{\sigma j}^*(\eta)\}|J|\mathrm{d}\eta\right)\mathrm{d}\xi\right)^{0.5} \tag{25.35}$$

The integration limits are $\xi_0 = 1$ and $\xi_1 \to \infty$, leading to

$$\int_{\xi_0}^{\xi_1} \xi^{\lambda_i+\lambda_j-1}\mathrm{d}\xi = \left[\frac{1}{\lambda_i+\lambda_j}\xi^{\lambda_i+\lambda_j}\right]_{\xi_0}^{\xi_1} = \frac{1}{\lambda_i+\lambda_j}(-1) \tag{25.36}$$

Thus, the same expression results for the unbounded medium.

Zero eigenvalues correspond to rigid body modes, in which both raw and recovered stresses are zero at all points. Consequently, any such modes present can be left out of the summation.

The energy norm of the stress error can be written for both bounded and unbounded domains as

$$\|\{e_\sigma^*\}\|_V = \left(\sum_{i=1}^{n} \sum_{j=1}^{n} \frac{c_i c_j}{-\lambda_i - \lambda_j} \int_S \{e_{\sigma i}^*(\eta)\}^T [D]^{-1} \{e_{\sigma j}^*(\eta)\} |J| \mathrm{d}\eta \right)^{0.5} \tag{25.37}$$

The boundary integral can be performed numerically, element by element, in the usual way. Integration in the radial direction has been performed analytically. Should a local error estimate be required, the integration limits for each mode in Equation (25.33) can be adjusted appropriately. The advantages of the scaled boundary finite element method are preserved, since discretisation is limited to the boundary only and no numerical integration is needed in the radial direction.

The energy norm of the recovered stresses (Equation (25.1)) can be computed in the same way as

$$\|\{\sigma^*\}\|_V = \left(\sum_{i=1}^{n} \sum_{j=1}^{n} \frac{c_i c_j}{-\lambda_i - \lambda_j} \int_S \{\sigma_i^*(\eta)\}^T [D]^{-1} \{\sigma_j^*(\eta)\} |J| \mathrm{d}\eta \right)^{0.5} \tag{25.38}$$

with the recovered modal stress at the boundary $\{\sigma_i^*(\eta)\}$ specified in Equation (25.24).

Equations (25.37) and (25.38) can be substituted into Equation (25.7) to determine the error estimator, η^*. This estimator is constructed in a manner that is totally consistent with its use in the finite element method, since the energy norms have been computed over the entire domain, both for bounded and unbounded problems. Consequently, the accuracy of the finite element and scaled boundary finite element models can be compared directly by using this estimator. This is in stark contrast with error indicators developed for the boundary element method, which are determined only on the boundary, and against which no direct comparison can be made.

The contribution of the individual scaled boundary finite elements to the global error is of interest for adaptivity. Breaking the boundary integral contained in Equation (25.37) into a summation of contributions from m individual elements, where m is the number of elements used to discretise the boundary

$$\|\{e_\sigma^*\}\|_V = \left(\sum_{i=1}^{n} \sum_{j=1}^{n} \frac{c_i c_j}{-\lambda_i - \lambda_j} \sum_{e=1}^{m} \int_{S_e} \{e_{\sigma i}^*(\eta)\}^T [D]^{-1} \{e_{\sigma j}^*(\eta)\} |J| \mathrm{d}\eta \right)^{0.5} \tag{25.39}$$

where S_e is the portion of the boundary modelled by element e. By rearranging the order of summation, the contribution of each element to the total error can be identified as

$$\|\{e_\sigma^*\}\|_V = \left(\sum_{e=1}^{m} \sum_{i=1}^{n} \sum_{j=1}^{n} \frac{c_i c_j}{-\lambda_i - \lambda_j} \int_{S_e} \{e_{\sigma i}^*(\eta)\}^T [D]^{-1} \{e_{\sigma j}^*(\eta)\} |J| \mathrm{d}\eta \right)^{0.5} = \left(\sum_{e=1}^{m} \|\{e^*{}_\sigma\}\|_e^2 \right)^{0.5} \tag{25.40}$$

In this equation $\|\{e_\sigma\}\|_e$ is the energy norm of the stress error for the region modelled by element e, and

$$\|\{e_\sigma^*\}\|_e = \left(\sum_{i=1}^{n} \sum_{j=1}^{n} \frac{c_i c_j}{-\lambda_i - \lambda_j} \int_{S_e} \{e_{\sigma i}^*(\eta)\}^T [D]^{-1} \{e_{\sigma j}^*(\eta)\} |J| \mathrm{d}\eta \right)^{0.5} \tag{25.41}$$

25.5 EXAMPLE FOR STRESS RECOVERY AND ERROR ESTIMATION

25.5.1 Infinite Plate with Central Circular Hole Under Uniaxial Tensile Stress

To illustrate the stress recovery procedure and to demonstrate the accuracy and convergence of the error estimator for analyses using the finite element and scaled boundary finite element method, a plate of infinite extent containing a central circular hole of radius a and subjected to a uniaxial stress field σ_0 is addressed (Figure 25.3). The problem is chosen because it has an analytical solution, and so the recovered stress field can be compared with the exact stress field. The true accuracy of the numerical solutions can also be determined. Plane strain conditions are assumed, and Poisson's ratio ν is taken as 0.3. The analytical solution for the stress field with components defined in the Cartesian coordinate system is [T4]

$$\sigma_x = \left(1 - \frac{a^2}{r^2} \left(\frac{3}{2} \cos 2\theta + \cos 4\theta \right) + \frac{3}{2} \frac{a^4}{r^4} \cos 4\theta \right) \sigma_0 \tag{25.42a}$$

$$\sigma_y = \left(-\frac{a^2}{r^2} \left(\frac{1}{2} \cos 2\theta - \cos 4\theta \right) - \frac{3}{2} \frac{a^4}{r^4} \cos 4\theta \right) \sigma_0 \tag{25.42b}$$

$$\tau_{xy} = \left(-\frac{a^2}{r^2} \left(\frac{1}{2} \sin 2\theta + \sin 4\theta \right) + \frac{3}{2} \frac{a^4}{r^4} \sin 4\theta \right) \sigma_0 \tag{25.42c}$$

and the analytical solution for the displacement field is

$$u_r = \frac{1+\nu}{4E} \frac{a}{r} \left((2-4\nu) \frac{r^2}{a^2} + 2 \left((4-4\nu) + 1 + \frac{r^2}{a^2} - \frac{a^2}{r^2} \right) \cos 2\theta \right) \sigma_0 a \tag{25.43a}$$

$$u_\theta = -\frac{1+\nu}{2E} \frac{a}{r} \left(\left(\frac{r^2}{a^2} + (2-4\nu) + \frac{a^2}{r^2} \right) \sin 2\theta \right) \sigma_0 a \tag{25.43b}$$

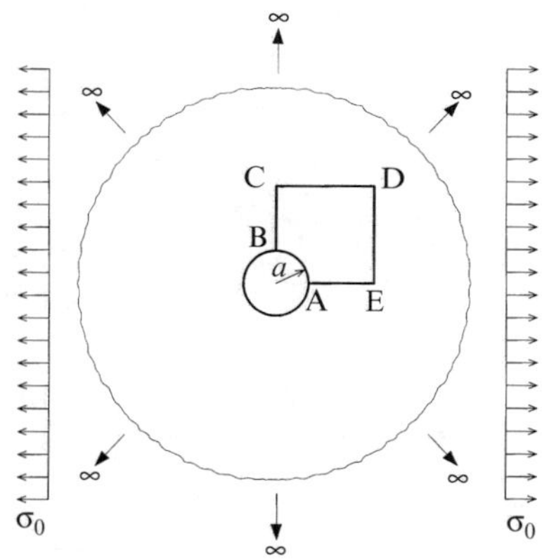

Figure 25.3 Infinite plate with central circular hole subjected to uniaxial tensile stress (Reproduced by permission of John Wiley & Sons Ltd)

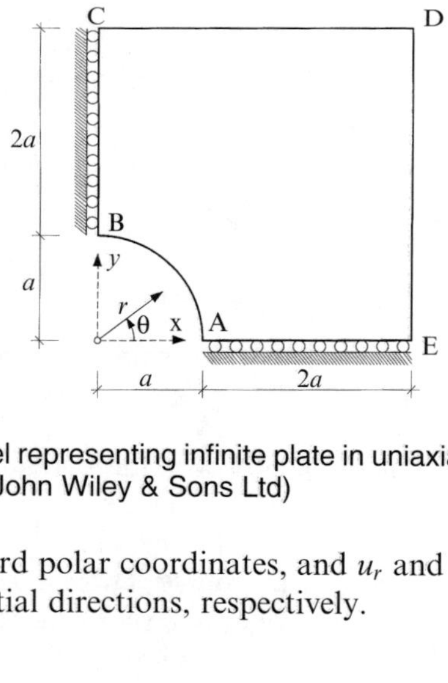

Figure 25.4 Bounded model representing infinite plate in uniaxial stress field (Reproduced by permission of John Wiley & Sons Ltd)

where (r, θ) are the standard polar coordinates, and u_r and u_θ are the displacements in the radial and circumferential directions, respectively.

25.5.2 Bounded Model

A bounded model of the problem is constructed by using symmetry and truncating the medium at a distance of $3a$ along each coordinate axis, as shown in Figures 25.3 and 25.4. The surface traction boundary conditions on the truncated faces *CD* and *DE* are obtained from the exact stress field for the infinite plate (Equation (25.42)).

Results obtained for three scaled boundary finite element meshes of increasing number of degrees of freedom will be presented. The meshes, denoted as coarse, intermediate and fine, are shown in Figure 25.5. Although only the boundary is discretised, each boundary element corresponds to a region of the domain. These regions are indicated in the figure. Three-noded quadratic line elements are used to discretise the boundary. This is not an ideal problem for the scaled boundary finite element method, since there is no obvious scaling centre that can be used to take advantage

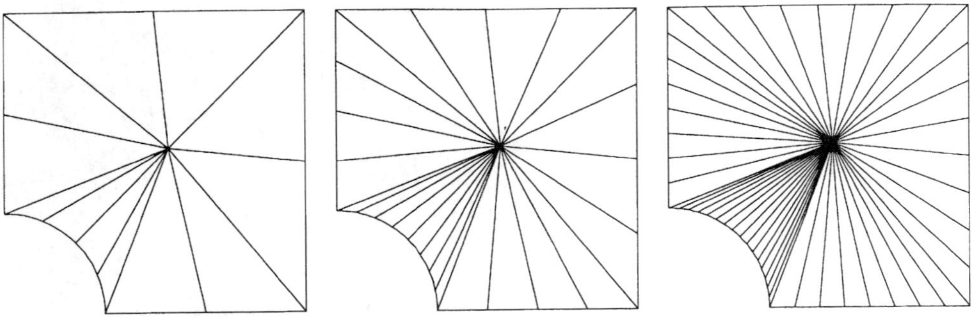

Figure 25.5 Coarse, intermediate and fine bounded scaled boundary finite element meshes (Reproduced by permission of John Wiley & Sons Ltd)

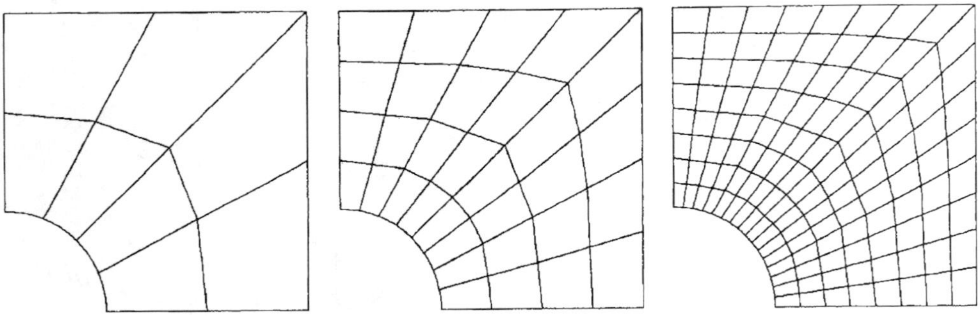

Figure 25.6 Coarse, intermediate and fine finite element meshes (Reproduced by permission of John Wiley & Sons Ltd)

of the unique properties of the method. However, the available analytical solution allows the performance of the error estimator to be evaluated for a bounded domain.

Results from a comparable sequence of three finite element meshes will also be presented. The meshes (also denoted as coarse, intermediate and fine) are shown in Figure 25.6. Eight-noded quadratic isoparametric elements are used. The boundary discretisation and boundary shape functions of the corresponding scaled boundary finite element and finite element meshes are identical.

The accuracy of the stress recovery will be discussed first. Although each stress component is recovered independently, the effectiveness of the recovery procedure will be illustrated here by looking at the effective stress, which for problems of plane strain is conventionally defined as

$$\bar{\sigma} = \sqrt{\frac{1}{2}\left((\sigma_x - \sigma_y)^2 + (\sigma_y - \sigma_z)^2 + (\sigma_z - \sigma_x)^2 + 6\tau_{xy}^2\right)} \tag{25.44}$$

The raw and recovered effective stresses non-dimensionalised by σ_0 for the coarse scaled boundary finite element mesh are presented in Plate 25.1 and for the intermediate finite element mesh in Plate 25.2. The effectiveness of the recovery procedure in obtaining a more useful stress field is clearly illustrated. The accuracy of the recovered stress field may be evaluated by examining the deviation of the recovered effective stress from the exact solution illustrated in Plate 25.3 for both the coarse and intermediate scaled boundary finite element meshes. The maximum absolute value of the non-dimensional effective stress present (corresponding to the stress concentration in Point *B*) is 8/3. This figure shows that, even with the coarse model, very accurate results are obtained over most of the mesh. Points of highest error occur where the boundary direction changes suddenly. As the mesh is refined, the errors reduce considerably. The recovery procedure is therefore seen to be highly effective.

The point-wise variation of the errors in the recovered effective stresses obtained by finite element analysis for the intermediate and fine meshes is presented in Plate 25.4 The superconvergent patch recovery technique is used to obtain the stresses. To allow ready comparison, the contours are restricted to the range used in Plate 25.3, although the intermediate finite element mesh contains larger errors near the hole. The errors in the coarse finite element mesh are so large that plotting at this scale is not appropriate.

On comparing the results of the intermediate finite element mesh in Plate 25.4 with those of the intermediate scaled boundary finite element mesh in Plate 25.3, the recovered finite element stresses are seen to be significantly less accurate than those obtained using the scaled boundary finite element method, even though the level of discretisation and the shape functions used around the boundary are identical. The intermediate scaled boundary mesh is seen to give more accurate results than the fine finite element mesh, which has twice the number of degrees of freedom on the boundary.

The error estimation is addressed next. The performance of the global error estimator is shown in Table 25.1 for the scaled boundary finite element method, and in Table 25.2 for the finite element method. These tables present the estimated global errors η^* (Equation (25.7)) obtained by using the recovery procedures described previously for the scaled boundary finite element method (Equations (25.37) and (25.38)) and for the finite element method. The actual global error η (Equation (25.4)) computed using the exact solution is also listed for each model, along with the number of degrees of freedom. The displacement in the x-direction (non-dimensionalised by $\sigma_0 a(1+\nu)(1-\nu)/E$) u_x at point A in Figure 25.4, and the recovered normal stress in

Table 25.1 Scaled boundary finite element error estimator for bounded model

Mesh	DOF	η^*(%)	η(%)	u_x	σ_x^*	Time (s)
Coarse	38	2.9	2.1	2.996	2.741	2.6
Intermediate	78	0.9	0.9	3.000	2.911	15.2
Fine	158	0.3	0.3	3.000	2.976	146.0

Table 25.2 Finite element error estimator for bounded model

Mesh	DOF	η^*(%)	η(%)	u_x	σ_x^*	Time (s)
Coarse	64	14.0	6.6	2.974	1.597	0.1
Intermediate	224	4.6	2.7	2.999	2.333	0.2
Fine	832	1.1	1.0	3.000	2.601	0.8

the x-direction (non dimensionalised by σ_0) σ_x^* at point B, are also tabulated to provide an indication to the point-wise convergence of the displacement and stress fields. The exact values of these two quantities can be obtained from Equations (25.43a) and (25.42a).

The amount of computer time required to perform the computations (on a 450 MHz Pentium III) is tabulated for information, although the main aim of this example is to illustrate the accuracy of the new error estimator. (The timings are obtained using rather inefficient general purpose routines to solve the quadratic eigenvalue problem in the scaled boundary finite element method, while the standard finite element timings are obtained using an efficient skyline solution scheme and optimal degree-of-freedom numbering. Significant improvements to the computational efficiency of the scaled boundary finite element method are expected in the future.)

Several key observations can be made:

1. The performance of the error estimator (in terms of its ability to predict the exact error) improves as the mesh is refined for both methods.
2. For the same level of boundary discretisation, the scaled boundary finite element method produces a lower energy norm error, and the accuracy of the error estimator is better.
3. The displacement converges rapidly, as is expected, since it is the basic function.
4. The error in the normal stress at the stress concentration (point B) converges less rapidly, and generally exceeds the global error (which is a weighted average over the domain). For the same global error level, the computed stress at the concentration point obtained by the scaled boundary finite element method is significantly more accurate than that obtained by the finite element method. This implies that the error is more evenly distributed across the mesh in the scaled boundary finite element solution.
5. The scaled boundary finite element method is not competitive in terms of computational time for this bounded domain model, as no advantage is taken of the special features of the method. Nevertheless, this example clearly demonstrates the accuracy of the error estimator, and the quality of the solutions obtained.

To be useful for adaptivity, the error estimator should also be able to approximate the contribution of the individual elements to the global error accurately. The contribution of each element to the error energy norm can be determined using Equation (25.41). This contribution is plotted for the coarse and intermediate scaled boundary finite element meshes in Figure 25.7. On comparing the element error energy norm contributions with the distribution of the point-wise error illustrated in Plate 25.3, it is clear that the distribution of the error energy norm correlates well with the point-wise error in the effective stress.

25.5.3 Unbounded Model

The error estimation procedure based on stress recovery can be applied to both bounded and unbounded models. Whereas the infinite plate has to be truncated to form a bounded model (and the exact solution used to provide boundary conditions) for the finite element method to be employed, the problem can be solved without truncation (and without reference to the exact solution) by using the special features of the scaled boundary finite element method. The resulting solution provides an approximation for the stress and displacement fields over the complete unbounded plate.

In this section, the plate is modelled as an unbounded medium with side-faces along each of the axes of symmetry and the scaling centre at the centre of the circular hole, as illustrated in Figure 25.8. Only the circumference of the hole is discretised, using three-noded quadratic scaled boundary finite elements. Three meshes, designated coarse, intermediate and fine (Figure 25.9) are used. For each unbounded mesh, the level of

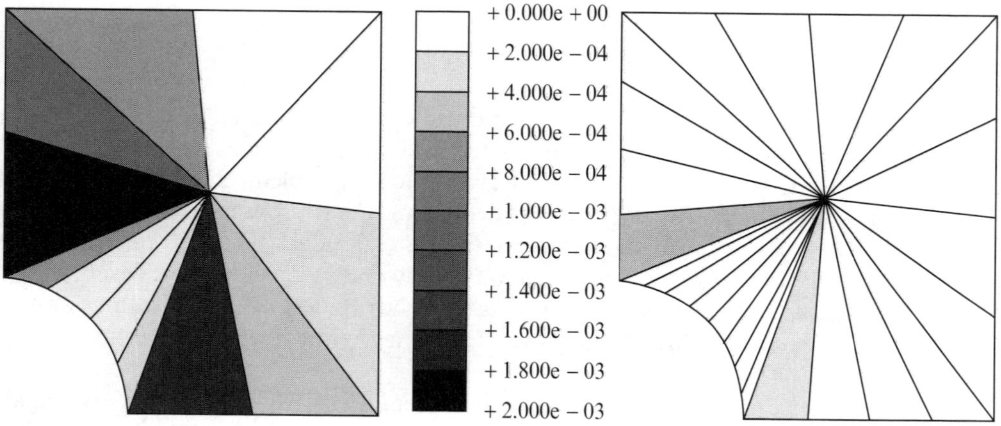

Figure 25.7 Contributions to error energy norm for coarse and intermediate meshes analysed by scaled boundary finite element method with bounded model (Reproduced by permission of John Wiley & Sons Ltd)

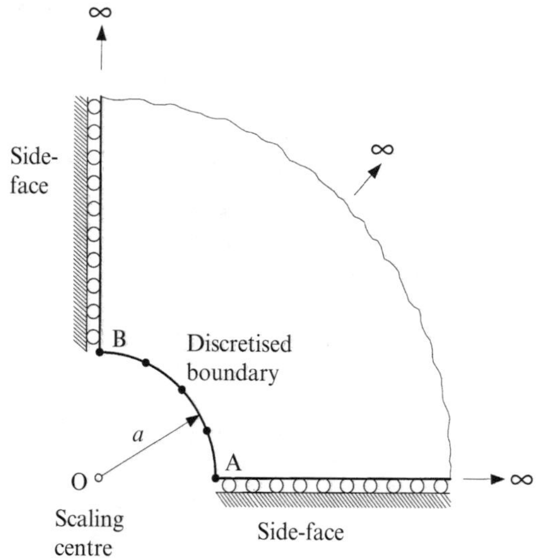

Figure 25.8 Unbounded model representing infinite plate in uniaxial stress field (Reproduced by permission of John Wiley & Sons Ltd)

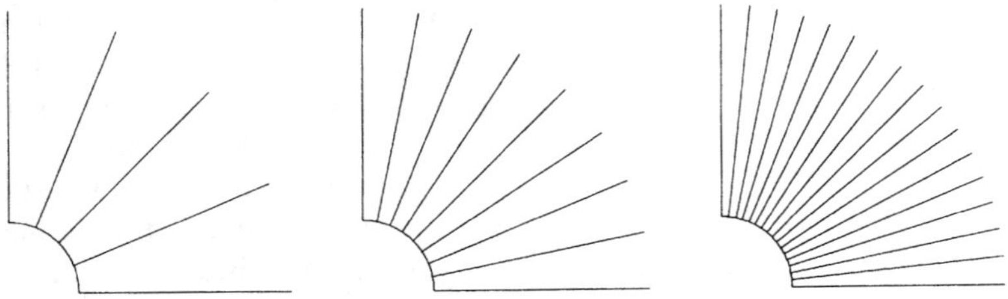

Figure 25.9 Coarse, intermediate and fine unbounded scaled boundary finite element meshes (Reproduced by permission of John Wiley & Sons Ltd)

discretisation over the circumference of the hole is identical to the corresponding bounded mesh (Figure 25.5).

The energy norm of the stress in the unbounded domain subjected to uni-axial stress is infinite (since away from the hole, the stress and strain in the x-direction are both approximately constant and non-zero). To overcome the problem, this creates with the error estimator, the deviation of the stress field from a uniform uni-axial stress field is computed. The energy norm of this 'deviation' stress field is finite, and the error estimator can be calculated in the usual way. However, this is a more severe error norm than that used in the bounded case, as the contribution of the smooth uni-axial stress field to the energy norm of the stress field (the divisor in the error estimator) is eliminated.

The performance of the error estimator is presented in Table 25.3. The estimated global errors η^* are presented, along with the actual global errors η (also determined for the 'deviation' stress field), the dimensionless displacement u_x at A, the normal stress σ_x^* in the x-direction at B, the computation time, and the number of degrees of freedom. The definitions of the dimensionless quantities and the applicable equations are the same as discussed in connection with Table 25.1 in section 25.5.2. Once again, the new error estimator is seen to perform extremely well. The increased severity of the error norm is illustrated by the convergence of the stress at point B, the error in which is less than the global error for each mesh. It is also illustrated by the point-wise variation of the effective stress error, which is plotted in Plate 25.5 using the same contour range as in Plates 25.3 and 25.4 for easy comparison. The point-wise errors are significantly smaller than those obtained by the bounded models, although the global error norms are comparable. The accuracy of the error estimator in predicting the distribution of error between the elements is maintained, as demonstrated by the correlation between the element contributions to the error energy norm (Figure 25.10) and the point-wise variation of effective stress error (Plate 25.5).

Table 25.3 Scaled boundary finite element error estimator for unbounded model

Mesh	DOF	η^*(%)	η(%)	u_x	σ_x^*	Time (s)
Coarse	16	5.2	3.5	2.998	3.125	0.1
Intermediate	32	1.0	1.0	2.999	3.026	0.8
Fine	64	0.2	0.2	3.000	3.000	5.7

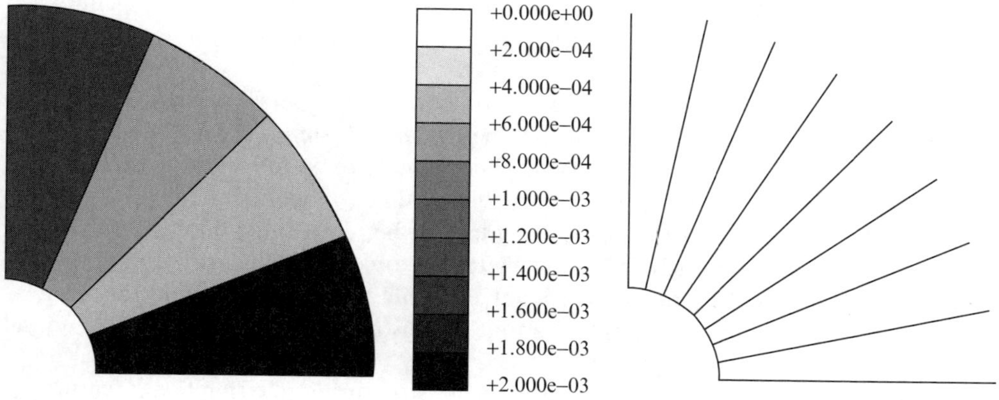

Figure 25.10 Contributions to error energy norm for coarse and intermediate meshes analysed by scaled boundary finite element method with unbounded model (Reproduced by permission of John Wiley & Sons Ltd)

The main aim of this example is to demonstrate the accuracy of the new error estimator. However, some comments on the performance of the scaled boundary finite element method are appropriate. Although the unbounded scaled boundary finite element model contains fewer degrees-of-freedom than for the corresponding bounded model, the accuracy attained is higher because better use is made of the semi-analytical nature of the method. In the unbounded model, the variation of displacement and stress along the axes of symmetry is obtained analytically, while in the bounded model the use of elements along these boundaries constrains the variation to the shape functions. The additional degrees of freedom therefore *reduce* the accuracy of the method, but increase the computational cost dramatically. The fine mesh of the finite element model and the intermediate mesh of the unbounded scaled boundary finite element model take the same amount of time to solve. However, the scaled boundary finite element mesh produces a more accurate solution, does not require the final solution to be known in advance (to provide boundary tractions), and produces an approximate solution for the complete unbounded domain.

25.6 H-HIERARCHICAL ADAPTIVITY

The accuracy of a discrete model (finite element, boundary element or scaled boundary finite element model) depends upon the level of discretisation; the higher the level of discretisation, the more accurate is the solution. An inadequate level of discretisation leads to inaccurate results. If an error estimator is available, the discrete model can be adapted automatically to meet some specified level of accuracy. This is clearly a desirable property for a numerical method.

The basic algorithm for adaptivity is very simple, and is illustrated in Figure 25.11. The process commences with a minimum description of the domain, material properties, boundary conditions, and a target level of accuracy. An initial coarse model is constructed from this information, and analysed. An estimate is then made of the error in the solution, and a refined model constructed from this information. The process is continued until the estimated error is less than the target error. The method is most efficient when the distribution of error is taken into account during the construction of the refined model.

In *h*-hierarchical methods, the shape functions from the coarser meshes are retained in the basis of the shape functions for the finer meshes. A simple means of accomplishing this is to subdivide any elements in which the error is too high at each step in the adaptive process, i.e. to apply substructuring. Each model contains all the nodes in the previous model, and the mesh structure can be represented efficiently by a tree structure. Such an approach is ideally suited to object-oriented programming and recursive techniques. For one-dimensional elements, a binary tree may be used to describe the refined mesh. A simple example is illustrated in Figure 25.12. The initial discrete model is denoted as parent.

As the *h*-hierarchical procedure is based on substructuring, the significant features of the scaled boundary finite element method discussed in Chapter 22 apply (potential singular behaviour with *a priori* unknown power of singularity at scaling centre with no discretisation on adjacent side-faces). For each subdivision the coefficient matrices have to be recalculated, as no intermediate results from the previous iteration can be used, in contrast to other *h*-adaptivity methods. However, this minor disadvantage is more than

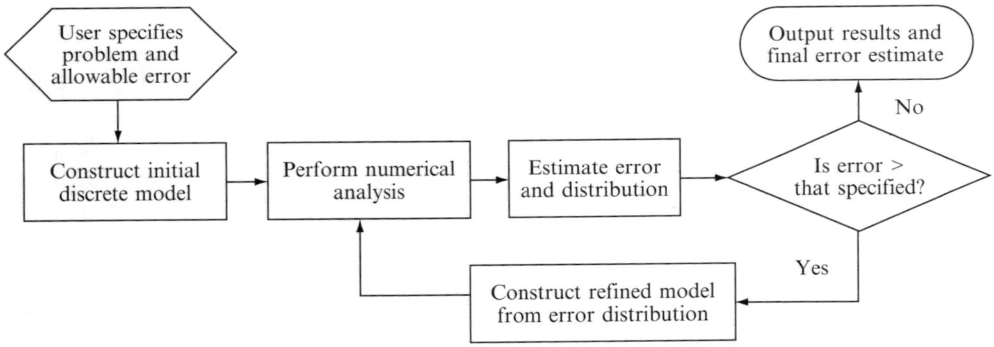

Figure 25.11 Algorithm for adaptive analysis (Reproduced by permission of John Wiley & Sons Ltd)

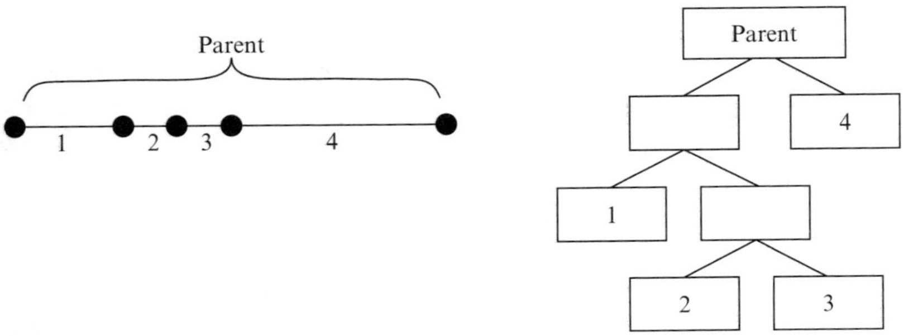

Figure 25.12 Binary tree representation of *h*-hierarchically subdivided two-noded line element (Reproduced by permission of John Wiley & Sons Ltd)

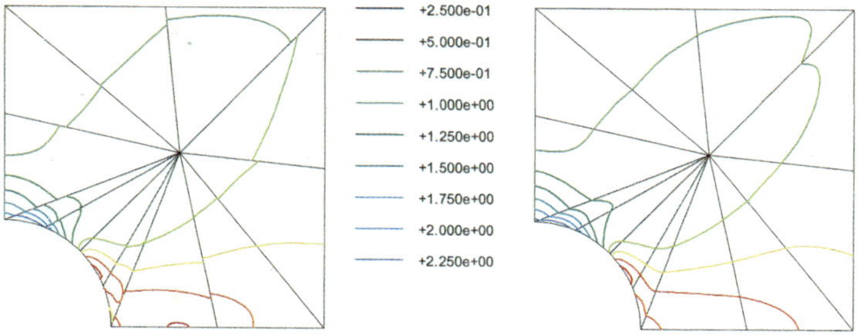

Plate 25.1 Raw (left) and recovered (right) stress as computed by scaled boundary finite element method for coarse mesh (Reproduced by permission of John Wiley & Sons, Ltd)

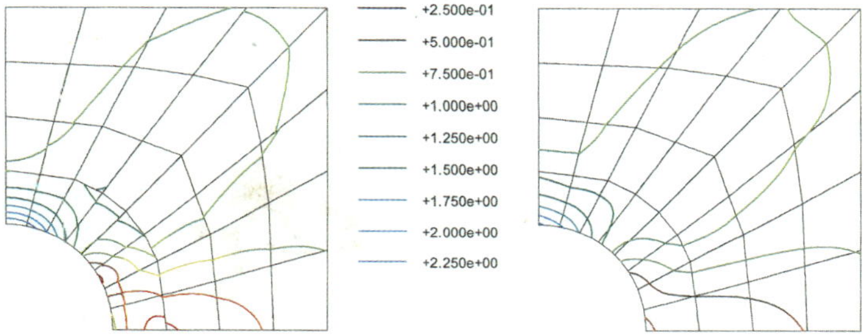

Plate 25.2 Raw (left) and recovered (right) stress as computed by finite element method for the intermediate mesh (Reproduced by permission of John Wiley & Sons, Ltd)

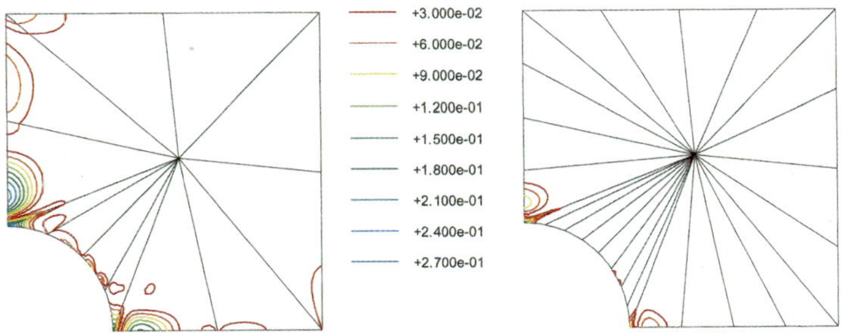

Plate 25.3 Point-wise variation of effective stress error for coarse and intermediate meshes of bounded model analysed by the scaled boundary finite element method (Reproduced by permission of John Wiley & Sons, Ltd)

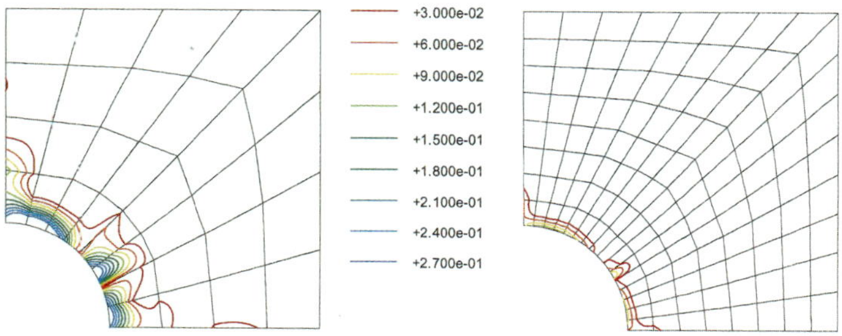

Plate 25.4 Point-wise variation of effective stress error for intermediate and fine meshes analysed by finite element method (Reproduced by permission of John Wiley & Sons, Ltd)

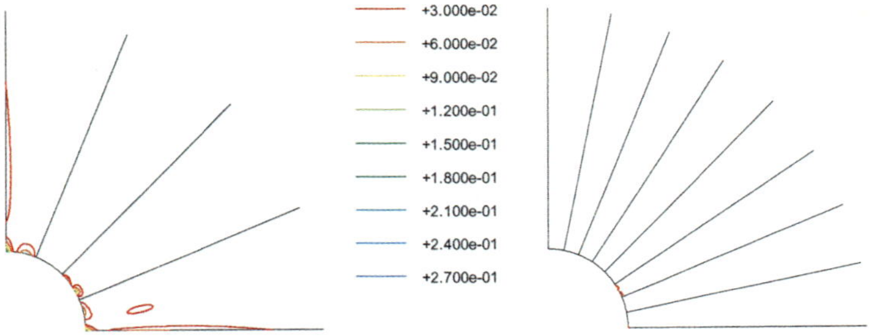

Plate 25.5 Point-wise variation of effective stress error for coarse and intermediate meshes of unbounded model analysed by scaled boundary finite element method (Reproduced by permission of John Wiley & Sons, Ltd)

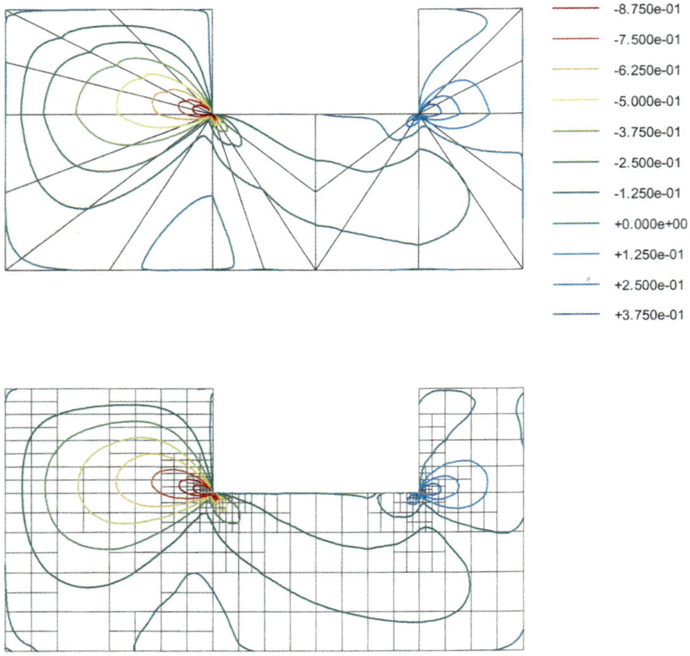

Plate 25.6 Shear stress at 2% target error: scaled boundary finite element method (top), finite element method (bottom) (Reproduced by permission of John Wiley & Sons, Ltd)

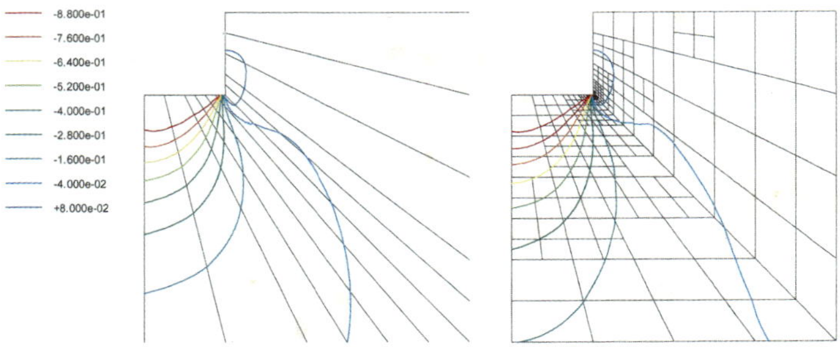

Plate 25.7 Vertical stress at 5% target error: scaled boundary finite element method (left), finite element method (right) (mesh a) (Reproduced by permission of John Wiley & Sons, Ltd)

compensated for by the striking advantage that a series of eigenvalue problems are solved, each of significantly smaller size than the eigenvalue problem resulting when the original domain is solved directly. As the major component of the computational effort is represented by the solution of the eigenvalue problem for which the number of operations increases much more than linearly with the number of unknowns, substructuring will, in general, lead to a more efficient procedure.

The *h*-hierarchical refinement strategy is discussed in the following. It is postulated that the optimum discretisation (mesh) is obtained when each element contributes equally to the error. If the target error in the relative energy norm (Equation (25.4)) is denoted by $\bar{\eta}$, the goal of the refinement strategy should be to ensure that for each element

$$\left\|\{e^*_\sigma\}\right\|_e \leq \bar{\eta}\sqrt{\frac{\|\sigma^*\|_V^2}{m}} \tag{25.45}$$

where $\|\sigma^*\|_V$ is the recovered stress over the entire domain, $\left\|\{e^*_\sigma\}\right\|_e$ is the contribution of the region modelled by element e to the estimated energy norm of the stress error and m is the number of elements in the discretisation (mesh).

An *h*-hierarchical refinement strategy consistent with this principle is to subdivide all elements that fail to meet this criterion at each step. Although the optimum mesh cannot (in general) be reached by this technique (since the level of discretisation varies in a stepwise rather than continuous manner), a reasonable approximation to the optimum can be obtained.

If the refinement is performed in two dimensions, a one-level rule is also employed, which ensures that adjacent elements do not differ in refinement level by more than one. This is necessary because the shape functions of the larger element constrain the deformations of the smaller elements. Such a rule is not necessary for the refinement of one-dimensional elements.

For problems of elastostatics, the scaled boundary finite element method obtains an approximate solution as the weighted summation of n modes. Equation (25.38) shows that the energy norm of the recovered stresses is obtained in the form

$$\|\{\sigma^*\}\|_V = \left(\sum_{i=1}^{n}\sum_{j=1}^{n}\frac{c_i c_j}{-\lambda_i-\lambda_j}\int_S \{\sigma^*_i(\eta)\}^T[D]^{-1}\{\sigma^*_j(\eta)\}|J|\mathrm{d}\eta\right)^{0.5} \tag{25.46}$$

and the contribution of each element to the estimated stress error energy norm is calculated as (Equation (25.41))

$$\left\|\{e^*_\sigma\}\right\|_e = \left(\sum_{i=1}^{n}\sum_{j=1}^{n}\frac{c_i c_j}{-\lambda_i-\lambda_j}\int_{S_e} \{e^*_{\sigma i}(\eta)\}^T[D]^{-1}\{e^*_\sigma(\eta)\}|J|\mathrm{d}\eta\right)^{0.5} \tag{25.47}$$

where $\{e^*_{\sigma i}(\eta)\}$ are the modal stress errors on the boundary (Equation (25.30)). Note that these equations even allow the contributions of elements with stress singularities to be evaluated. Equations (25.46) and (25.47) can be substituted directly into Equation

(25.45) to provide a refinement criterium. The h-hierarchical refinement strategy described above can then be applied.

25.7 EXAMPLES FOR ADAPTIVITY

A bounded medium with stress singularities and an unbounded medium are analysed using the scaled boundary finite element method with the h-hierarchical adaptive procedure. The problems are also calculated using a similar h-hierarchical adaptive finite element procedure, and the computational efforts required to reach a prescribed error level are compared. The numerical procedures used in the scaled boundary finite element method are far from optimum at this stage: a general purpose eigenvalue routine is used to obtain the eigenvalues and eigenvectors of the Hamiltonian matrix corresponding to the quadratic eigenvalue problem, with no advantage taken of the Hamiltonian properties. Complex arithmetic is used throughout the stress recovery procedure and error computation, although savings could be made by combining the complex conjugate modes to remove the imaginary components. On the other hand, in the finite element analyses the stiffness matrix is stored in a skyline form with the degrees of freedom renumbered. Full advantage is taken of the symmetry of the matrix, and a fast active column solver used to obtain the solution. Consequently, the computer times are biased in favour of the finite element method.

25.7.1 Plate with Two Square Holes Under Uniaxial Tensile Stress

To demonstrate the advantages of substructuring, a plate with two square holes subjected to uniaxial tensile stress σ_0 is analysed (Figure 25.13a). At the re-entrant corners stress singularities are present. Poisson's ratio ν equal 0.3.

Even after taking advantage of biaxial symmetry (Figure 25.13b), no location for the scaling centre exists from which the entire boundary is visible. Substructuring into two subdomains is thus performed. Additional degrees of freedom are present on the common interface. For each subdomain, the scaling centre O is chosen at the re-entrant corner, and the adjacent side-faces, which are the internal faces of the total model, are not discretised. The initial discretisation of the remaining boundary for the scaled boundary finite element method is shown in Figure 25.13b. Each subdomain contains six three-node line elements, and the analysed total model has 23 nodes.

An analysis with finite elements is also performed. For each subdomain, the initial mesh has three eight-node quadratic finite elements. On the boundary, the shape functions of the finite elements are the same as those of the scaled boundary finite elements.

The problem is analysed using both the adaptive scaled boundary finite element method and the adaptive finite element method for various target errors $\bar{\eta}$. The corresponding final error estimator η^* is plotted versus the computation time in Figure 25.14. Note that both axes are logarithmic. For comparison, the problem is also analysed by a series of uniformly refined meshes for both methods.

The adaptive and uniform meshes of the scaled boundary finite element method give very similar rates of convergence, as both capture the singularity rigorously. On the

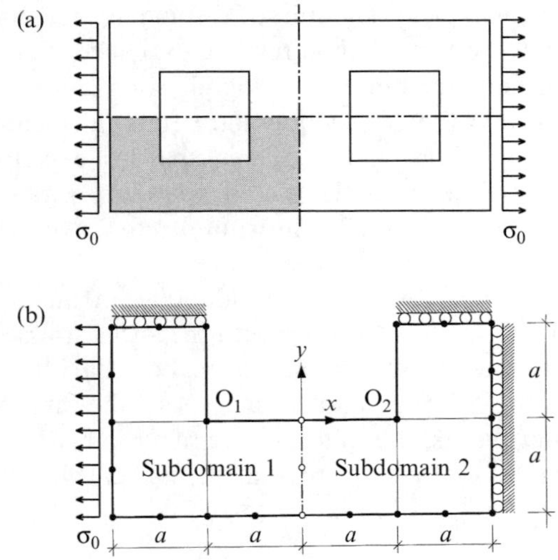

Figure 25.13 Plate with two square holes subjected to uniaxial tensile stress. (a) Geometry and loading, (b) model with two subdomains (Reproduced by permission of John Wiley & Sons Ltd)

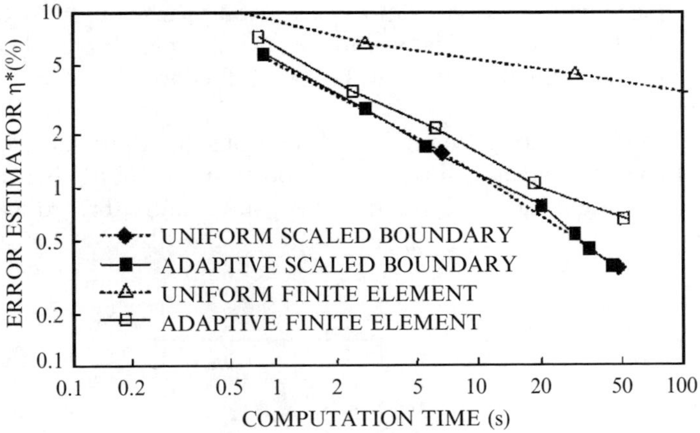

Figure 25.14 Accuracy versus computational effort for uniform and adaptive refinements (Reproduced by permission of John Wiley & Sons Ltd)

discretised circumferential boundary there is no singularity present in this example, which means that the details of the discretisation are not important. However, the adaptive procedure is still very useful from a practical point of view, as it allows the user to specify a very coarse initial mesh and a target error, and not be concerned about the appropriate element size to employ in a uniform mesh.

The difference between the convergence rates of the uniformly and adaptively refined finite element models is much more significant, due to the presence of a stress singularity in the discretised domain.

The convergence rate of the adaptive scaled boundary finite element method is remarkably similar to that of the adaptive finite element method, although the computation time is somewhat less. The shift of the line representing the convergence of the scaled boundary finite element method relative to that representing the convergence of the adaptive finite element method on the log plot of Figure 25.14 indicates that (to achieve a set level of error) the scaled boundary finite element method requires an average 40% less time. This also indicates that the use of substructuring does not significantly reduce the efficiency of the scaled boundary finite element method. As already mentioned, the use of more efficient routines to solve the eigenvalue problem can be expected to improve this ratio in the future.

The final meshes attained by the two adaptive methods for the 2% target error level are shown in Plate 25.6, together with the shear stresses, τ_{xy} (non-dimensionalised by σ_0). The contour range for the scaled boundary finite element results is truncated, as infinite stresses are computed at each of the interior corners. However, away from these singular points, the solutions are remarkably similar despite the much smaller number of degrees of freedom present in the scaled boundary finite element mesh.

25.7.2 Cylindrical Foundation Embedded in Half-Space

As an unbounded medium a cylindrical foundation of radius a embedded with depth $e = a$ in a half-space with Poisson's ratio $\nu = 0.25$ and shear modulus G is examined (Figure 25.15). The flexible baseplate is loaded with a pressure p. The wall of the cylinder rigidly restrains the half-space in the radial direction, but allows movement in the vertical direction.

Advantage is taken of the axisymmetry of the problem. Although the displacement field has only two components in a plane, with potential discontinuities of the stresses in this plane in a numerical analysis, the problem remains three dimensional. The

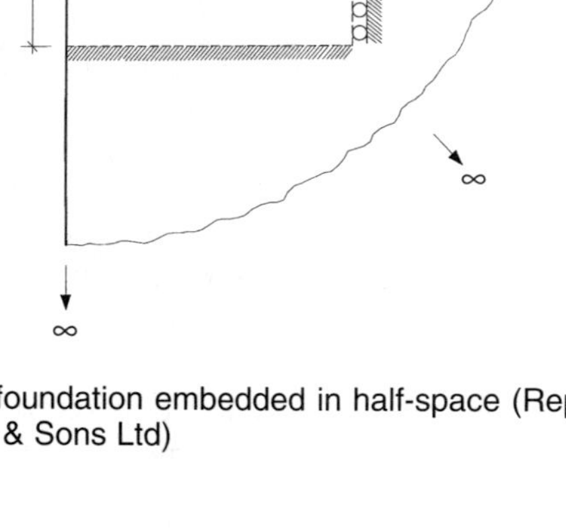

Figure 25.15 Cylindrical foundation embedded in half-space (Reproduced by permission of John Wiley & Sons Ltd)

formulation of the error estimation in section 25.4 applicable to a two-dimensional case must thus be modified slightly. The scaled boundary finite element equation in displacement for $s = 3$ applies with the integrations in the circumferential direction around the axis of axi-symmetry in the coefficient matrices $[E^0]$, $[E^1]$, $[E^2]$, performed analytically. Certain minor changes occur in the formulation. For instance, comparing the displacements in the radial direction in the two- and three-dimensional cases (Equation (16.15)), it follows that the same expression arise as for the two-dimensional case if the eigenvalues λ_i are replaced by $\lambda_i - 0.5$ for the unbounded medium. The modifications are straightforward. For details, References [W9] and, performed somewhat differently, [D5] can be consulted.

When the finite element method is applied to such problems, the semi-infinite nature of the domain presents a difficulty. The most common approach is to truncate the mesh at an artificial boundary considered to be sufficiently far from the foundation that the effect of applying rigid restraints will be negligible. The distance of the boundary from the origin O is designated as b in Figure 25.15. Other methods, such as infinite elements, can also be used. However, the error estimator cannot easily be applied when such methods are used, and consequently the truncation approach is used here. The base of the truncated mesh is fixed in both directions, while the circumferential artificial boundary is rigidly restrained in the radial direction, but unrestrained in the vertical direction.

Five initial finite element meshes are constructed using eight-noded quadratic elements. The meshes are truncated at distances b from the origin O varying between $4a$ and $64a$. The elements are increased in size in accordance with the distance from the origin, as indicated in Figures 25.16 a–e.

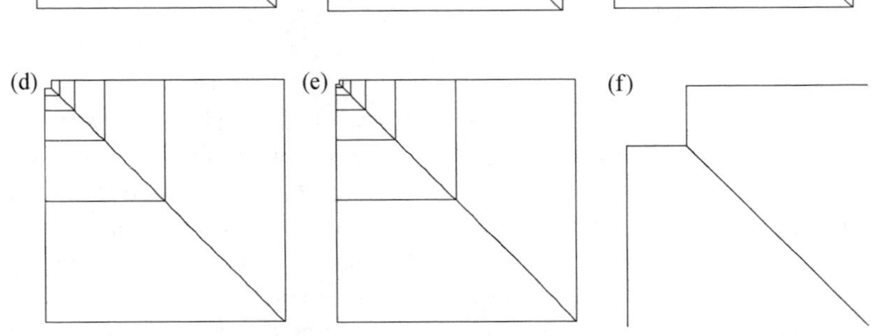

Figure 25.16 (a)–(e) Initial finite element meshes (scale varies), (f) initial scaled boundary finite element mesh (Reproduced by permission of John Wiley & Sons Ltd)

Table 25.4 Computational efficiency and accuracy of adaptive finite element analyses (meshes a to e) and adaptive scaled boundary finite element analysis (mesh f) (Reproduced by permission of John Wiley & Sons Ltd)

Mesh	Truncation b	DOF	Time	η^*	Iterations	Displacement
a	$4a$	1411	4979	5.00	7	−0.2636
b	$8a$	1597	5675	4.98	7	−0.2994
c	$16a$	1875	6833	4.84	7	−0.3162
d	$32a$	2089	8317	4.78	7	−0.3244
e	$64a$	2325	9396	4.75	7	−0.3284
f	∞	34	2998	4.99	4	−0.3318

An initial scaled boundary finite element mesh consisting of two three-noded quadratic elements along the baseplate and wall in Figure 25.15 is used. All meshes therefore have the same initial discretisation level along this interface. The initial scaled boundary finite element mesh is shown in Figure 25.16(f). The scaling centre O is selected at the origin. Only the interface is discretised. No spatial discretisation of the free surface or along the axis of symmetry is introduced.

Adaptive analyses are performed for each mesh using the h-hierarchical procedures described earlier, with a target stress error energy norm $\bar{\eta}$ of 5%. The results, including the achieved error estimator η^* and the number of iterations (subdivisions), are presented in Table 25.4. The vertical displacement is calculated at the centre of the baseplate and is non-dimensionalised by dividing by pa/G.

The baseplate displacements indicate dramatically the effect of truncating the solution domain. Although the target error is met for each mesh, the footing displacement varies considerably. Since the error within the domains has been restricted to the desired level, the differences in the displacements are due to the domain truncation. As the distance to the artificial boundary is increased, the finite element solution tends towards the scaled boundary finite element solution, in which the domain has not been truncated. However, the (unoptimised) scaled boundary finite element method takes 40% less time than the finite element mesh with the most severe truncation, 70% less time than the largest mesh analysed, and produces a solution which is more accurate than could be achieved with any truncated finite element mesh refined to the same level of error. Only 34 degrees of freedom are required to produce the solution, as opposed to between 1411 and 2325 for the finite element solutions, and so the memory requirements are also significantly less.

The accuracy of the stresses may be verified by comparing the stress contour plots. Close agreement is found in the vicinity of the footing, with differences only becoming apparent near the vertical truncation of the finite element model. The most severe indication of this is in the first finite element mesh a, where the truncation occurs at $4a$. The vertical stress contours recovered from this mesh and from the scaled boundary finite element mesh are presented in Plate 25.7. Distortion of the stress bulbs near the artificial boundary in the finite element solution is evident, while the agreement between the two methods near the footing is excellent. All the other finite element meshes exhibit improved agreement with the scaled boundary finite element results, due to the reduced influence of the domain truncation.

These results indicate that the scaled boundary finite element method error estimator and adaptive procedure work well even for unbounded media. A similar level of accuracy to the finite element method is achieved when the same target error is specified, but at lower cost (both in terms of time and in terms of memory), and without the error introduced by truncating the medium at an arbitrary distance.

26 Concluding Remarks

It is demonstrated in the preceding chapters that, based on a weighted-residual approach, the scaled boundary finite element method transforms in the circumferential directions the *partial* differential equations to *ordinary* differential equations, which can be solved *analytically*. Effectively, the governing partial differential equations are weakened in the circumferential directions in a finite element manner, but remain strong in the radial direction. Only the boundary is discretised with curved surface finite elements. The scaled boundary finite element method is thus a *semi-analytical* procedure for solving partial differential equations.

To a large extent, the challenging differential equations of elastodynamics in the frequency and time domains, as well as in statics, are addressed, and to a lesser extent, those of diffusion. It must be emphasised that the novel computational procedure can, in principle, be applied to all linear partial differential equations which can be elliptic, parabolic and hyperbolic. This includes, for instance, the differential equations governing electromagnetism. Thus, many problems in engineering, physical science and applied mathematics can be solved based on the scaled boundary finite element method.

In Chapter 2 the scaled boundary finite element method is introduced, and in Chapter 14 conclusions based on the Model Problem of Part I are stated. The derivation, solution procedure and the properties are discussed. In particular, Table 14.1 presents the advantages of the scaled boundary finite element method compared to those of the finite element and boundary element methods. As these items are not repeated in this chapter, the reader should go back to and re-study Chapters 2 and 14. In the following, only the additional advantages and features which could not be discussed based on the Model Problem are addressed.

The following aspects examined in Part II are mentioned.

- *Incompressibility* (section 21.2)
  In contrast to the finite element method, Poisson's ratio = 0.5 is enforced analytically in the scaled boundary finite element method. (This is even more straightforward in the boundary element method, as only the analytical fundamental solution is affected.)
- *Material Properties Variation in Radial Direction* (section 21.1)
  The elasticity matrix and mass density can vary as power functions of the radial coordinate. The coefficient matrices are calculated with the material properties at the boundary. The scaled boundary finite element equation in dynamic stiffness and in displacement for statics have the same form as for the homogeneous case, with only

two scalar coefficients being affected. As in the circumferential direction an arbitrary variation of the material properties is possible, more general variations can be calculated, e.g. described by power functions in the direction of a Cartesian coordinate.

- *Substructuring* (Chapter 22)
  For a bounded medium the restrictions on geometry and material variation enforced by scaling can be completely avoided by substructuring. In addition, in many cases, a more efficient procedure results, as a series of eigenvalue problems, each of modest size, is calculated instead of one eigenvalue problem of large size. For dynamic problems, a sufficient number of degrees of freedom in the interior of the domain can be generated. By selecting similar boundaries applying scaling, no additional eigenvalue problems are solved (section 17.1). The salient features at the scaling centre, i.e. that a singularity of *a priori* unknown power can analytically be calculated, and that on the adjacent straight segments of the boundary, the side-faces, the boundary conditions are satisfied exactly without any discretisation, will also lead to substructuring into subdomains, each with its own scaling centre. This permits problems with multiple singularities and boundary segments with exact enforcement of boundary conditions to be processed. For an unbounded medium, the use of substructuring is limited.
- *Two-dimensional Layered Unbounded Medium and Three-dimensional Prismatic Unbounded Medium* (sections 21.4 and 21.5)
  For these two cases, the scaling centre is located at infinity, which permits the formulation to be streamlined. The scaled boundary finite element equation in displacement has constant coefficients, and in the scaled boundary finite element equation in dynamic stiffness, the derivative term vanishes, which yields a quadratic eigenvalue problem for each frequency.
- *Reduced Set of Base Functions* (section 21.3)
  To be able to reduce the size of the matrices, the degrees of freedom on the boundary are represented by a set of base functions and corresponding amplitudes. This permits realistic dynamic unbounded medium–structure interaction analyses with many degrees of freedom to be performed. The set of base function can be chosen as the first few eigenvectors corresponding to the lowest eigenvalues of the static problem.
- *Dynamic Stress-Intensity Factor* (section 23.6)
  Transient dynamic analysis of fracture mechanics problems can be addressed. The bounded medium is substructured into subdomains, introducing some internal degrees of freedom to be able to represent the response up to a sufficiently high frequency. The static-stiffness and mass matrices of each subdomain are calculated based on the scaled boundary finite element method. For the subdomain containing the crack, the scaling centre is chosen at the crack tip, and the adjacent straight crackfaces are not discretised. The dynamic response is then calculated directly in the time domain using a standard integration scheme. The dynamic stress-intensity factor (as a function of time) then follows analytically directly based on its definition. The same advantages as for the analysis of static stress-intensity factors thus apply.
- *Error Estimation and Adaptivity* (Chapter 25)
  Stress recovery and error estimation need only be performed numerically on the boundary. All calculations in the radial direction for bounded and unbounded media are analytical and do not introduce any discretisation. Scaling ensures that

the recovered stresses are superconvergent throughout the domain. The recovery-type error estimator for the scaled boundary finite element method is in the same form as for finite elements, allowing for a direct comparison between the accuracies of the two numerical methods. The same applies for adaptive procedures, which permit an analysis with a prescribed target error to be carried out. The h-hierarchical adaptive refinement strategy is developed.

The accuracy and computational efficiency of the scaled boundary finite element method is examined next. For statics, the following conclusions are reached (Chapter 25):

- The stress recovery and error estimation technique work well with stress singularities and unbounded media, in contrast to their finite element counterparts.
- For the same level of boundary discretisation, even when no stress singularity or unbounded medium is present, the scaled boundary finite element method yields a smaller global error estimator.
- For the same global error estimator, the computed stresses at the singularity point or stress concentration point obtained by the scaled boundary finite element method are significantly more accurate than those obtained by the finite element method. The displacements and stresses in other points are in close agreement.
- The scaled boundary finite element estimator is shown to predict the true error more closely than the equivalent finite element error estimator.
- For the same specified accuracy, in situations involving stress singularities and unbounded media, the h-hierarchical adaptivity procedure yields smaller computer run times for the scaled boundary finite element method, and significantly fewer degrees of freedom, as a measure of memory requirement, than for the finite element method.

In a nutshell, it can be stated that for problems with stress singularities (or concentrations) and for unbounded media, the scaled boundary finite element method performs much better than the finite element method from all points of view.

The scaled boundary finite element method, of course, also exhibits certain disadvantages. Where there is light, there is shadow! These restrictive properties are listed as follows:

- *Geometry and Material Variation*
  The novel method is based on scaling (similarity). For a bounded medium several substructures (Chapter 22), each with its own scaling centre, can be introduced, which yields additional boundaries between adjacent substructures. For each substructure, its boundary must be visible from its scaling centre. This is a very powerful extension permitting, for instance, multiple cracks with stress singularities at their tips (Figure 22.3) to be analysed rigorously.
  For an unbounded medium, substructuring is limited (Figure 22.4). An approximate representation of the unbounded medium satisfying scaling (similarity) is illustrated in Figure 26.1 in an analysis of a dynamic unbounded medium–structure interaction problem. Strongly inclined parallel interfaces between parts of a half-plane consisting of different materials are present. By moving the structure–medium interface (boundary) outwards, an approximate model satisfying scaling is constructed. The further away the structure–medium interface is chosen, the better the approximation becomes.

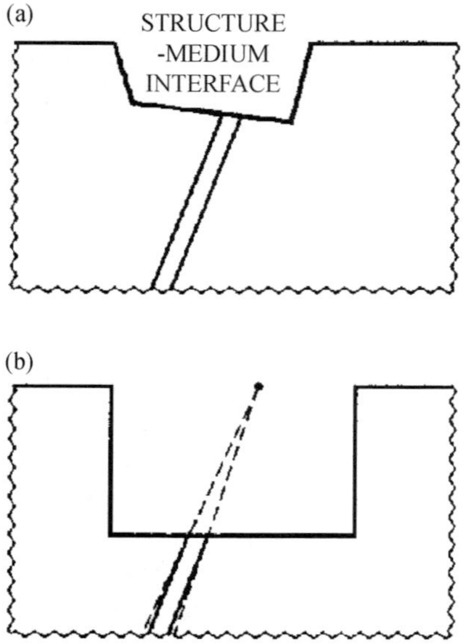

Figure 26.1 (a) Original dynamic unbounded medium–structure interaction problem with inclined material interfaces; (b) approximate representation enforcing scaling of unbounded medium with structure–medium interface moved outwards (Reproduced by permission of John Wiley & Sons Ltd)

The procedure rigorously calculates the dynamic system with the dashed interfaces up to infinity. This approach should be compared with that of truncating the discretisation of the interfaces extending to infinity, as in the boundary element method (working with the fundamental solution of the full plane), where it is not clear what dynamic system is actually analysed from the truncation point up to infinity. Thus, in this case a bounded subdomain in the near field of the unbounded medium and an unbounded subdomain in the far field are introduced, and the boundary between the bounded and unbounded subdomains is moved further away from the scaling centre. In addition, the physical problem is slightly changed to be able to work with one scaling centre.

- *Eigenvalue Problem*
  In the solution of statics (Chapter 16) or dynamics (section 20.1), an eigenvalue problem must be solved, in contrast to the finite element and boundary element methods. This means that the scaled boundary finite element is hardly competitive for standard bounded media with smooth stress variations. However, in the presence of stress singularities, the solution of the eigenvalue problem determines the power of the stress singularities, leading to high accuracy without any special measures. This is not the case for the finite element and boundary element methods, which use pre-determined polynomials to interpolate the displacements. A very large number of elements or special techniques, which, for example, incorporate the power of the singularities that must thus be known *a priori*, are necessary. For the unbounded medium, solution of the eigenvalue problem permits the radiation condition to be also satisfied exactly (Chapter 18) in complex situations (anisotropy, incompressible soil, presence of free surfaces and material interfaces), and to be consistent with the discretisation in the circumferential

directions. The eigenvalue problem can be avoided by constructing the high-frequency limit of the dynamic stiffness of the medium using impedances.

- *Unit-impulse Response*
  In a boundary element method working in the time domain, a transient excitation can be processed directly. This is not possible in the scaled boundary finite element method, where unit-impulse response matrices are calculated first. A transient analysis then involves convolution integrals (Chapter 19) with a computational effort that is proportional to the square of the number of time steps. These can be avoided by using a rational approximation and corresponding realisation. For instance, the dynamic behaviour of the unbounded medium is then represented by a recursive evaluation of the interaction forces, or by a spring-dashpot-mass model leading to a solution of linear differential equations, both with a computational effort which is proportional to the number of time steps.
- *Spatial Discretisation in Circumferential Directions*
  The boundary (structure–medium interface) is discretised spatially with surface finite elements, leading to a sufficient number of nodes to represent the shortest wave length in the circumferential directions (in the radial direction, an analytical formulation is present without any discretisation). For a surface parallel to the structure–soil interface in the unbounded medium with a constant radial coordinate, the same number of nodes will be present due to scaling. Although the distance between the nodes has increased according to the scaling law, the accuracy does not deteriorate. This is remarkable!

It is worth mentioning again those problems for which the scaled boundary finite element method is ideally suited, and is far superior to any other numerical procedure. Only applications where the salient features of the novel method result in significant advantages are discussed. Border-line cases with minor improvements are not addressed. Elastodynamics and as an important special case statics are examined, addressing the requirements on geometry and material as well as the ease with which the final differential equations can be solved. First bounded media and second unbounded media are addressed, both for statics and then for dynamics. For a *bounded medium with stress concentrations*, the scaled boundary finite element method is very powerful. The scaling centre coincides with the singular point. The power of the stress singularities is determined as part of the analytical solution, and does thus not have to be known before the analysis. No restrictions on geometry and material behaviour exist when using substructuring. For each stress singularity a subdomain is selected. Crack faces are not discretised. For dynamics, further subdomains are, in general, introduced, leading to static-stiffness and mass matrices. In statics, the governing differential equations are solved analytically (Euler–Cauchy), leading to a semi-analytical procedure. In dynamics, the assembled property matrices are solved numerically directly in the time domain (e.g. Newmark algorithm). The stress singularity follows based on its definition from the analytical expression in the subdomain with the singularity at the crack tip also in dynamics. The other subdomains only affect the displacements in the nodes of the subdomain with the singularity. Examples are described in sections 23.1–23.5 and 25.7.1 for statics, and in section 23.6 for dynamics. For an *unbounded medium with one scaling centre* (at the intersection of the extensions of the side-faces and material interfaces, if any), the scaled boundary finite element method is very attractive. Free and fixed surfaces, as well as material interfaces with extensions passing through the scaling centre, are not discretised.

The boundary condition at infinity (radiation condition in dynamics) is satisfied exactly. For statics, the governing differential equation are solved analytically (Euler–Cauchy). For dynamics, the high-frequency asymptotic expansion for the dynamic stiffness is determined analytically, or constructed directly from impedances, combined with a numerical procedure in the lower frequency range or equivalently in the time domain. An example is described in section 24.1. Summarising, the superiority of the scaled boundary finite element method manifests itself overwhelmingly in two types of analyses: in the presence of singularities in statics, and even in dynamics, yielding an analytical expression; and for unbounded media leading in statics to an analytical solution throughout the domain satisfying the boundary condition at infinity exactly, and in dynamics to a solution satisfying the radiation condition at infinity, exactly expressed as a high-frequency asymptotic expansion for the dynamic stiffness. The radial coordinate with its origin, the scaling centre, plays a key role. The limit of the radial coordinate towards zero describes the singular behaviour, if any, and for an unbounded medium the other limit towards infinity permits the boundary condition at infinity (radiation condition) to be incorporated into the solution.

It is appropriate to recall the philosophical steps that lead to construction of the scaled boundary finite element method. It is desirable to solve the partial differential equations with the boundary conditions *analytically*. In contrast to a numerical procedure, leading typically to a system of equations with results in distinct points throughout the domain, in general, this leads to a closed-form expression of the results, permitting an in-depth understanding of the physical behaviour. In addition, results in selective specific points can easily be determined. For the general case, such an analytical solution cannot be determined. The possibility of developing an analytical solution, which should be as simple as possible, is significantly increased if the partial differential equations are transformed into ordinary differential equations. A transformation of the partial differential equations from the Cartesian coordinate system to the scaled boundary coordinates is first performed. The latter consists of the radial coordinate and circumferential coordinates on the boundary. The radial coordinate is kept as the independent variable of the ordinary differential equations. At its origin, the singular behaviour can be calculated analytically, and for the other limit of radial coordinate towards infinity, the boundary condition at infinity (radiation condition) can be enforced analytically. The unknown functions on the boundary (and corresponding scaled surfaces) are discretised *numerically* by applying a weighted residual approach with shape functions, as in *finite elements*. This leads, as in the *boundary element* method, to a reduction of the spatial dimension by one, and eliminates the circumferential coordinates. The boundary conditions are also converted into distinct values of the functions and their derivatives, calculated for the radial coordinate corresponding to the boundary. The key feature of the scaled boundary finite element method, as discussed in section 15.1, consists of interpolating the functions $\{u(\xi, \eta, \zeta)\}$ with the same shape functions $[N(\eta, \zeta)]$ of the boundary with the circumferential coordinates η, ζ for all surfaces with a constant radial coordinate ξ as (Equation (15.3))

$$\{u(\xi, \eta, \zeta)\} = [N(\eta, \zeta)]\{u(\xi)\} \tag{26.1}$$

where $\{u(\xi)\}$ represent the functions of the radial coordinate, the dependent variables, determined analytically by solving the ordinary differential equations.

Finally, a suggestion for future research will be provided. No verification has been performed. This incomplete list can be broken down into topics addressing applications in other areas, extending the formulation and improving the efficiency.

Probably the most promising activity is the derivation of the formulation and development of the solution procedures for other problems. First results [W4] are available to analyse the dynamics of a two-phase material, e.g. saturated poroelastic soil (Biot's theory). As unbounded domains are often encountered in wave propagation in electromagnetics, this subject could be addressed. This is just one example!

Turning to the extensions of the formulation, the concept of scaling could possibly be extended, covering not only with respect to a single fixed point (the scaling centre), but also with respect to points moving, for example, on an axis or a curve. This means that for the scaling of an element on the boundary, the scaling centre would move over a small segment of the locus of scaling centres. In addition, to model an unbounded medium with, for instance; two fixed scaling centres each governing part of the domain, an element performing the transition between the two scaling centres would be helpful. In this case, the corresponding scaled boundary finite element equation in displacement could be derived. However, the variable coefficients of the latter would be so complicated that it is to be expected that an analytical solution of this second-order differential equation, even for the simplest cases, would hardly be possible. A formulation addressing the 2½-dimensional problem would also be welcome, in which the geometry is two-dimensional but the loading is three-dimensional.

As far as the efficiency of the solution procedures is concerned, a lot of work remains to be done. Real progress should result by incorporating concepts which are standard in linear system and control theory, where the same relationships arise (e.g. solving the algebraic Riccati equation). When solving the eigenvalue problem in statics, using the Hamiltonian property of the coefficient matrix would result in significant computational savings in an operation which is crucial for running times. In dynamics, the scaled boundary finite element equation in displacement should be solved in the radial coordinate analytically exactly, incorporating the radiation condition and not using the high-frequency asymptotic expansion of the dynamic stiffness to determine the boundary condition. The procedure for a scalar using second-kind Hankel functions should be generalised to the matrix case!

The finite element and boundary element methods have been developed during a long period to a high level of sophistication which has led to significant know-how and experience. The latter should be applied by scientists and engineers to address the scaled boundary finite element method, where real possibilities still exist. Significant progress in the novel computational procedure could then be expected to occur in a comparatively short time. In particular, PhD students and their advisers will discover that the scaled boundary finite element method is a real Eldorado to define meaningful thesis topics.

References

[B1] Bartels, R.H. and Stewart, J.L. (1972) Solution of the matrix equation AX + XB = C. *Communications ACM*, **15**, 820–826.

[B2] Bathe, K.J. (1995) *Finite Element Procedures*, Prentice-Hall, Englewood Cliffs, NJ.

[B3] Becker, E.B., Carey, G.F. and Oden, J.T. (1981) *Finite Elements – an Introduction*, Prentice-Hall, Englewood Cliffs, NJ.

[B4] Brebbia, C.A. and Dominguez, J. (1992) *Boundary Elements: an Introductory Course*, 2nd edition, Computational Mechanics Publications, Southampton.

[C1] Chen, D.H. and Harada, K. (1996) Stress singularities for crack normal to and terminating at bimaterial interface on orthotropic half-plates. *International Journal of Fracture*, **81**, 147–162.

[C2] Chen, Y. (1975) Numerical computation of dynamic stress intensity factors by a Lagrangian finite difference method (the Hemp code). *Engineering Fracture Mechanics*, **7**, 653–660.

[C3] Cho, B.C., Lee, K.R., Choy, Y.S. and Yuuki, R. (1992) Determination of stress intensity factors and boundary element analysis for interface cracks in dissimilar anisotropic materials. *Engineering Fracture Mechanics*, **43**, 603–614.

[D1] Dasgupta, G. (1979) Impedance matrices for unbounded media. *Proceedings Seventh Canadian Congress of Applied Mechanics*, Sherbrooke, 891–892.

[D2] Dasgupta, G. (1982) A finite element formulation for unbounded homogeneous continua. *Journal of Applied Mechanics*, **49**, 136–140.

[D3] Deeks, A. and Wolf, J.P. (2002) Stress recovery and error estimation for the scaled boundary finite-element method. *International Journal for Numerical Methods in Engineering*, **54**, 557–583.

[D4] Deeks, A. and Wolf, J.P. (2002) An *h*-hierarchical adaptive procedure for the scaled boundary finite-element method. *International Journal for Numerical Methods in Engineering*, **54**, 585–605.

[D5] Deeks, A. and Wolf, J.P. (2002) A virtual work derivation of the scaled boundary finite element method for elastostatics. *Computational Mechanics*, **28**, 489–504.

[D6] Deeks, A. and Wolf, J.P. (2002) Semi-analytical elastostatic analysis of unbounded two-dimensional domains. *Numerical and Analytical Methods in Geomechanics*, **26**, 1031–1057.

[D7] Doherty, J.P. and Deeks, A. (2002) Scaled boundary finite-element analysis of a non-homogeneous elastic half-space. *International Journal for Numerical Methods in Engineering* (in press).

[F1] Friedrich, K. and Schmid, G. (2000) Personal communication.

[G1] Gantmacher, F.R. (1977) *The Theory of Matrices*, 2, Chelsea, New York.

[G2] Golub, G.H. and Van Loan (1996) *Matrix Computations*, John Hopkins University Press, Baltimore.

[K1] Kausel, E., Roësset, J.M. and Waas, G. (1975) Dynamic analysis of footings on layered media. *Journal of Engineering Mechanics*, **101**, 679–693.

[L1] Laub, A.J. (1979) A Schur method for solving algebraic Riccati equations. *IEEE Transactions Automatic Control*, **AC-24**, 913–921.

[L2] Lekhnitskii, S.G. (1968) *Anisotropic Plates*, Gordon and Breach, London.

[L3] Lysmer, J. (1970) Lumped mass method for Rayleigh waves. *Bulletin of the Seismological Society of America*, **60**, 89–104.

[M1] Miyazaki, N., Ikeda, T., Soda, T. and Munakata, T. (1993) Stress intensity factor analysis of interface crack using boundary element method – applications of contour integral method. *Engineering Fracture Mechanics*, **45**, 599–610.

[P1] Patel, R.V., Lin, Z. and Misra, P. (1994) Computation of stable invariant subspaces of Hamiltonian matrices. *Journal of Matrix Analysis and Applications*, **15**, 284–298.

[P2] Paris, F. and Cañas, J. (1997) *Boundary Element Method: Fundamentals and Applications*, Oxford University Press, Oxford.

[P3] Press, W.H., Flannery, B.P., Teukolsky, S.A. and Vetterling, W.T. (1988) *Numerical Recepies*, Cambridge University Press, Cambridge.

[R1] Roësset, J.M. and Scaletti, H. (1979) Boundary matrices for semi-infinite problems. *Proceedings Third Engineering Mechanics Speciality Conference*, ASCE, Austin, TX, 384–387.

[S1] Silvester, P.P., Lowther, D.A., Carpenter, C.J. and Wyatt, E.A. (1977) Exterior finite elements for 2-dimensional field problems with open boundaries. *Proceedings of the Institution of Electrical Engineers*, **124**, 1267–1270.

[S2] Sommerfeld, A. (1949) *Partial Differential Equations in Physics*, Academic Press, New York, NY.

[S3] Song, Ch. and Wolf, J.P. (1995) Consistent infinitesimal finite element cell method: out-of-plane motion. *Journal of Engineering Mechanics*, **121**, 613–619.

[S4] Song, Ch. and Wolf, J.P. (1996) Consistent infinitesimal finite element cell method for diffusion equation in unbounded medium. *Computer Methods in Applied Mechanics and Engineering*, **132**, 319–334.

[S5] Song, Ch. and Wolf, J.P. (1996) Consistent infinitesimal finite element cell method: three-dimensional vector wave equation. *International Journal for Numerical Methods in Engineering*, **39**, 2189–2208.

[S6] Song, Ch. and Wolf, J.P. (1997) Consistent infinitesimal finite element cell method for incompressible unbounded medium. *Communications in Numerical Methods in Engineering*, **13**, 21–32.

[S7] Song, Ch. and Wolf, J.P. (1997) The scaled boundary finite element method – alias consistent infinitesimal finite element cell method – for elastodynamics. *Computer Methods in Applied Mechanics and Engineering*, **147**, 329–355.

[S8] Song, Ch. and Wolf, J.P. (1998) The scaled boundary finite element method: analytical solution in frequency domain. *Computer Methods in Applied Mechanics and Engineering*, **164**, 249–264.

[S9] Song, Ch. and Wolf, J.P. (1999) The scaled boundary finite element method – alias consistent infinitesimal finite element cell method – for diffusion. *International Journal for Numerical Methods in Engineering*, **45**, 1403–1431.

[S10] Song, Ch. and Wolf, J.P. (1999) Body loads in scaled boundary finite element method. *Computer Methods in Applied Mechanics and Engineering*, **180**, 117–135.

[S11] Song, Ch. and Wolf, J.P. (2000) The scaled boundary finite-element method – a primer: solution procedures. *Computers and Structures*, **78**, 211–225.

[S12] Song, Ch. and Wolf, J.P. (2001) Semi-analytical evaluation of dynamic stress intensity factors. *Proceedings First Asian-Pacific Congress on Computational Mechanics* (APCOM'01), Valliappan, S. and Khalili, N., eds., Sydney, 2, 1041–1046.

[S13] Song, Ch. and Wolf, J.P. (2002) Semi-analytical representation of stress singularities as occurring in cracks in anisotropic multi-materials with the scaled boundary finite element method. *Computers and Structures*, **80**, 183–197.

[T1] Tan, C.L. and Gao, Y.L. (1992) Boundary element analysis of plane anisotropic bodies with stress concentrations and cracks. *Composite Structures*, **20**, 17–28.

[T2] Thatcher, R.W. (1975) Singularities in the solution of Laplace's equation in two dimensions. *Journal of the Institute of Mathematics and Its Applications*, **16**, 303–319.

[T3] Thatcher, R.W. (1978) On the finite element method for unbounded regions. *SIAM Journal on Numerical Analysis*, **15**, 466–477.

[T4] Timoshenko, S. and Goodier, J.N. (1951) *Theory of Elasticity*, 2nd edition, McGraw-Hill, New York.

[W1] Waas, G. (1972) *Linear Two-Dimensional Analysis of Soil Dynamics Problems in Semi-Infinite Layered Media*. PhD dissertation, University of California, Berkeley, CA.

[W2] Wolf, J.P. (1994) *Foundation Vibration Analysis Using Simple Physical Models*, Prentice-Hall, Englewood Cliffs, NJ.

[W3] Wolf, J.P. (2002) Response of unbounded soil in scaled boundary finite element method. *Earthquake Engineering and Structural Dynamics*, **31**, 15–32.

[W4] Wolf, J.P. and Huot, F.G. (2001) On modelling unbounded saturated poroelastic soil with the scaled boundary finite element method. *Proceedings First Asian-Pacific Congress on Computational Mechanics (APCOM'01)*, Valliappan, S. and Khalili, N., eds., Sydney, 2, 1047–1056.

[W5] Wolf, J.P. and Moussaoui, F. (2001) Far-field displacements of soil in scaled boundary finite element method. *Proceedings 10th International Conference of the International Association for Computer Methods and Advances in Geomechanics*, Tucson, AZ, Vol. 1, 101–108.

[W6] Wolf, J.P. and Song, Ch. (1994) Dynamic-stiffness matrix of unbounded soil by finite-element multi-cell cloning. *Earthquake Engineering and Structural Dynamics*, **23**, 233–250.

[W7] Wolf, J.P. and Song, Ch. (1995) Consistent infinitesimal finite element cell method: in-plane motion. *Computer Methods in Applied Mechanics and Engineering*, **123**, 355–370.

[W8] Wolf, J.P. and Song, Ch. (1996) Static stiffness of unbounded soil by finite-element method. *Journal of Geotechnical Engineering*, **122**, 267–273.

[W9] Wolf, J.P. and Song, Ch. (1996) *Finite-Element Modelling of Unbounded Media*, John Wiley and Sons, Chichester. (Program SIMILAR downloaded from ftp://ftp.wiley.co.uk/pub/books/wolf/ and http://lchpc25.epfl.ch/ as well as http://www.civeng.unsw.edu.au/staff/song.c/sbfem/SIMILAR/ and http://www.civil.uwa.edu.au/~deeks/sbfem/).

[W10] Wolf, J.P. and Song, Ch. (1996) Consistent infinitesimal finite element cell method: three-dimensional scalar wave equation. *Journal of Applied Mechanics*, **63**, 650–654.

[W11] Wolf, J.P. and Song, Ch. (1996) Consistent infinitesimal finite element cell method in frequency domain. *Earthquake Engineering and Structural Dynamics*, **25**, 1307–1327.

[W12] Wolf, J.P. and Song, Ch. (1998) Unit-impulse response of unbounded medium by scaled boundary finite-element method. *Computer Methods in Applied Mechanics and Engineering*, **159**, 355–367.

[W13] Wolf, J.P. and Song, Ch. (2000) The scaled boundary finite-element method – a primer: derivations. *Computers and Structures*, **78**, 191–210.

[W14] Wolf, J.P. and Song, Ch. (2001) The scaled boundary finite element method – a fundamental-solution-less boundary-element method. *Computer Methods in Applied Mechanics and Engineering*, **190**, 5551–5568.

[W15] Wolf, J.P. and Song, Ch. (2002) Some cornerstones of dynamic soil-structure interaction. *Engineering Structures*, **34**, 13–28.

[W16] Wolf, J.P. and Song, Ch. (2002) The semi-analytical fundamental-solution-less scaled boundary finite-element method to model unbounded soil. *Euromech Colloquium 414. Boundary Element Methods for Soil/Structure Interaction*, Hall, W.S. and Oliveto, G., eds., Kluwer Academic.

[W17] Wolf, J.P. and Weber, B. (1982) On calculating the dynamic-stiffness matrix of the unbounded soil by cloning. *International Symposium on Numerical Models in Geotechnics*, Dungar, R. *et al.*, eds., Zürich, 486–494.

[Y1] Ying, L.-a. (1978) The infinite similar element method for calculating stress intensity factors. *Scientia Sinica*, **21**, 19–43 (also presented in Chinese at the National Conference on Fracture Mechanics, Nanning, 1974).

[Y2] Ying, L.-a. (1995) *Infinite Element Methods*, Friedr. Vieweg + Sohn (also published in Chinese by Peking University Press, Beijing, 1992).

[Y3] Yosibashi, Z. (1997) On solutions of two-dimensional linear elastostatic and heat-transfer problems in the vicinity of singular points. *Journal of Solids and Structures*, **34**, 243–274.

[Y4] Yuuki, R. and Cho, S.B. (1989) Efficient boundary element analysis of stress intensity factors for interface cracks in dissimilar materials. *Engineering Fracture Mechanics*, **34**, 179–188.

[Z1] Zhang, J.J., Tan, C.L. and Afagh, F.F. (1997) Treatment of body-force volume integrals in BEM by exact transformation for 2-d anisotropic elasticity. *International Journal for Numerical Methods in Engineering*, **40**, 89–109.

[Z2] Zhang, Z., Yao, Z.H. and Du, Q.H. (1994) Boundary element method for determining stress intensity factors of bimaterial interface crack. *Theory and Applications of Boundary Element Methods, Proceedings of Sixth China–Japan Symposium on BEM*, Shanghai, 315–320, International Academic Publishers.

[Z3] Zienkiewicz, O.C. and Zhu, J.Z. (1987) A simple error procedure and adaptive procedure for practical engineering analysis. *International Journal for Numerical Methods in Engineering*, **24**, 337–357.

[Z4] Zienkiewicz, O.C. and Zhu, J.Z. (1992) The superconvergent patch recovery and a posteriori error estimates, Part 1: the recovery technique, Part 2: error estimates and adaptivity. *International Journal for Numerical Methods in Engineering*, **33**, 1331–1364, 1365–1382.

Index